THE EVOLUTIONARY HISTORY OF NEMATODES –

as revealed in stone, amber and mummies

THE EVOLUTIONARY HISTORY OF NEMATODES – as revealed in stone, amber and mummies

George O. Poinar Jr

David J. Hunt and Roland N. Perry (Series Editors)

NEMATOLOGY MONOGRAPHS AND PERSPECTIVES
VOLUME 9

BRILL
LEIDEN-BOSTON
2011

This book is printed on acid-free paper.

Library of Congress Cataloging-in-Publication Data

Library of Congress Cataloging-in-Publication Data are available from the publisher

ISBN: 978 90 04 17521 1

Contents

Dedication

I dedicate this book to my wife, Roberta, for her support, encouragement, discussions and understanding over many years.

In memory of earlier workers

B.G. Chitwood, A.A. Paramonov and K.I. Skrjabin, whose works and ideas provided a basis for the study of nematode evolution.

B.G. Chitwood

A.A. Paramonov

K.I. Skrjabin

Nematology Monographs & Perspectives, 2011, Vol. 9, xvii-xviii

Preface

The mysteries surrounding the origin and evolution of the nematodes have resulted in spirited debates over the past 100 years. Since my college days at Cornell under the tutelage of Bill Mai, I have been fascinated by the remarkable diversity of nematodes and their apparent unrelatedness to other invertebrates. There are so many unanswered and perhaps unanswerable questions about these rarely seen invertebrates. Perhaps they occurred in Precambrian ooze at the bottom of the seas. Some must have parasitised trilobites, eurypterids and other soft-bodied Cambrian invertebrates, like those discovered in the Burgess Shale. Maybe they were already parasitising the first terrestrial amphibians that emerged from Devonian swamps. We known the dinosaurs suffered from roundworms but we still do not know when and from where humans first acquired nematode parasites.

After leaving Ithaca, I was determined to learn more about nematode evolution and continued my training with renowned nematologists around the world. These included J.B. Goodey and his associates at Rothamsted, A. Chabaud and his group in Paris, and Oostenbrink and his team in Wageningen. An introduction to entomogenous nematodes was provided by H.E. Welch in Belleville, Ontario, and then by E.S. Kirjanova and I.A. Rubstov in St Petersburg. During these periods, I had the good fortune to discuss problems of nematode evolution with other eminent nematologists, including W.G. Inglis in London, K.I. Skrjabin, A.A. Paramonov and S.L. Blinova-Lazarevskaja in Moscow, G.A. Kakuliya and I. Eliava in Tbilisi, R.Ph. Dollfus in Paris, and E.C. Dougherty and E. Hansen at Berkeley.

My interest in palaeontology began with J.W. Wells at Cornell, continued later at Berkeley with the late Wyatt Durham and Joe Peck, and is ongoing with Art Boucot at Oregon State University. The late Joe Peck introduced me to my first fossil nematodes in Mexican amber in 1975. From then on, I have spent the better part of my life tracking nematodes through time while visiting amber deposits around the world.

While the fossil records of nematodes do not do justice to their abundance and biodiversity, all reports of these invertebrates from sedimentary deposits, coprolites, amber and mummies over the past 150 years are included in the present work, including descriptions of 31 new fossil genera and 66 new fossil species from amber. Because of the rarity of fossils, we will never know more than a miniscule fraction of the nematode species that lived in the past.

Still, it is surprising what fossils can tell us. The remains of nematodes are used here to reveal their evolutionary history, a topic previously addressed mainly through the analyses of morphological and behavioural characters of living forms. Fossils provide minimum dates for the occurrence of nematode lineages at various locations and time periods in the earth's history. They also tell us when nematodes had established symbiotic and parasitic associations with plants and animals. Last, but not least, fossils of both nematodes and their hosts can be used to calibrate molecular clocks.

Taxonomic placement of the fossils described herein is based on my experience with extant nematodes, and some decisions will undoubtedly be debated or challenged. Aside from morphological characters, I have used the principle of behaviour fixity to assign nematode fossils to both extant as well as collective fossil genera. This concept states that the behaviour, ecology and climatic preferences of fossil organisms will be similar to those of their present-day descendants at the generic and, in some cases, the family level. Hopefully, this work will provide a foundation for further studies in paleonematology and will show how, in conjunction with morphological and molecular studies, fossils can provide insights into the origin and evolution of nematodes.

Acknowledgements

I thank Roberta Poinar for reading the entire manuscript and offering suggestions. I am also grateful to the following individuals for commenting on various sections of the manuscript or providing information, photographs or fossil material: Scott Anderson, Art Boucot, Alex Brown, Ron Buckley, Ron Cauble, Andrew Cholewinski, Pat Craig, G. Brovad, Richard Halsey, A.J. Helden, Raymond Jacobson, Dan McAuley, L.R. Miller, Michael Oschin, Sigitas Podenas, Hendrik Poinar, Ru Smith, E. Tarasco, Jim Work and Jorg Wunderlich. I also thank Greg Poinar for Adobe Photoshop assistance, G.A. Kakulia for supplying the photograph of B.G. Chitwood, I.S. Eliava for providing the photograph of A.A. Paramonov, and K.I. Skrjabin for personally providing his photograph (on April 3, 1969).

All fossil specimens in amber illustrated here are deposited in the Poinar amber collection maintained at Oregon State University except for the specimens shown in the following figures: Figure 39 and Plate 6, Ron Buckley collection; Figures 157 and 158, Ron Cauble collection; Figure 24, Raif Milki collection; Figure 25 and Plate 3, Aftim Acra collection; Figure 97, Andrew Cholewinski collection; Figures 148 and 149, Ru Smith collection; Figures 243, 244, 245, 248, 249, 250, 251, 252 and 253, University of California Museum of Paleontology; Figures 213, 214 and Plate 26, Staatlichs Museum, Stuttgart, and Figure 63, Zoological Museum in Copenhagen. The location of specimens depicted in Figures 47, 48, 49, 50, 60, 61 and 71 is unknown.

Nematology Monographs & Perspectives, 2011, Vol. 9, 1-5

Introduction

Nematodes are one of the most abundant groups of invertebrates on the face of the earth and probably have maintained this position since their origin, which could well have been in the Precambrian. They rival the Arthropoda in biodiversity and species abundance. Their numbers are estimated to range from 1.5 billion in the upper 20 millimetres of an acre of marine beach sand and 16 million in a square metre of intertidal area to 3 billion in a single acre of soil and 380 million in a square metre of leaf litter (Chitwood, 1950; Teal & Wieser, 1966; Wasilewska, 1979). Even in the sand-covered hydrogen sulphide 'black zone' beneath the ocean floors, where toxic products of anaerobic decomposition accumulate, nematodes can reach populations of 1000 cm^{-2} (Fenchel & Riedl, 1970). Aside from those living in aquatic and terrestrial environments, there are multitudes that parasitise plants and animals. While some 20 000 nematode species have been described, estimates for species diversity range from 100 000 to 10 million (Poinar, 1983; Malakhov, 1994; Andrássy, 2005).

Of all the Ecdysoma, or pseudocoelomate Bilateria as they were originally called (Hyman, 1951), nematodes are by far the most successful. Their structure, physiology, diverse reproductive patterns and adaptability have resulted in their invasion of more habitats on land, or in fresh and salt water than any other multicellular group of animals, including insects. While geneticists have produced a wide range of mutant rhabditids, especially *Caenorhabditis* spp., by artificial means, natural mutations of nematodes appear to be rare, indicating that their basic body plan is highly canalised (strongly conserved). However, a high degree of morphological variability of the buccal cavity and secondary reproductive structures shows a disruption of the canalising mechanism. Occasionally, aberrant forms are produced in culture as a result of teratogenic components from foreign bacteria (Poinar, 1988) (Fig. 1) although such abnormalities have not been reported from natural populations.

Nematodes combine several survival strategies. Free-living microbotrophs can be considered r-strategists (Smith, 1996), which are characterised by small body size, rapid growth, high reproductive rates and relatively short lives. As r-strategists, they are more adaptable and better

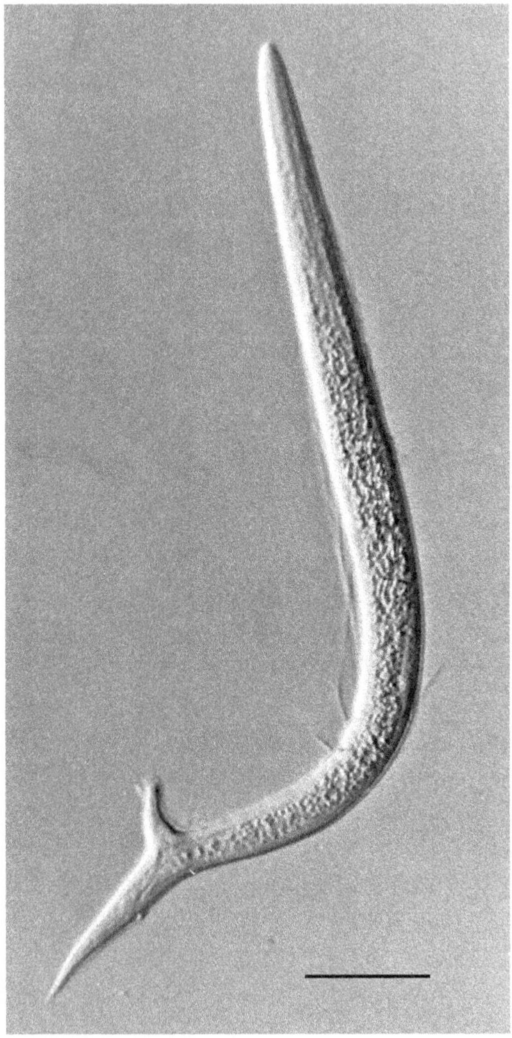

Fig. 1. *Abnormal forked tail on a cultured infective juvenile of* Heterorhabditis bacteriophora *Poinar, 1976, caused by the teratogenic effects of a foreign bacterium (Poinar, 1988). (Scale bar = 64 μm.)*

able to survive in the environment under adverse conditions. However, many of the parasitic lineages can be considered K-strategists, which have slow development rates, larger body sizes, low productive rates and long lives. Nematodes are also ectothermal and can develop under a wide set of temperatures, ranging from frozen tundra to hot springs.

Two handicaps that nematodes have had to overcome are their need for moisture (some consider all nematodes to be aquatic since they require a moisture film for development) and their reduced powers of locomotion. While most free-living nematodes can swim and/or crawl, long distance dispersal depends on abiotic and biotic agents. The most common biotic agents for dispersing nematodes are invertebrates, especially arthropods, and vertebrates. The ability of many terrestrial nematodes to form resistant stages that can withstand complete or partial desiccation has provided unique survival strategies that have solved, in part, their need for moisture and contributed to their success. Nematodes can enter various types of diapause and quiescence triggered either by endogenous or environmental factors (Evans & Perry, 1976). These dauer stages (dauerlarven) can survive temperature extremes, osmotic fluctuations, low oxygen levels and the absence of food. Their ability to attach themselves to adult stages of insects has solved the problem of dispersal for many. Those parasitic inside animals are of course distributed by their hosts. Abiotic agents such as water and wind also play a role in dispersing nematodes and, in the past, vicarience events were important in moving nematode lineages around the globe.

However, even without resistant stages, nematodes are able to thrive in adverse habitats like deep ocean trenches, hot springs, deserts, rotting cadavers, or within the bodies of a wide range of warm-blooded vertebrates. Unlike insects, nematodes flourish in many marine habitats, not only as free-living microbotrophs but also as parasites of seaweeds, invertebrates and vertebrates.

The diverse reproductive methods of nematodes are also a key to their success (Poinar & Hansen, 1984). Aside from the normal amphimictic or bisexual condition, nematodes undergo several types of autotoky (reproduction without males), including hermaphroditism, pseudogamy and parthenogenesis. Some nematodes practice heterogamy (with sexual and autotokous life cycles) and switch from amphimixis to hermaphroditism during alternating life cycles. A highly unusual reproductive strategy employed by a few nematodes is the production of male hermaphrodites. In these rare cases, after mating with a female, the male produces eggs within its body, fertilises them and then dies when the newborn break through his body to exit.

While nematodes are abundant today and certainly were so in the past, their fossil record is minuscule. Their small size, soft bodies and concealed habits contribute to the rarity of nematode fossils. Certainly, many

Fig. 2. *Location of nematode fossils discussed in the present work. 1.* Palaeonema phyticum *Poinar, Kerp & Hass, 2008 from Rhynie, Scotland (Devonian). 2.* Nemavermes mackeei *Schram, 1973 from Illinois, USA (Carboniferous). 3.* Eophasma jurassica *Arduini, Pinna & Teruzzi, 1983 from Italy (Jurassic). 4.* Captivonema cretacea *Boström* in *Manun, Bose, Sayer & Boström, 1994 from Norway (Lower Cretaceous). 5. Lebanese amber fossils from Lebanon (Lower Cretaceous). 6. Burmese amber fossils from northern Myanmar (Lower Cretaceous). 7.* Ascarites *spp. from Belgium (Lower Cretaceous). 8.* Palaeoenoploides alessandrellianus *Mattavelli & Bracchi, 2008 from Italy (Upper Cretaceous). 9. Baltic amber fossils from northern Europe (Eocene). 10. Dominican amber fossils from the Dominican Republic (Miocene-Eocene). 11. Mexican amber fossils from Chiapas, Mexico (Miocene-Oligocene). 12.* Heydenius tabani *in Germany (Pliocene).*

more nematode fossils exist than is realised. It is just that they are difficult to locate, hard to recognise, and few scientists bother to look for them.

The first fossil nematode was described by von Heyden (1860) when he discovered the mermithid *Heydenius antiquus* (von Heyden, 1860) emerging from a beetle in a deposit of brown coal (lignite) in Germany. This was followed by descriptions of free-living and mermithid nematodes in Baltic amber (von Duisburg, 1862; Menge, 1866, 1872). It was a period of some 60 years before the next description of a fossil nematode appeared (Taylor, 1935). While the subject was reviewed by Dollfus (1950) and Roger (1965), a number of years passed before additional descriptions appeared (Poinar, 1977a).

This monograph summarises all previous descriptions of fossil and subfossil nematodes and describes new taxa from amber deposits around the world (Fig. 2) thus establishing a total of 49 genera and 91 species of fossil or subfossil nematodes. This is a far cry from the 2271 extant nematode genera tabulated some 26 years ago (Anderson, 1984), but it is a beginning. Nematode taxa followed by an asterisk (*) after the generic or specific name are fossil lineages new to science that are described in Appendix I of the present work.

Nematology Monographs & Perspectives, 2011, Vol. 9, 7-11

Types of preservation

Amber

Amber is fossilised tree resin that preserves a wide range of organisms, from microbes to vertebrates (Poinar, 1992). While most trees produce resins, only a few of these have the properties to resist decay and become fossilised. One genus of trees whose resin does become fossilised is *Agathis* Salisbury, 1807, a tropical conifer belonging to the Araucariaceae. Resin from this family of trees forms the bulk of Cretaceous amber as well as some early Tertiary amber (Lambert & Poinar, 2002). Amber from Lebanon, Burma (Poinar *et al.*, 2007a) and the Baltic area originated from resin produced by members of the Araucariaceae. By contrast, Tertiary amber from Mexico and the Dominican Republic originated from leguminous trees of the genus *Hymenaea* L., 1753 (Poinar, 1991a; Poinar & Brown, 2002). The bulk of organisms trapped in amber represent forms that lived in, on, or under the bark of the resin-producing tree. An interesting property of amber is that it frequently entrapped several organisms together in the same piece. These inclusions are microcosms that reveal various types of associations that would rarely be preserved with other types of fossilisation.

While amber is one of the best media for preserving nematodes, the degree of preservation depended on how quickly water left the tissues and was replaced by fixatives. The combination of inert dehydration and fixation by natural plant products was mainly responsible for the superb preservation of organic inclusions in amber (Poinar & Hess, 1985). The desiccated condition of internal tissues is revealed by transmission electron micrographs of nematodes preserved in amber. With adults of *Oligaphelenchoides atrebora* Poinar, 1977 in Mexican amber, only a layer of shrunken tissue adhering to the body wall remained (Fig. 3). While the contents of the eggs of *O. atrebora* were desiccated (Fig. 4), large sections of the chitin-protein complex of the shell remained intact (Fig. 5).

Studying nematodes in amber usually requires re-shaping the amber in order to align the nematodes so they can be viewed under the microscope (see Appendix II). While it is preferable to examine specimens under a

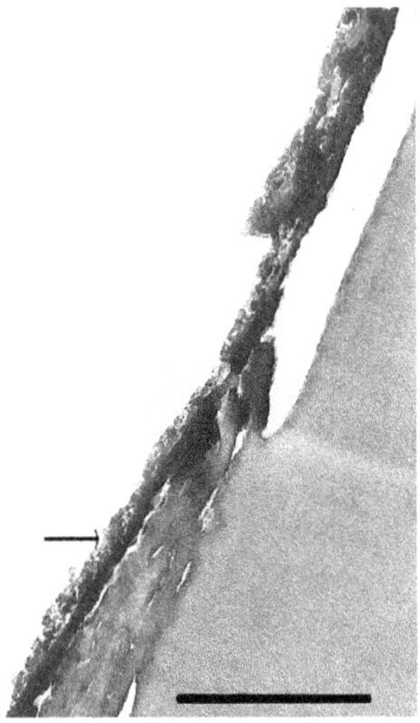

Fig. 3. *Transmission electron microscope section of an adult* Oligaphelenchoides atrebora *Poinar, 1977, in Mexican amber. The internal tissue (arrow) is condensed against the body wall (photo by Roberta Poinar). (Scale bar = 2.3 μm.)*

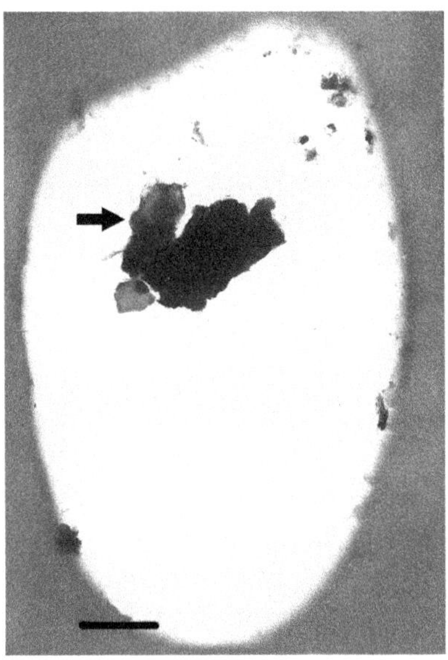

Fig. 4. *Transmission electron microscope view through an egg of* Oligaphelenchoides atrebora *Poinar, 1977, in Mexican amber showing the condensed internal tissue (arrow) (photo by Roberta Poinar). (Scale bar = 2.8 μm.)*

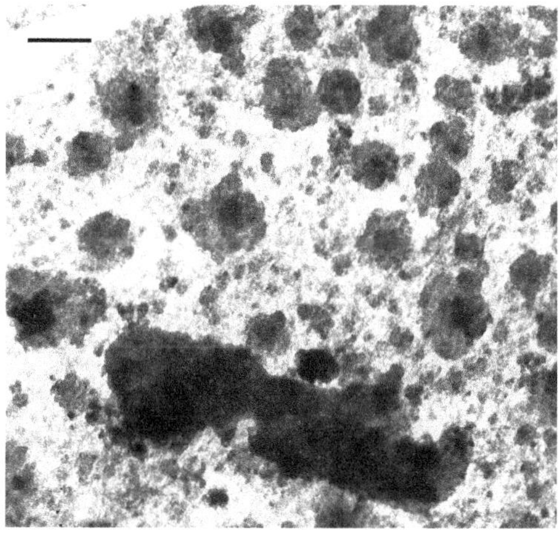

Fig. 5. *Transmission electron microscope photograph of the surface of the chitin-protein complex of the eggshell of* Oligaphelenchoides atrebora, *Poinar, 1977 in Mexican amber (photo by Roberta Poinar). (Scale bar = 2.2 μm.)*

high power dry microscope objective, this is often not possible when the nematodes are adjacent to insect hosts or plant tissue.

Dating of amber is somewhat controversial depending on the methods used by various researchers. For instance, Dominican amber was dated at 20-15 mya (million years ago) based on foraminifera (Iturralde-Vincent & MacPhee, 1996), but 45-30 mya based on coccoliths (Cepek *in* Schlee, 1990). Mexican amber is estimated to be 22-26 mya (Poinar, 1992), Baltic amber 40-50 mya (Larsson, 1978), Burmese amber 97-110 mya (Cruickshank & Ko, 2003) and Lebanese amber 130-135 mya (Poinar & Milki, 2001). A major difficulty in dating amber lies in the fact that it is usually secondarily deposited in sedimentary rocks so the actual age is older than determined, but just how much older is difficult to say. Geological ages mentioned in this work are taken from Granstein *et al.* (2004).

Rock deposits

Nematode remains in sedimentary deposits mostly occur as compression fossils with only the body outline and external ornamentation evident. However, in some fine-grained sedimentary deposits known as Lagerstätten, where conditions were especially favourable for preservation, sclerotised parts such as onchia, spicules and gubernacula remain and, occasionally, even details of the alimentary and reproductive systems can be detected.

In some instances, as with the Early Devonian Rhynie Cherts in Scotland, silica and other dissolved minerals impregnated the fossil-containing rocks. This permineralisation resulted in an astonishing degree of preservation that allowed us to examine cellular features of the Rhynie Chert nematode, *Palaeonema phyticum* Poinar, Kerp & Hass, 2008. The silica in these *ca* 400 million-year-old deposits originated from hot springs and geysers similar to those occurring today in the Yellowstone National Park in North America.

Trace fossils

Trace fossils are fossils showing the activity of an organism. Trace fossils attributed to nematodes are trails or tracks of their movements, which are frequently sinusoidal wave progressions formed when movement occurred on high viscosity surfaces. Sinuous, sinusoidal or travelling wave trails possibly made by nematodes have been recorded in Cambrian through Tertiary deposits. It is possible that trails made by the Carboniferous-Triassic form genus *Cochlichnus* Hitchcock, 1858 and the Palaeozoic-Cenozoic form genus *Gordia* Emmonds, 1844 were the result of nematode activity (Hantzschel, 1975). Eocene sinusoidal tracks in the Green River Formation, Utah, USA, have been attributed to nematodes (Moussa, 1969) but it is difficult to be certain. These fossils are discussed further under the geological periods in which they occur.

Coprolites

Analysis of lithified coprolites dating back millions of years (Fig. 6) and mummified dung of more recent ages (Fig. 7) can reveal the presence of vertebrate-parasitic nematodes. Ancient dung samples are one of the best resources for establishing early records of human nematode parasites, a subject discussed later in this work. Of course, paleoparasitologists who study parasites in human and other mummified dung must distinguish between true nematode parasites and those that entered the dung after it was deposited.

Fig. 6. *Lithified dinosaur coprolite. Ascarid eggs have been recovered from similar Early Cretaceous dinosaur coprolites in Belgium. (Scale bar = 6 mm.)*

Fig. 7. *Mummified shasta sloth dung. (Scale bar = 23 mm.)*

Nematology Monographs & Perspectives, 2011, Vol. 9, 13-14

Systematic treatment of fossil nematodes

Extant nematodes that are immature or lacking diagnostic characters can be described in collective group genera, as defined in Article 42.2.1 of the XV International Code of Zoological Nomenclature. If and when additional material with diagnostic characters is obtained, the species can be placed in an extant or new genus. Collective genera for extant nematodes are: *Agamomermis* Stiles, 1903 for immature mermithids, *Agamofilaria* Stiles, 1908 for immature Filarioidea in vertebrate tissues, *Microfilaria* Cobbold, 1880 for blood stages of Filarioidea, *Agamospirura* Henry & Sisoff, 1913 for immature Spiruroidea, and *Agamascaris* Steiner, 1924 for immature Ascaridomorpha.

Due to the inability to see diagnostic characters adequately, it is often impossible to assign a fossil nematode to an extant genus or family. This is why fossil collective genera are erected under the same guidelines as extant collective genera. Such collective genera have been nominated for many of the fossils proposed in the present work. Previously described fossil nematode collective genera are *Heydenius* Taylor, 1935 for Tertiary Mermithidae, *Cretacimermis* Poinar, 2001 for Cretaceous Mermithidae, *Oligoplectus* Taylor, 1935 for members of the family Plectidae and *Vetus* Taylor, 1935 for free-living nematodes not assignable to any known family.

Collective genera, with their respective species, are erected for taxonomic convenience and can be considered independent of the nomenclatural classification system used for extant forms. The assignment of rank is a matter of judgement, and while some of the genera and species can be accommodated within the present classification others have obscure relationships. Nevertheless, collective genera indicate the presence of certain lineages and associations at a particular time and place in the past. While type species are not required for collective genera, they are included in the present work since some of the obscure features may be elucidated with new techniques or additional material may appear.

Many of the descriptions presented here are based on morphological characters (body shape, length, body diameter, tail structure, *etc.*) of dauer juveniles. Behavioural features (*e.g.*, hosts, carriers, food source, *etc.*) are also used to relate fossils to modern taxa based on the principle of behavioural fixity, a feature that has been widely demonstrated with a range of fossil organisms (Boucot, 1990; Boucot & Poinar, 2010). It infers that the behaviour, ecology and climatic preferences of fossil organisms will be similar to those of their present-day descendants at the generic, and, in some cases, family, level. Systematic treatment of higher nematode categories in Appendix I and in the tables basically follows that of De Ley and Blaxter (2002, 2004) and Eyualem-Abebe *et al.* (2006).

Nematology Monographs & Perspectives, 2011, Vol. 9, 15-18

Nematodes from the Precambrian

We know that bacteria have existed for at least 3.5 billion years, which places them in the Archaean Eon of the Precambrian (Table 1). Protozoa have a fossil record extending back some 2 billion years to the Proterozoic Eon of the Precambrian. The fossil record of fungi, another source of nutrition for nematodes, begins in the Ordovician, some 460 mya (Taylor & Taylor, 1993), although molecular estimates

Table 1. *Geologic time table.*

Period/Epoch	Age (mya)	Nematode fossils and sources
Holocene	0.0115-present	Colombian copal
Pleistocene	1.81-0.0115	Earliest human nematodes
TERTIARY (CENOZOIC)		
Pliocene	5.33-1.81	
Miocene	23.03-5.33	Mexican and Dominican amber
Oligocene	33.9-23.03	Mexican and Dominican amber
Eocene	55.8-33.9	Baltic amber
Palaeocene	65.5-55.8	
MESOZOIC		
Cretaceous	145.5-65.5	Burmese and Lebanese amber
		Captivonema cretacea
		Palaeoenoploides alessandrellianus
Jurassic	199.6-145.5	*Eophasma jurassica*
Triassic	251.0-199.6	
PALAEOZOIC		
Permian	299.0-251.0	
Carboniferous	359.2-299.0	*Nemavermes mackeei*
Devonian	416.0-359.2	*Palaeonema phyticum*
Silurian	443.7-416.0	
Ordovician	488.3-443.7	
Cambrian	542.0-488.3	
PROTEROZOIC	2500-542.0	
ARCHEAN	2500 and earlier	

Fig. 8. *The Precambrian nematode-like fossil* Grypania spiralis *(Walcott, 1899) in Walter* et al., *1976 (their Figure 10 on Plate 2, p. 876). This species was originally described as* Helminthoidichnites spiralis *by Walcott (1899). (Scale bar = 3 mm.)*

Fig. 9. *The Precambrian nematode-like fossil* Tawuia dalensis *Hofmann & Aitken, 1979 (their Figure 13E, p. 161). (Scale bar = 4 mm.)*

of fungal phylogeny place the origin of terrestrial fungi (Ascomycota, Basidiomycota and Glomales) in the Proterozoic (Redecker *et al.*, 2000).

The big question remains – when did the first nematodes appear? If nematodes existed in the Proterozoic, some 0.542 to 2.5 billion years

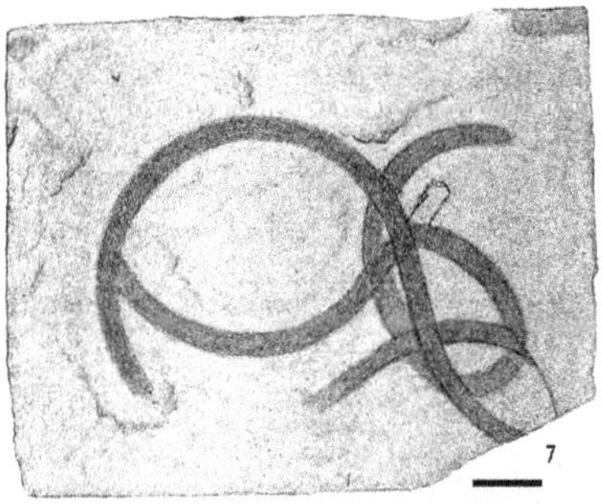

Fig. 10. *The Precambrian nematode-like fossil* Helminthoidichnites meeki *Walcott, 1899 (his Figure 7 on Plate 24, p. 240). (Scale bar = 3.3 mm.)*

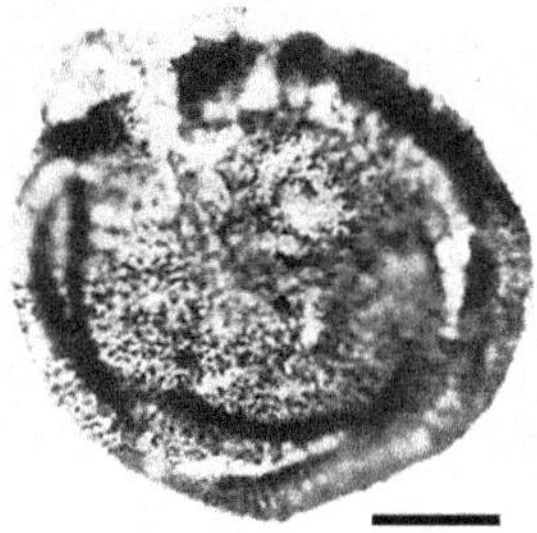

Fig. 11. *The Precambrian fossil* Trachysphaeridium laufeldi *Vidal, 1976 resembles a nematode egg (his Figure 21B on p. 37). (Scale bar = 8 μm.)*

ago (Table 1), there would have been an ample supply of microbes to sustain them, but would they have left a fossil record? If nematodes did originate in the marine environment, then there could be remains of them in Precambrian sediments.

Some elongate, marine, worm-like objects from the Proterozoic that lack appendages and body segments could possibly be nematodes. One of these is *Helminthoidichnites spiralis* Walcott, 1899, which was later re-described as *Grypania spiralis* by Walter *et al.* (1976) (Fig. 8). Another is *Tawuia dalensis* Hofman & Aitken, 1979, which reached up to 77 mm in length and 6 mm in diam. (Fig. 9). This species had a circular mouth opening and a central core that could represent the alimentary tract (Hofman & Aitken, 1979).

The large *Helminthoidichnites meeki* Walcott, 1899 (Fig. 10) from the Greyson Shale in Montana was 55 mm long and 2 mm in diam. with a uniformly curved axis (Walcott, 1899; Walter *et al.*, 1976). The spherical fossil, *Trachysphaeridium laufeldi* Vidal, 1976, resembles nematode eggs (Fig. 11). While these fossils are intriguing, there is not enough detail present to assign them definitely to the Nematoda.

Nematology Monographs & Perspectives, 2011, Vol. 9, 19-25

Nematodes from the Palaeozoic

Much continental area was covered by the sea during the Cambrian, which began 542 mya, and marine nematodes were probably widespread, feeding on bacteria, fungi and protozoa among decomposing organic matter along the bottom of the sea or in brackish inlets.

Predaceous nematodes, as well as those that established symbiotic associations with bottom-dwelling invertebrates, were probably also present. Taking the well known fauna of the Burgess Shale (530 mya) as an example of marine life occurring during that period, possible nematode habitats could have included the following: the gill chambers of the small, snail-like *Opabinia regalis* Walcott, 1912 or the tiny *Marrella splendens* Walcott, 1912 or among the body setae of the polychaete worm, *Canadia spinosa* Walcott, 1911. Nematodes could have even inhabited the burrows of the priapulid worms, *Louisella pedunculata* Walcott, 1911 and *Ottora prolifica* Walcott, 1911 (a portion of the gut of *O. prolifica* shows what could be a coiled nematode) (Whittington, 1985, see Fig. 4.21).

Additional habitats could have been in the gut or on the body surface of slow-moving, bottom-feeding detritivores such as trilobites (Fig. 12), the gregarious small crustacean *Canadaspis ovalis* (Walcott, 1912), or the slug-like *Amiskwia sagittiformis* Walcott, 1911. Some nematodes may have developed in the egg nests of the above groups. Members of the widespread nematode family Monhysteridae are not only free-living in the sea, but also occur on the gills, under the body plates and in the body cavities of various crustaceans (Poinar *et al.*, 2009). Some forms even appear to burrow into the shell of horseshoe crabs (Leibovitz & Lewbart, 2004), organisms that have a fossil record extending back to the Silurian.

The diverse habitats found during the Cambrian (marine, brackish water, freshwater, terrestrial) probably led to nematode diversification and associations were established with other organisms. Certainly by the Ordovician, when the first terrestrial plants appeared, nematodes were exploring the land, and by the Late Silurian, they were forming symbiotic and parasitic associations with the earliest known terrestrial

Fig. 12. *Nematodes could have lived between the segments or in the alimentary tract of Cambrian trilobites similar to these Devonian* Phacops megalomanicus *(Struve, 1982).*

invertebrates (arachnids, eurypterids and myriapods) (Jeram *et al.*, 1990).

Reproducing populations of *Palaeonema phyticum* Poinar, Kerp & Hass, 2008 in the stomatal chambers of the land plant, *Aglaophyton major* (Kidson & Lang, 1920) preserved in the Rhynie Chert are evidence that nematodes had formed trophic associations with plants by the Lower Devonian (396 mya) (Figs 13-16; Plate 1). These early plant parasites reproduced within substomatal cavities, intercellular spaces and cavities containing disrupted cortical tissue, as well as inside the cortex cells of *A. major*. The nematodes probably entered the stomatal openings much as present-day plant-parasitic nematodes enter the stomata of angiosperms. In order to obtain nourishment, *Palaeonema* used its buccal teeth to crush cortical plant cells.

The environment around *Palaeonema* was quite harsh with a number of hot springs and geysers. Periodic floods resulted in hot mineral water inundating the minute *Aglaophyton* plants with their nematode fauna. Aside from these physical conditions, *Palaeonema* had to contend

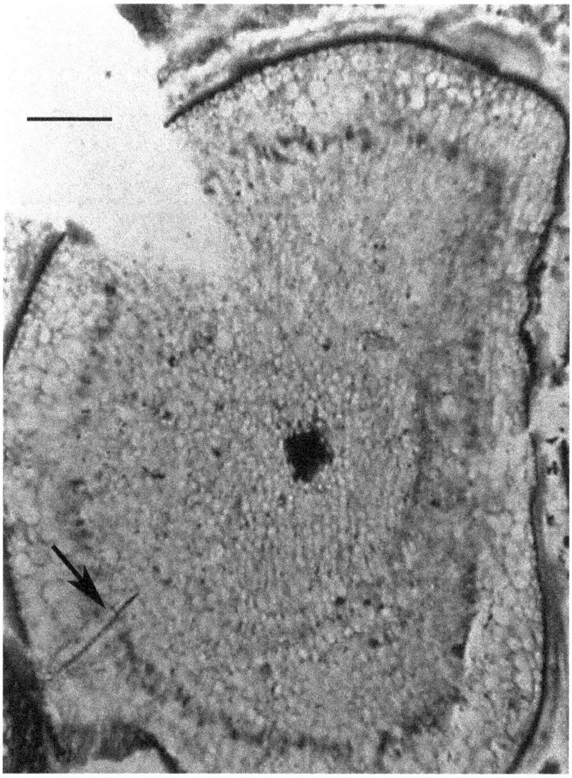

Fig. 13. *Holotype female* Palaeonema phyticum *Poinar, Kerp & Hass, 2008 (arrow) in the 'stem' of the primitive land plant* Aglaophyton major *in Rhynie Chert, Aberdeen, Scotland. (Scale bar = 460 μm.)*

with an assortment of predators, including mites, springtails and spider-like trigonotarbids. With these predators sharing the same habitat, it is unlikely that *Palaeonema* would have survived for long as a free-living microbotroph.

The type of parasitism demonstrated by *Palaeonema*, *i.e.*, using the buccal teeth to disrupt the plant tissues, still occurs today in some nematodes of the families Panagrolaimidae and Cephalobidae. These dyssaprobes utilise their sclerotised stoma mechanically to enter healthy plants and feed on root, stem and leaf tissue (Christie, 1933; Filipjev & Schuurmans Stekhoven, 1959; Paramonov, 1962). *Palaeonema* represents the earliest known herbivore, since previously the assemblage of animals at all Silurian-Devonian terrestrial sites consisted of predators

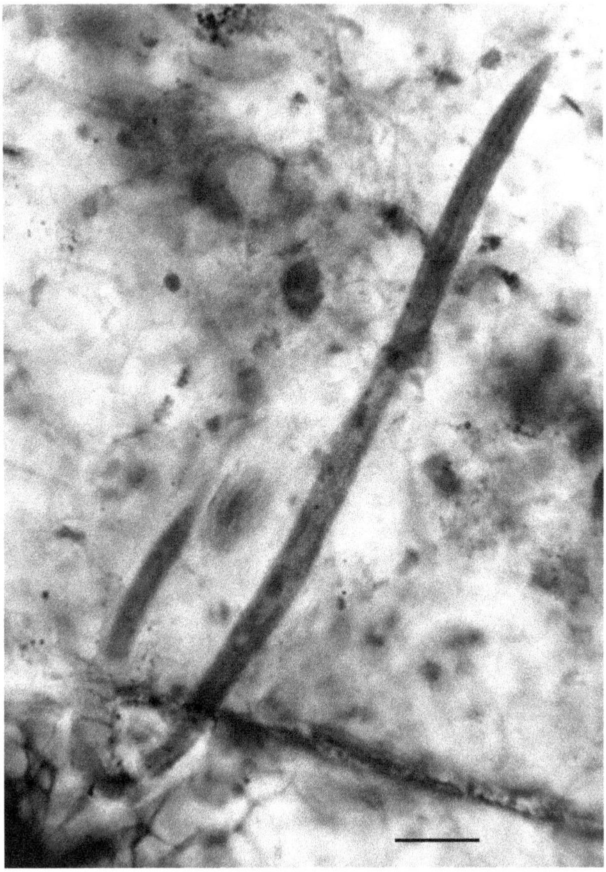

Fig. 14. *Holotype female* Palaeonema phyticum *Poinar, Kerp & Hass, 2008 surrounded by cortical cells of* Aglaophyton major *in Rhynie Chert, Aberdeen, Scotland (see also Plate 1). (Scale bar = 59 μm.)*

and detritivores, and those trophic systems were considered to pre-date the evolution of herbivory (Selden & Nudds, 2004).

The first definitive fossil marine nematodes date from the uppermost Mississippian of the Carboniferous (320 mya). The large bodied *Nemavermes mackeei* Schram, 1973 (Schram, 1973, 1979) (Figs 17, 18) lived in an offshore marine environment now known as the Mazon Creek Formation in northern Illinois, USA. The fossils were part of the Essex biota, which consisted predominately of marine organisms (Nudds & Selden, 2008). Originally the area was a swamp forest that was periodically inundated by the sea and maybe also by brackish waters from

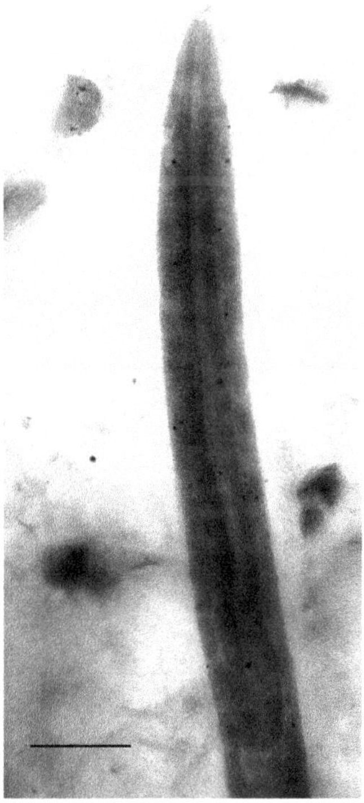

Fig. 15. *Anterior portion of holotype female* Palaeonema phyticum *Poinar, Kerp & Hass, 2008 in Rhynie Chert, Aberdeen, Scotland. (Scale bar = 12 μm.)*

estuaries. This resulted in a layer of marine fossils overlying coal seams containing terrestrial biota.

The large size of *Nemavermes* Schram, 1973 (100-140 mm long and 5-6 mm in diam.) and the long sparse setae positioned over its entire body are features found in a few extant marine nematodes, but none matches the description of *Nemavermes*. A separate family is established for this fossil in the present work.

The elongate processes (Fig. 18) protruding from the front end of *Nemavermes* are curious. While it was suggested that they were labial papillae, they are much longer than any known labial papillae and may have been used to probe the bottom layers for food. The detritus-filled gut of several specimens indicates that *Nemavermes* was a deposit feeder. The genus had a wide distribution and also occurred in Middle Pennsylvanian deposits some 310 mya (Schram, 1973). It survived for at

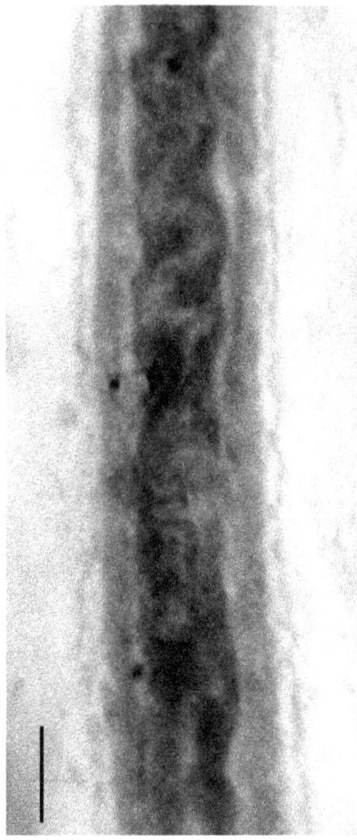

Fig. 16. *Mid-body portion of a juvenile* Palaeonema phyticum *Poinar, Kerp & Hass, 2008 in Rhynie Chert, Aberdeen, Scotland. (Scale bar = 7 μm.)*

least 10 million years in near-shore marine environments with possible freshwater inflows.

Fig. 17. *The Middle Pennsylvanian marine nematode* Nemavermes mackeei *Schram, 1973 from the Mazon Creek coal beds in northern Illinois (photo courtesy of J. Schram). (Scale bar = 6 mm.)*

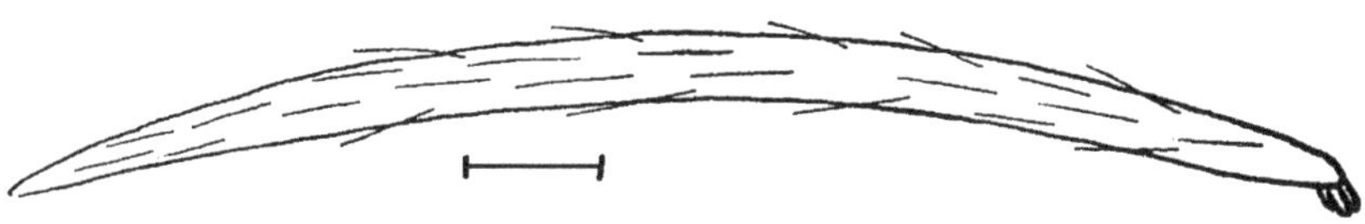

Fig. 18. *Drawing of the Middle Pennsylvanian marine nematode* Nemavermes mackeei *Schram, 1973 from the Mazon Creek coal beds in northern Illinois (drawing courtesy of J. Schram). (Scale bar = 12 mm.)*

Nematology Monographs & Perspectives, 2011, Vol. 9, 27-52

Nematodes from the Mesozoic

Jurassic nematodes

A marine detritivore from a Jurassic deposit in Italy is *Eophasma jurasicum* Arduini, Pinna & Teruzzi, 1983 (Fig. 19; Plate 2A). The six specimens of this species were preserved as compression fossils in sedimentary deposits dating back 160 mya. Although smaller than *Nemavermes*, it is still a large species with a length of 35-55 mm and a diam. of 1.2-2.0 mm. With a dome-shaped head, the fairly elongate body gradually tapers to a pointed flexible tail. The cuticle is coarsely annulated and the head and tail regions are covered with short setae.

Fig. 19. *The Jurassic* Eophasma jurasicum *Arduini, Pinna & Teruzzi, 1983 from Como, Italy (see also Plate 2A). (Scale bar = 4 mm.)*

Denticles were noted in the buccal chamber of some specimens and a valve-like structure separated the short pharynx from the dark-stained intestine. Based on its coarse cuticular annulations the specimen was tentatively placed in the Desmoscolecida. *Eophasma* was probably a free-living selective detritivore that also included small invertebrates in its diet. It shared its habitat with algae, cephalopods, echinoderms, polychaetes, crustaceans and fish (Arduini *et al.*, 1983).

Cretaceous nematodes

Cretaceous nematodes have been recovered from both aquatic (marine and freshwater) and terrestrial environments. In a freshwater swamp in Spitsbergen, Norway, some 120 mya in the Early Cretaceous, an aquatic oligochaete was forming an egg sac (cocoon) from the glandular cells of its clitellum. As the secretions were covering the egg sac, they impacted a small co-habiting nematode and both were fossilised together in a lignitic deposit. The enoplid nematode was described as *Captivonema cretaceae* Boström *in* Manun, Bose, Sayer & Boström, 1994. It was fairly small, measuring approximately 500 μm in length and 20 μm in diam. (Figs 20, 21; Plate 2B). Tapering setae up to 25 μm long covered most of its body. The cuticle was finely annulated and in places showed longitudinal striations similar to the lateral lines of extant nematodes. A gut was clearly visible, although the anus and details of the mouth were not evident. *Captivonema* was assumed to be a detritivore feeding on decomposing plant matter in the swamp.

Some 80 mya in the Late Cretaceous, shallow marine basins formed over portions of the Tethyan Sea that covered much of the Arabian Peninsula. In these shallow areas, fine-grained sediment settled over expired crustaceans, annelid polychaetes, ammonites and octopi. The resulting fossiliferous limestone beds then became exposed in north-west Lebanon between Beirut and the coastal city of Jbail. Among this assemblage were 172 specimens of a predatory nematode ranging from 8 to 46 mm in length and 0.3 to 2.3 mm in diam. (Figs 22, 23). This nematode was described as *Palaeoenoploides alessandrellianus* Mattavelli & Bracchi, 2008 and placed in the Enoplidae. What is interesting about this fossil is the shape of the buccal armature (Fig. 23). The disarticulated buccal teeth are composed of sclerotised elements ranging between 700 and 800 μm in length. It appears that the living

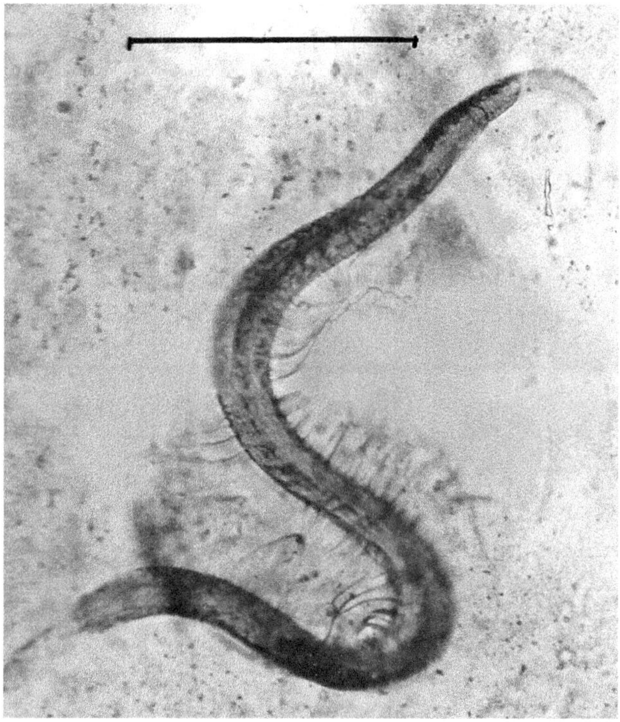

Fig. 20. *The Early Cretaceous* Captivonema cretacea *Boström* in *Manun, Bose, Sayer & Boström, 1994 from a freshwater swamp in Spitsbergen, Norway (photo courtesy of Svein B. Manun) (see also Plate 2B). (Scale bar = 120 μm.)*

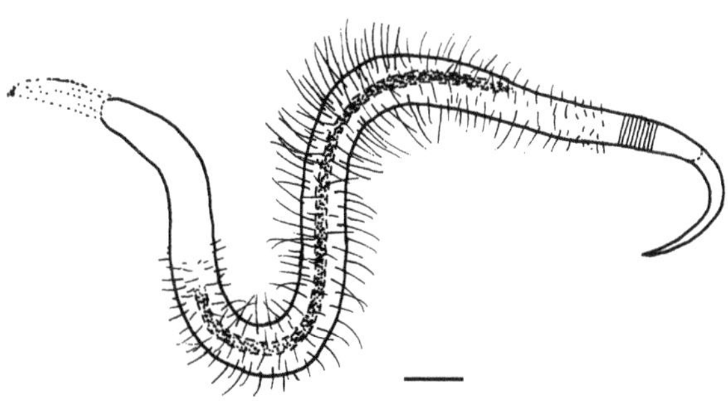

Fig. 21. *Drawing of the Early Cretaceous* Captivonema cretacea *Boström* in *Manun, Bose, Sayer & Boström, 1994 from a freshwater swamp in Spitsbergen, Norway (drawing courtesy of Svein B. Manun). (Scale bar = 20 μm.)*

Fig. 22. *The Late Cretaceous predatory enoplid* Palaeoenoploides alessandrellianus *Mattavelli & Bracchi, 2008 from marine deposits in Lebanon. Specimen in the collections of the Museo di Storia Naturale di Milano (photo courtesy of Anna Alessandrello). (Scale bar = 1 mm.)*

animal possessed three large teeth, one dorsal with a double hook and two ventral, each with a single hook. The spicules on 18 specimens were also preserved. Based on the large, sturdy buccal teeth, it is apparent that *P. alessandrellianus* was a predator. However, as these nematodes moved through the debris on the bottom of the sea in search of food, they also became prey since their remains were found inside the gut of a polychaete worm, as well as in coprolites of fish and crustaceans.

Lebanese amber nematodes

Cretaceous terrestrial and freshwater nematodes occur in amber. The oldest amber deposit in which nematodes have been found is Lebanese or Middle East amber ranging from 130 to 135 million years in age (Poinar & Milki, 2001). The resin originated from araucarians in a forest that stretched from Lebanon and Israel through Jordan all the way to the Persian Gulf. When the forest existed, that area of the Arabian Peninsula was still attached to Africa as part of the super-continent Gondwanaland.

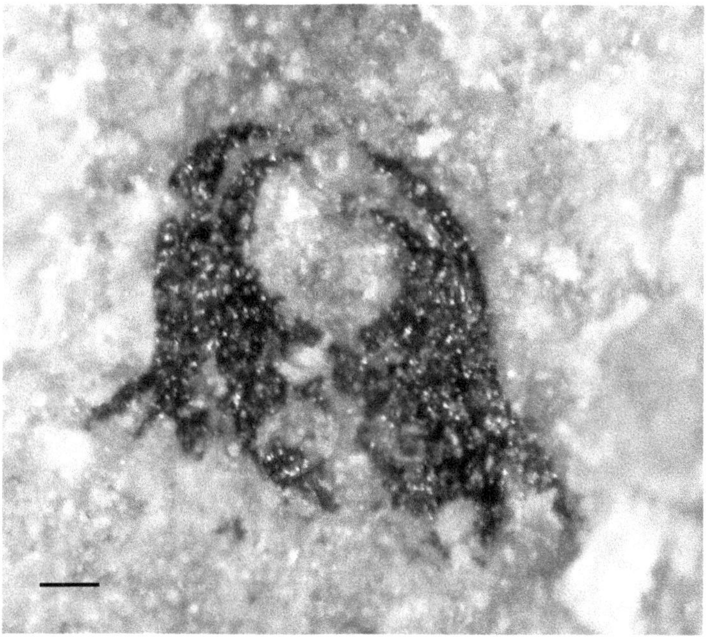

Fig. 23. *Three large, one double and two single hooked teeth of the marine nematode* Palaeoenoploides alessandrellianus *Mattavelli & Bracchi, 2008. Specimen in the collections of the Museo di Storia Naturale di Milano (photo courtesy of Anna Alessandrello). (Scale bar = 123 μm.)*

While India, Antarctica and Australia had already separated, Africa and South America were still contiguous. Thus all life forms in Middle East amber are endemic to the southern continents.

While a pair of nematodes (Fig. 24), described herein as *Vetus libani**, are the only free-living nematodes from these deposits, the most interesting nematode fossil is *Cretacimermis libani* (Poinar, Acra & Acra, 1994), the oldest representative of the family Mermithidae (Poinar *et al.*, 1994) (Fig. 25; Plate 3). There can be no doubt of its parasitic status since the single specimen is coiled up in the body cavity of the dipteran host, an adult midge of the family Chironomidae. This host-parasite association has persisted up to the present. Chironomid midges commonly undergo their larval development in aquatic habitats and *C. libani* probably infected the larval stage of the midge and then was carried through the pupal and into the adult stage, similar to the life cycle of the extant midge parasite, *Gastromermis anisotis* Poinar & Poinar, 2003.

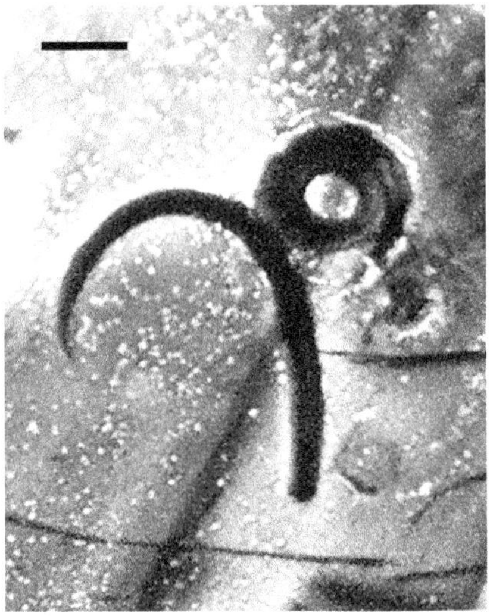

Fig. 24. *Two specimens of the free-living nematode* Vetus libani* *in Lower Cretaceous Lebanese amber (Milki amber collection, Beirut, Lebanon). (Scale bar = 35 μm.)*

BURMESE AMBER NEMATODES

Some 30-35 million years after Lebanese amber had formed, a Laurasian amber deposit formed in Myanmar (Burma). This amber was also produced from resin supplied by araucarian trees, but the life forms it contained were quite different from the Middle East deposit. Early Cretaceous Burmese amber (97-110 mya) contained a wide range of plant remains, including several angiosperms such as bamboos, dogwoods and rosaids, as well as cedars, dawn redwoods, sequoias, junipers and ferns. Insect inclusions in Burmese amber included blood-sucking insects vectoring vertebrate pathogens. There was a turnover of the insect fauna at that time with the disappearance of the established 'primitive' groups and the appearance of modern forms (Poinar & Poinar, 2008). A range of nematodes occurs in Burmese amber, including free-living microbotrophs, stylet-bearing mycetophages and invertebrate associates, with parasites of snails, midges, biting midges, craneflies and beetles.

Two dauer juveniles (Fig. 26) associated with a winged kalotermitid termite show that nematode-termite associations have existed for at least 100 million years. The Burmese amber termite that harboured the nematodes was a winged adult that had been attacked by a predator and

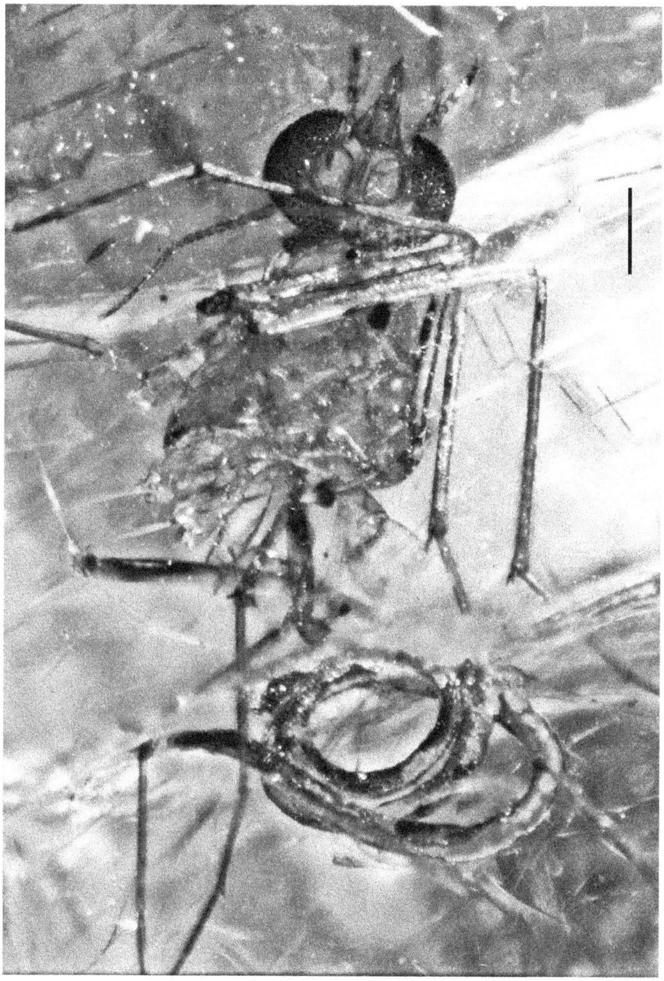

Fig. 25. Cretacimermis libani *(Poinar, Acra & Acra, 1994) inside a chironomid midge in Cretaceous Lebanese amber (Acra amber collection, Beirut, Lebanon) (see also Plate 3). (Scale bar = 34 μm.)*

lost a portion of its abdomen. This action exposed a number of mutualist protists that had been living in the termite's gut (Poinar, 2009a). It could not be determined whether the nematodes escaped from the gut or other internal organs, or had an external phoretic association with the termite. However, both nematodes were ensheathed, thus confirming their dauer state.

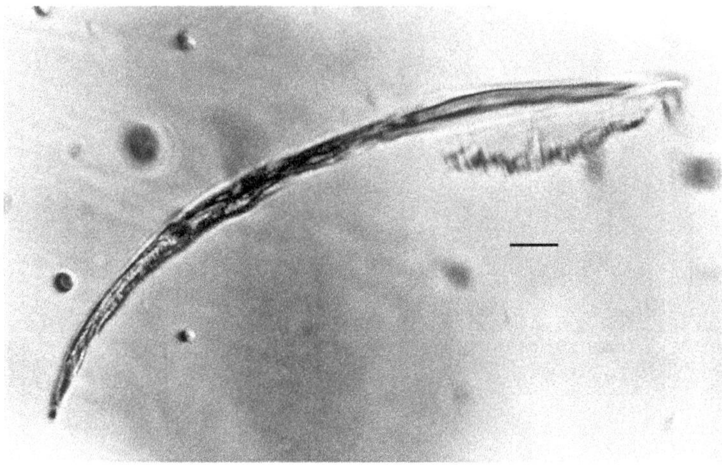

Fig. 26. *Dauer stage of* Cretodiplogaster termitophilus* *associated with the termite,* Kalotermes burmensis *Poinar, 2009, in Early Cretaceous Burmese amber (accession no. B-N-5). (Scale bar = 14 μm.)*

All extant nematodes that form dauer stages associated with termites belong to the Rhabditidae or Diplogastridae. In general, dauer stages of diplogastrids are smaller than those of the rhabditids, and the small size (195-215 μm) of these specimens, which are described herein as *Cretodiplogaster termitophilus**, falls within the range of extant dauer juveniles of diplogastrid termitophilous nematodes (Table 2). In extant termites, dauer stages of diplogastrids may occur in the mandibular glands and trachea (Fig. 27) or between the abdominal segments (Miller, 1989).

The adults of *C. termitophilus** probably occurred in the termite nests where they fed on bacteria and other microorganisms, and the two dauer climbed on the termite just before it became airborne. The nematodes would have continued their development when the termite established a new colony or remained on the carrier until it died and then fed on microbes associated with the cadaver.

As resin flows over the bark surface, it often collects small invertebrates that are feeding in or under the bark. This is probably how a small population of fungal-feeding aphelenchoidids, together with their fungal food source, became entrapped in Burmese amber (Figs 28, 29). Only 11 specimens were preserved, although these nematodes, described herein as *Cretaciaphelenchoides burmensis**, probably represented only a small

Table 2. *Dauer stages of Rhabditidae and Diplogastridae associated with ants and termites (all measurements in μm).*

Species	Length	Diam.	Reference
Diplogastridae			
Chroniodiplogaster aerivora (Cobb)	284-397	12-19	Poinar (1990a)
Chroniodiplogaster formosiana Poinar	247-293	12-17	Poinar *et al.* (2006a)
Pseudodiplogasteroides sp.	250-288	11-15	Miller (1989) (Fig. 27)
Pristionchus uniformis Fed. & Stan. (= *Isakis* sp.)	215-297		Fedorko & Stanuszek (1971)
*Filotermes minutus**	137-153	6-12	Present work (Fig. 124)
*Cretodiplogaster termitophilus**	195-215	9	Present work (Fig. 26)
Rhabditidae			
Pellioditis dolichura Schneider (= *Rhabditis janeti* Duthier)	360-480		Wahab (1962)
*Prescafilotermes dominicanus**	378-513	16-22	Present work (Fig. 128)
*Prescafilotermes elongatus**	654	29	Present work (Fig. 130)
Rhabditid in *Zootermopsis*	450	16	Present work (Fig. 131)
Rhabditis ossicula (Massey)	425	40	Massey (1971)
Rhabditonema fastidiosum (Massey)	538	25	Massey (1971)

portion of the original population. They ranged in length from 330 to 475 μm and in diam. from 11 to 17 μm. The males had tightly coiled tails while those of the females were nearly straight. Some resistant juvenile stages were also present, suggesting that this species had habits similar to some extant members of the family that are carried from tree to tree by wood-boring insects (Hunt, 1993). They represent the earliest fossil record of the Aphelenchoidea.

Aside from insects, there were other types of invertebrates associated with the Burmese resin-bearing araucarian trees. One of several snails in Burmese amber was associated with juvenile nematodes ranging from 410 to 473 μm in length and 21 to 26 μm in diam. and pre-infective stages 473 to 578 μm in length and 13 to 18 μm in diam. (Figs 30-32). Some of the latter were adjacent to, and probably emerging from, the mantle cavity of the snail (Fig. 32; Plate 4).

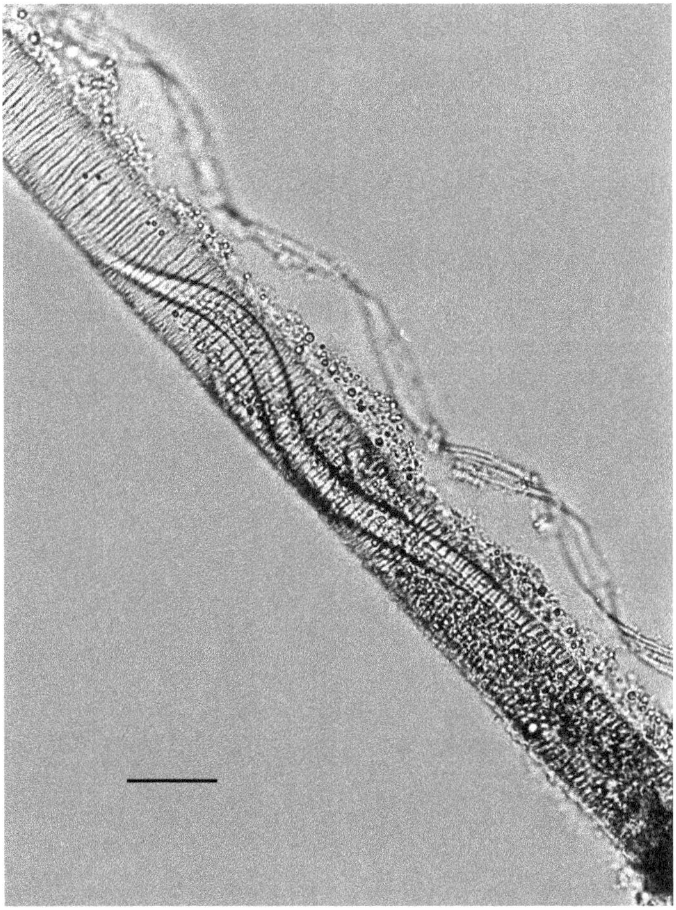

Fig. 27. *Extant diplogastrid dauer juvenile in trachea of* Mastotermes darwinensis *Froggatt in Australia (photo courtesy of L.R. Miller). (Scale bar = 39 μm.)*

Nematodes have various types of associations with terrestrial snails. Some use snails as carriers (internal and external phoresis) while others are facultative or obligate parasites. Nematode genera involved in these associations are *Alloionema* Schneider, 1859, *Aphelenchus* Bastian, 1865, *Aphelenchoides* Fischer, 1894, *Caenorhabditis* Osche, 1952, *Cephalobus* Bastian, 1865, *Cosmocercoides* Baylis, 1927, *Rhabditis* Dujardin, 1845, *Phasmarhabditis* Schneider, 1859 and *Plectus* Bastian, 1865 (Mengert, 1953; Coupland, 1995). Vertebrate-parasitic nematodes in the families Angiostrongyidae, Filaroididae and Skryabingylidae also

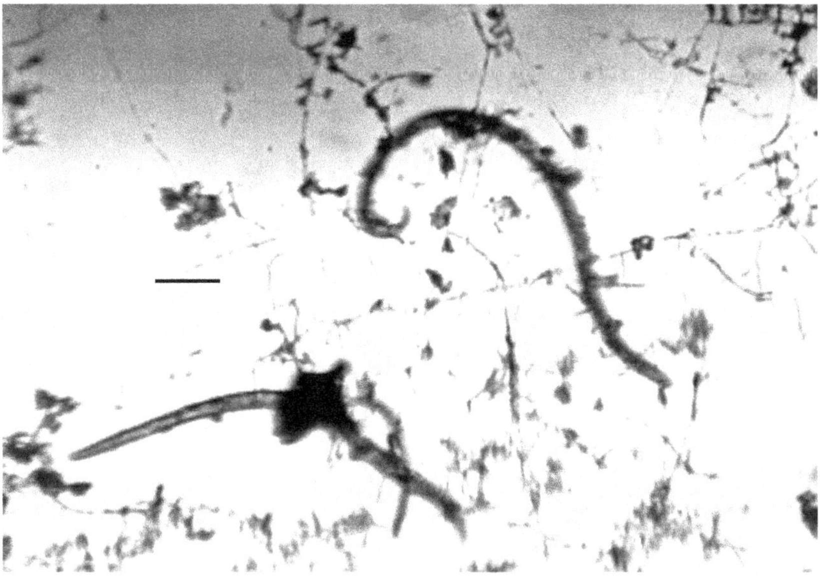

Fig. 28. *Group of the mycetophagous* Cretaciaphelenchoides burmensis* *in Early Cretaceous Burmese amber (accession no. B-N-7). (Scale bar = 56 μm.)*

utilise snails as intermediate hosts (Anderson, 2002), although the nematodes involved in these associations are usually juvenile stages encysted within the snail tissues. Based on the structure and habits of the juveniles and pre-infectives, the nematodes were placed in the family Cosmocercidae and described as *Palaeocosmocerca burmanica**.

The fossil species closely resembles the extant snail parasite *Cosmocercoides dukae* Anderson, 1960. The adults of *C. dukae* live in the intestinal tract of the snail and second-stage juveniles and third-stage infectives occur in the mantle cavity. Little is known about the nourishment of *C. dukae* but, since the pharynx is typical of microbotrophic nematodes, it is presumed to feed on microbes. A sibling species of *C. dukae* occurs in the rectum of frogs and toads (Harwood, 1930; Vanderburgh & Anderson, 1987), which shows a phylogenetic link between invertebrate and vertebrate nematode parasites of this lineage. It is difficult to determine whether the molluscs or amphibians were the original host of *Cosmocercoides* spp. although, if it were the latter, the presence of *P. burmanica** would suggest that this nematode lineage was already established in amphibians in the Early Cretaceous.

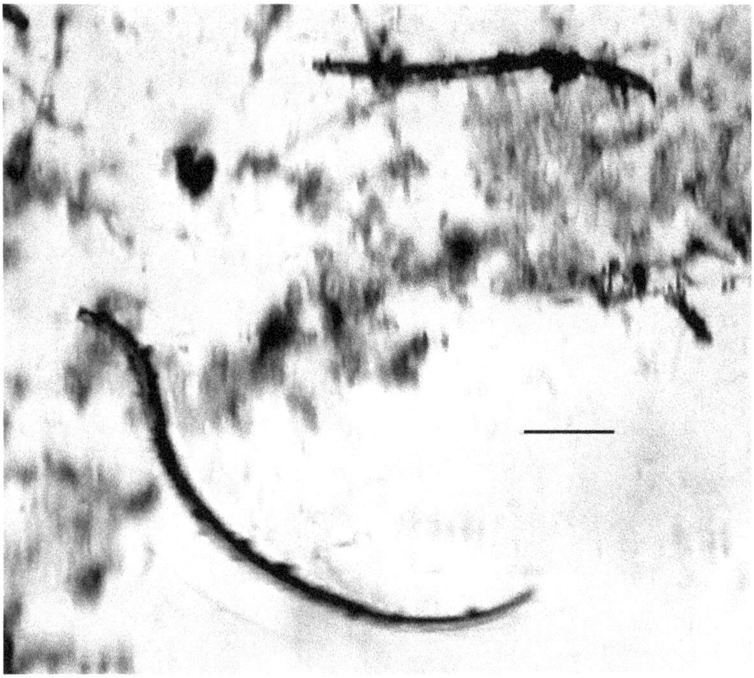

Fig. 29. *Two specimens of* Cretaciaphelenchoides burmensis* *in Early Cretaceous Burmese amber (accession no. B-N-7). (Scale bar = 56 μm.)*

Oxyurids are probably the first nematodes to have established parasitic associations with terrestrial animals, beginning with invertebrates and then expanding their host range to vertebrates. While the fossil record of oxyurids could, as parasites of millipedes, extend back to the Silurian, exactly when they first invaded insects is unknown. Most insect hosts are those in which either the adult or larval stages are detritivores or omnivores and thus have an opportunity to ingest the nematode eggs, which are deposited in the faecal matter of infected hosts. Extant craneflies are common hosts of thelastomatids (Poinar, 1975) and the parasite's eggs are presumably acquired by cranefly larvae.

A pre-adult male thelastomatid was discovered adjacent to an adult female cranefly in Burmese amber (Figs 33, 34; Plate 5). The nematode, described as *Paleothelastoma tipulae**, was 376 μm in length and 25 μm in diam. and possessed cuticular annulations in the anterior half of the body (Fig. 34). The pharynx was characteristic of those found in extant thelastomatids. The cranefly host was a species of *Gonomyia* Osten-

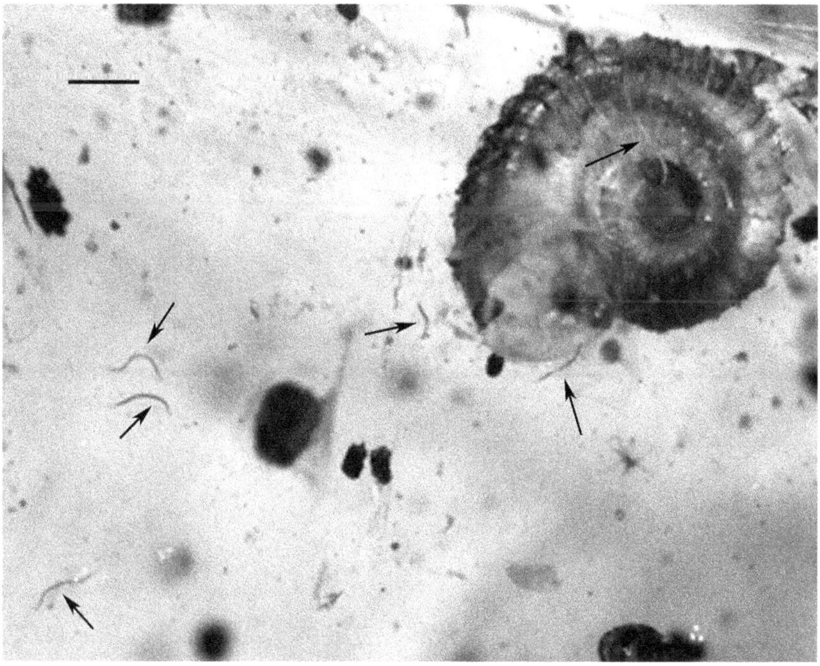

Fig. 30. *Several specimens (arrows) of the coscomercid* Palaeocosmocerca burmanica* *associated with a terrestrial snail in Early Cretaceous Burmese amber (accession no. B-N-6) (see also Plate 4). (Scale bar = 526 μm.)*

Sacken, 1869, the larvae of which are detritivores that develop in sandy or rich organic soil near water sources.

Three extant genera of thelastomatids occur in larval craneflies (Leibensperger, 1960; Dale, 1965; Jarry, 1969; Vojtková, 1971) but nematodes have never been collected from adult flies. However, this fossil shows how thelastomatids can be distributed by adult hosts. A detailed search may reveal oxyurids in extant adult craneflies, since these nematodes occur in adult hydrophilid and passalid beetles (Jarry, 1964; Skrjabin *et al.*, 1966).

Discovering the fossil remains of entomopathogenic nematodes (species of *Steinernema* Travassos, 1927 and *Heterorhabditis* Poinar, 1976) would appear to be nearly impossible. That is because both the infective and developmental stages, as well as the typical hosts which are the immature stages (larvae, pupae) of holometabolous insects, normally live in the soil. In addition, the insects die within 48 h of becoming infected and therefore have little chance of coming into contact with

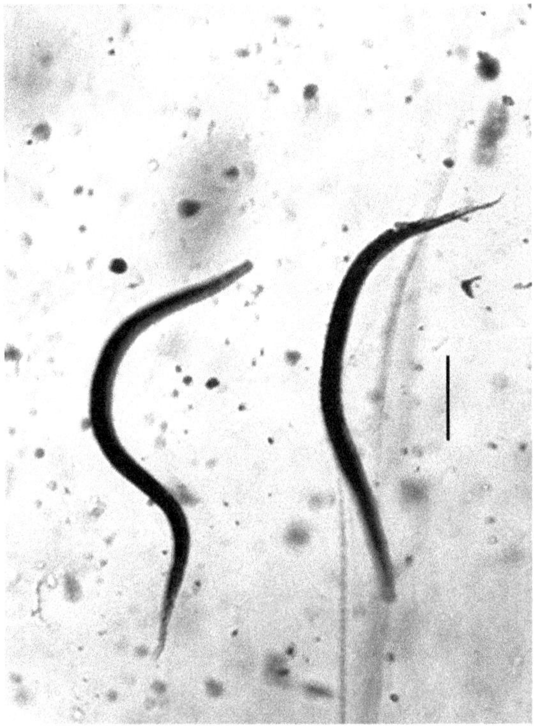

Fig. 31. *Two specimens of* Palaeocosmocerca burmanica* *in Early Cretaceous Burmese amber (accession no. B-N-6). (Scale bar = 88 μm.)*

any tree resin. Also, as soon as a new generation of infective stages is formed, the host cadaver is subject to rapid decomposition.

However, by a strange sequence of events, a beetle living in the Burmese amber forest became infected with a heterorhabditid and was then entrapped in resin. Most of the infective stages probably had already left the host but three were captured in resin. The infective stages were very similar to those of present-day heterorhabditids (Figs 35, 36). They were enclosed by their second-stage cuticles, which contained longitudinal striations and possessed a minute tooth at the tip of their heads. A similar tooth is used in extant heterorhabditids for entering the body cavity of potential hosts. The fossil specimens ranged from 485 to 533 μm in length, which is similar to the length (479-573 μm) of infectives of *Heterorhabditis indicus* Poinar, Karunakar & David, 1992. Further evidence that the fossil species, described herein as *Proheterorhabditis burmanicus**, belonged to the Heterorhabditidae was

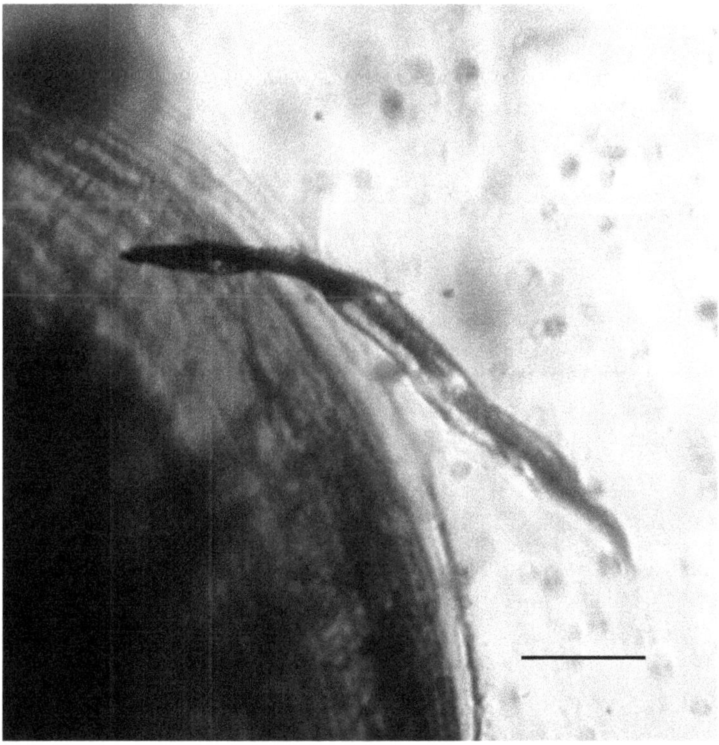

Fig. 32. *Specimen of* Palaeocosmocerca burmanica* *adjacent to the mantle cavity of a terrestrial snail in Early Cretaceous Burmese amber (accession no. B-N-6). (Scale bar = 111 μm.)*

the presence of rod-shaped bacteria ranging from 5.1 to 6.6 μm in length in the body cavity of the adjacent beetle (Fig. 37). These bacterial cells were similar in length to those of *Photorhabdus luminescens* (Thomas & Poinar, 1979), the symbiotic bacterium associated with extant species of *Heterorhabditis* (Poinar, 1979a) and which measure 5.2 (3.1-9.8) μm in size.

A possible scenario leading to the preservation of these fossils is that one or more infective stages of *Proheterorhabditis** entered the gut of the beetle when it was in the final stages of pupal development, either in soil or detritus at the base of the tree or possibly in a rotten area on the trunk. The beetle emerged and flew into a pool of resin at about the same time as the nematodes invaded the host's haemocoel and released their symbiotic bacteria. The beetle remained on the surface

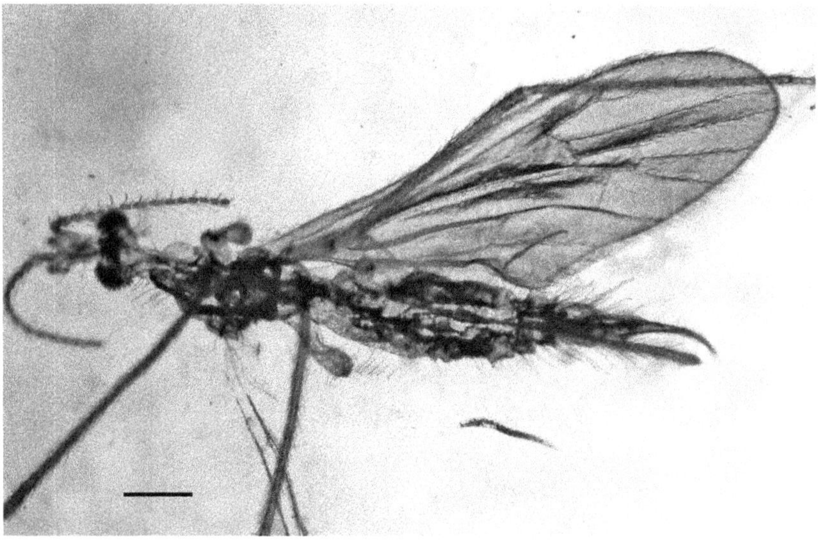

Fig. 33. *The thelastomatid* Paleothelastoma tipulae* *adjacent to its tipulid host in Early Cretaceous Burmese amber (accession no. B-N-4) (see also Plate 5). (Scale bar = 251 μm.)*

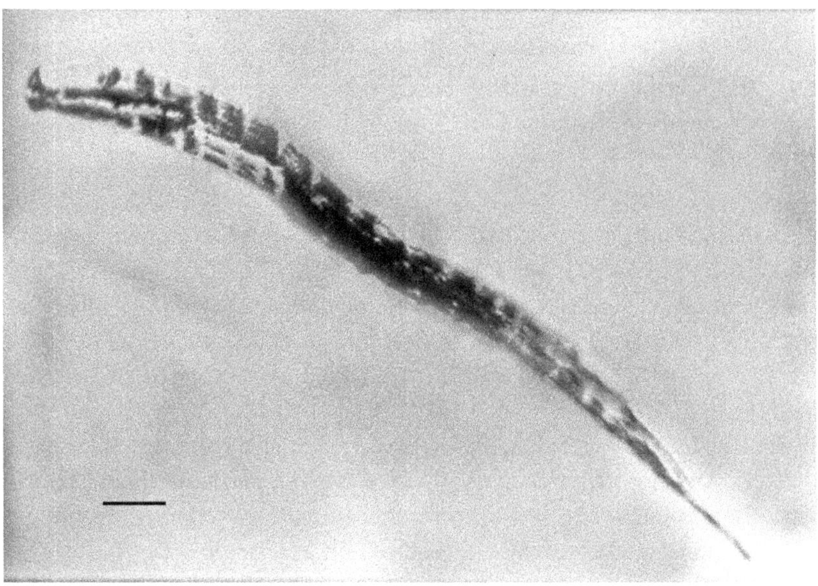

Fig. 34. *Detail of the thelastomatid* Paleothelastoma tipulae* *in Early Cretaceous Burmese amber (accession no. B-N-4). (Scale bar = 27 μm.)*

Fig. 35. *Specimen of* Proheterorhabditis burmanicus* *(arrow) adjacent to its beetle host in Early Cretaceous Burmese amber (accession no. B-N-8). (Scale bar = 152 μm.)*

of the resin for several days, long enough for the completion of at least one hermaphroditic generation, which produced a limited number of infective juveniles. A subsequent resin flow covered the beetle and entrapped the three nematodes as they were exiting the host. The beetle host has some characteristics of rove beetles, which develop in damp habitats such as under bark or among moss and fungi (Peterson, 1957). Both adult and larval stages of rove beetles are susceptible to infection by *Heterorhabditis* species today (Georgis *et al.*, 1991). Parasitism of adult hosts is an important means of distributing entomopathogenic nematodes (Poinar, 1979; Klein, 1993).

Several mermithid nematodes occur in Burmese amber. One is *Cretacimermis chironomae**, represented by two specimens associated with an adult chironomid midge (Grimaldi *et al.*, 2002). One of the

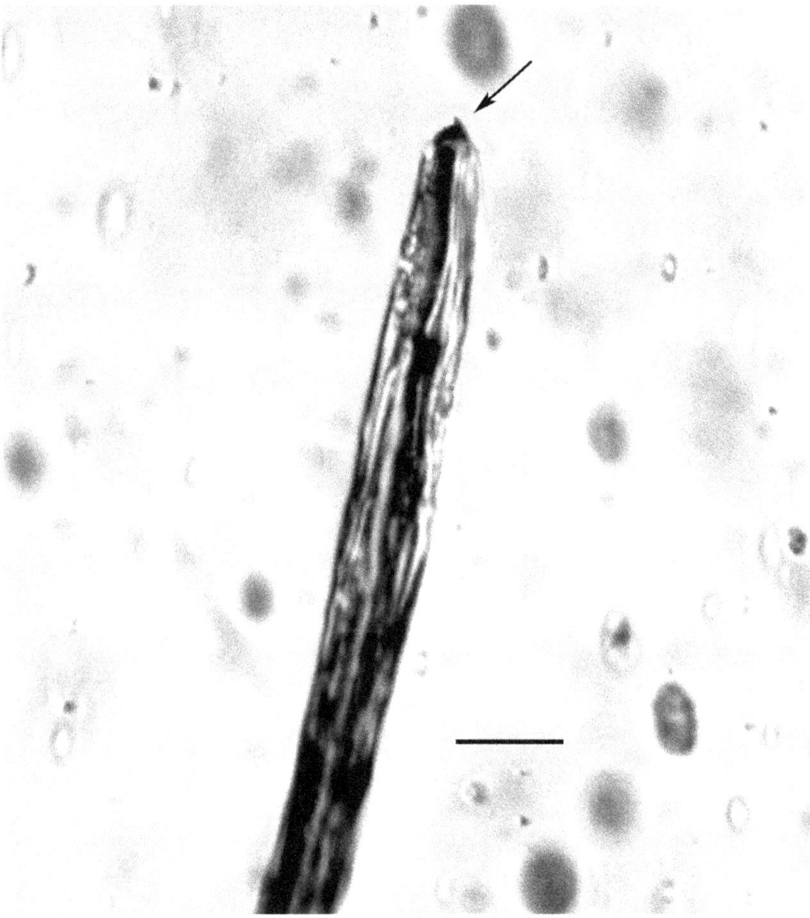

Fig. 36. *Detail of anterior portion of* Proheterorhabditis burmanicus* *in Early Cretaceous Burmese amber, showing tooth at tip of head (arrow) (accession no. B-N-8). (Scale bar = 25 μm.)*

nematodes had already emerged from the host, while the tail portion of the other is still inside the midge. These specimens, together with the older species from Lebanese amber, *C. libani*, show that mermithid parasitism of the Chironomidae occurred in Laurasia and Gondwanaland by the Early Cretaceous and possibly also existed when the continents were still contiguous as Pangaea, some 220 mya (see section on nematode origins). The life cycle of *C. chironomae** was probably very similar to that of the mermithids that infect chironomid midges

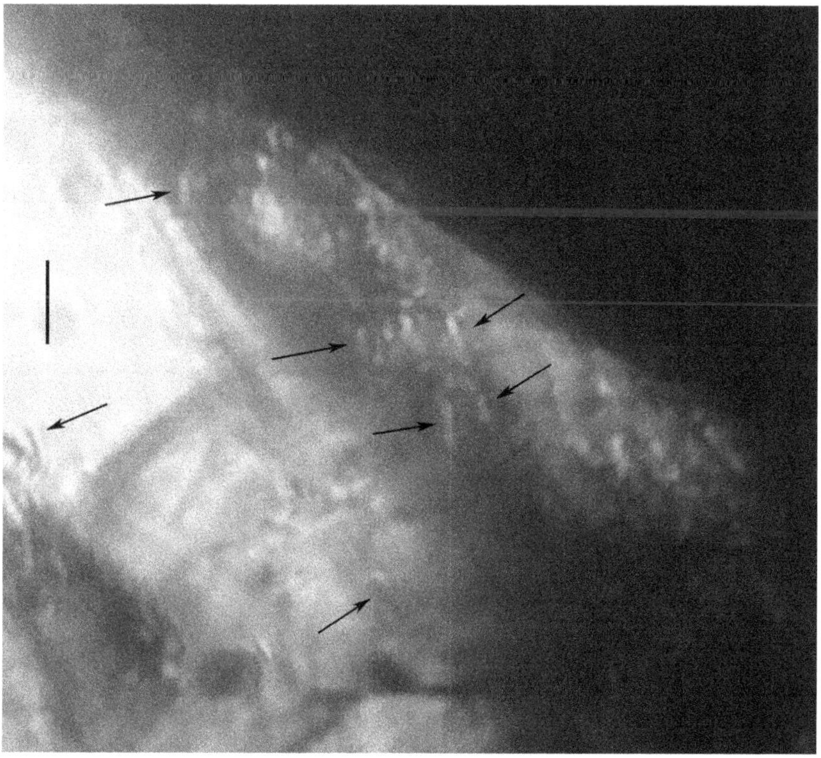

Fig. 37. *Rod-shaped bacteria (arrows) in the body cavity of the beetle host of* Proheterorhabditis burmanicus* *similar in size and shape to extant species of* Photorhabdus *Thomas & Poinar (accession no. B-N-8). (Scale bar = 10 μm.)*

today. Infection is initiated when the pre-parasitic juvenile mermithids penetrate through the body wall of host larvae. There is a period of rapid growth and the parasites usually exit from the host larvae. However, if the pre-parasites enter late instar larvae, they are often carried through the pupal stage and emerge from the adult midges, as occurs in the extant species, *Gastromermis anisotis* (Fig. 38). Once carried back to the aquatic environment by the host, the post-parasites of *C. chironomae** would have moulted, mated and deposited their eggs in debris at the bottom of the water source. Pre-parasitic juvenile mermithids emerging from eggs would have infected host larvae and the cycle would continue.

The mermithid *Cretacimermis protus* Poinar & Buckley, 2006 parasitised biting midges (Ceratopogonidae) in Burmese amber (Fig. 39; Plate 6). This species is represented by several specimens, including one

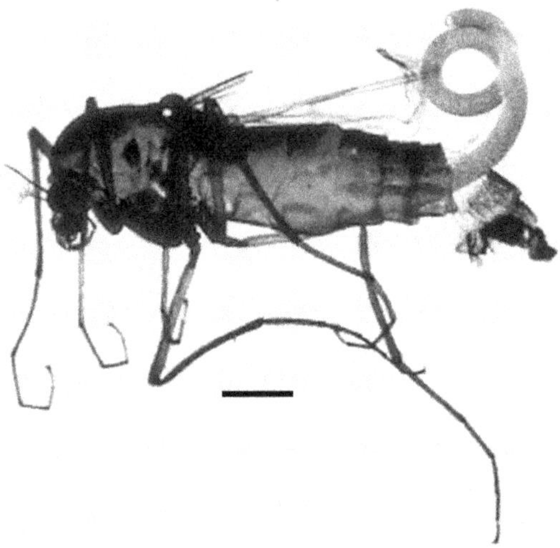

Fig. 38. *The extant mermithid* Gastromermis anisotis* *Poinar & Poinar, 2003 emerging from the chironomid* Cricotopus nostocicola *Wirth has a life cycle that probably was very similar to that of* Cretacimermis chironomae* *in Early Cretaceous Burmese amber. (Scale bar = 77 μm.)*

that had partially emerged from an adult biting midge of the genus *Atriculicoides* Remm, 1976 (Fig. 39), two additional specimens in another piece of Burmese amber adjacent to an adult biting midge of the genus *Leptoconops* Skuse, 1889 (Poinar & Monteys, 2008), and two individuals partially emerged from an unidentified biting midge (Fig. 40). The single mermithid from *Atriculicoides* Kemm, 1976 is 4.3 mm in length, suggesting it is a female, while the two specimens from *Leptoconops* Skuse, 1889 are 2.5 mm and 2.7 mm in length, respectively. Both of these latter specimens are probably males, which are usually smaller than females. The size ranges of *C. protus* compares favourably with those of the extant biting midge mermithid parasite, *Heleidomermis cataloniensis* Poinar & Monteys, 2008, whose females average 4.2 mm and males 2.2 mm in length. The life cycle of *Heleidomermis* Rubstov, 1970 is similar to that of *Gastromermis anisotis*, although the former genus possesses accelerated development and the final moult to the adult stage occurs in the host rather than in the environment (Poinar & Poinar, 2003). Surrounding cuticles and a male genital opening indicates that *C. protus* also possessed accelerated development, suggesting that this species belongs to the same lineage as the recent genus *Heleidomermis* (Fig. 41).

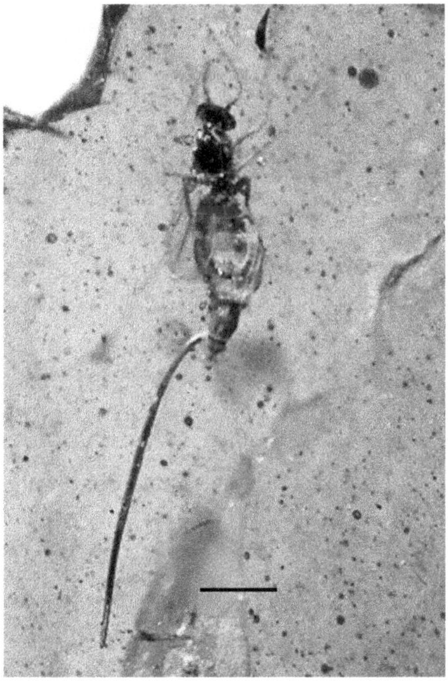

Fig. 39. *The mermithid* Cretacimermis protus *Poinar & Buckley, 2006 emerging from a biting midge in Early Cretaceous Burmese amber (Ron Buckley amber collection) (see also Plate 6). (Scale bar = 1 mm.)*

Further details on the ecology of nematodes in the Burmese amber forest can be provided by the hosts of these mermithids. Extant species of *Leptoconops* feed on the blood of mammals, birds and reptiles, and the larvae live in sand along coastal and inland beaches or in moist and often saline soil in dry locations. While the genus *Atriculicoides* is now extinct, it is likely that the females fed on vertebrate blood and the larvae occurred along the edges of water sources (Szadziewski & Poinar, 2005).

NEMATODES FROM CRETACEOUS COPROLITES

Medical parasitologists are quite familiar with the technique of analysing human faecal material to find traces of alimentary tract parasites. Thus, recovered helminth stages can indicate the cause of gasterointestinal diseases. Similar methods of extraction can be used to identify parasites in fossilised dung, or coprolites.

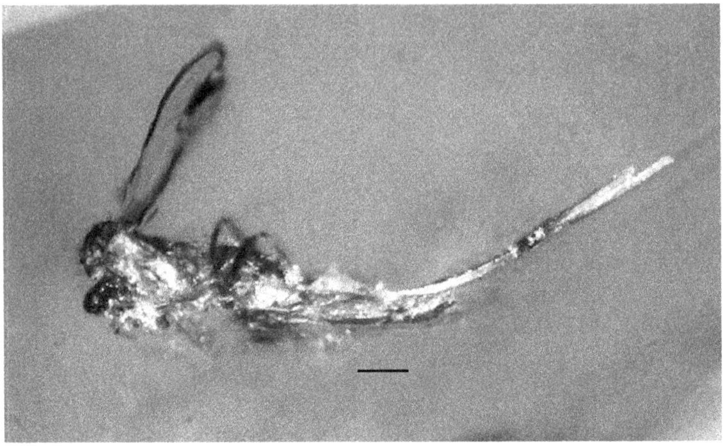

Fig. 40. *Two additional specimens of* Cretacimermis protus *Poinar & Buckley, 2006 emerging from a biting midge in Early Cretaceous Burmese amber (accession no. B-N-3). (Scale bar = 114 μm.)*

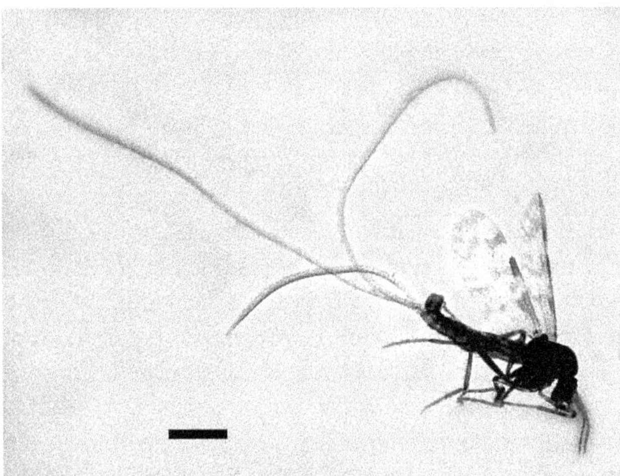

Fig. 41. *Mermithids (*Heleidomermis *sp.) emerging from an extant biting midge in Europe (specimen courtesy of Eustacio Tarasco). (Scale bar = 600 μm.)*

The analysis of a dinosaur coprolite from the Early Cretaceous Iguanodon shaft in Belgium revealed the presence of several ascarid eggs, these representing the oldest record of the Ascarididae and terrestrial nematode parasites of vertebrates. One egg that still had its outer mammillated surface was described as *Ascarites priscus* Poinar &

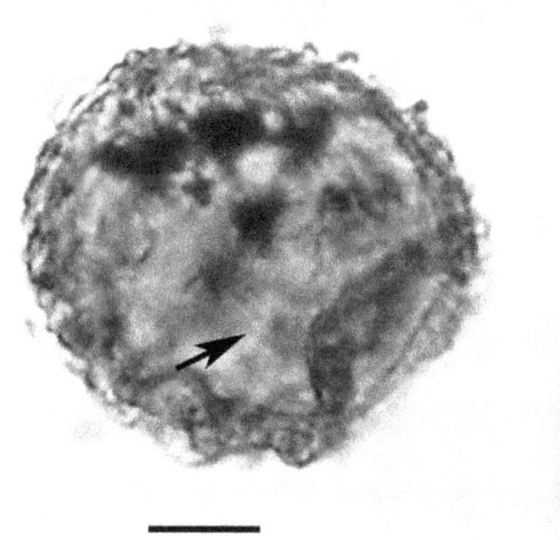

Fig. 42. *Egg of the early Cretaceous* Ascarites priscus *Poinar & Boucot, 2006 in a coprolite that was assigned to the predatory dinosaur* Megalosaurus *from the Bernissant Wealden Iguanodon locality in Belgium. Arrow shows developing juvenile. (Slide No. Y2 852 (2) deposited in Zoology Department, Oregon State University). (Scale bar = 11 μm.)*

Boucot, 2006 (Fig. 42). A second egg lacked this outer shell layer and was described as *Ascarites gerus* Poinar & Boucot, 2006 (Fig. 43). The eggs, which contain developing juvenile nematodes, are within the size range of those from extant species in the Ascarididae (Fig. 44). Between 50 and 100 species of ascarid occur in reptiles today, and, while those that parasitised dinosaurs are certainly extinct, they probably belonged to lineages found in reptiles and birds today.

Nematodes associated with Cretaceous foraminifera

A report of apparent nematode bore holes on extant foraminifera tests has been used to imply the presence of Cretaceous nematodes in the marine environment. Sliter (1971) reported borings up to 14 μm in diam. in the tests of the foraminifera *Rosalina globularis* d'Orbigny, 1826 and *Bolivina doniezi* Cushman & Wickenden, 1929 collected off southern California and maintained in the laboratory (Fig. 45). He noted that some of the foraminifera with borings contained internal nematodes. While the actual boring activity by nematodes was not observed, Sliter assumed

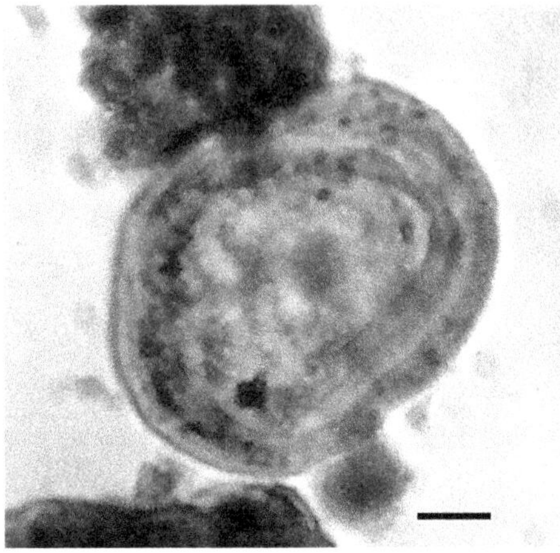

Fig. 43. *Egg of the early Cretaceous* Ascarites gerus *Poinar & Boucot, 2006 in a dinosaur coprolite that was assigned to the predatory dinosaur* Megalosaurus *from the Bernissant Wealden Iguanodon locality in Belgium (Slide No. Y2 852 (2) deposited in Zoology Department, Oregon State University). (Scale bar = 9 μm.)*

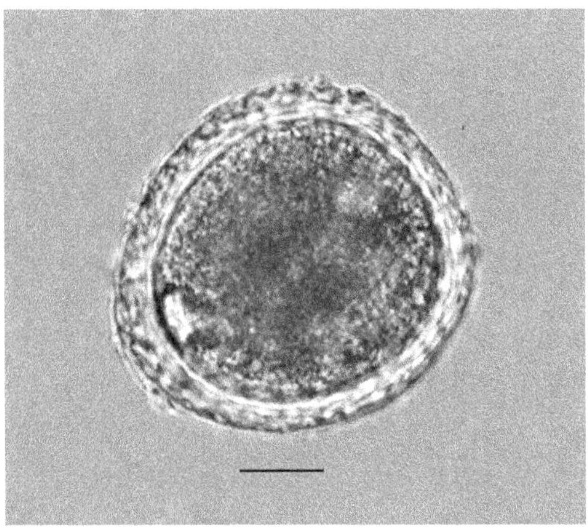

Fig. 44. *Extant egg of the human white stomach roundworm,* Ascaris lumbricoides *(L.). Note mammillated outer surface (see also Plate 32B). (Scale bar = 18 μm.)*

that the holes were made by nematodes in order to enter the foram cell. Identical borings occur in bathyal and neritic foraminifera from Cretaceous deposits (Fig. 46) (Sliter, 1971). If the extant borings were

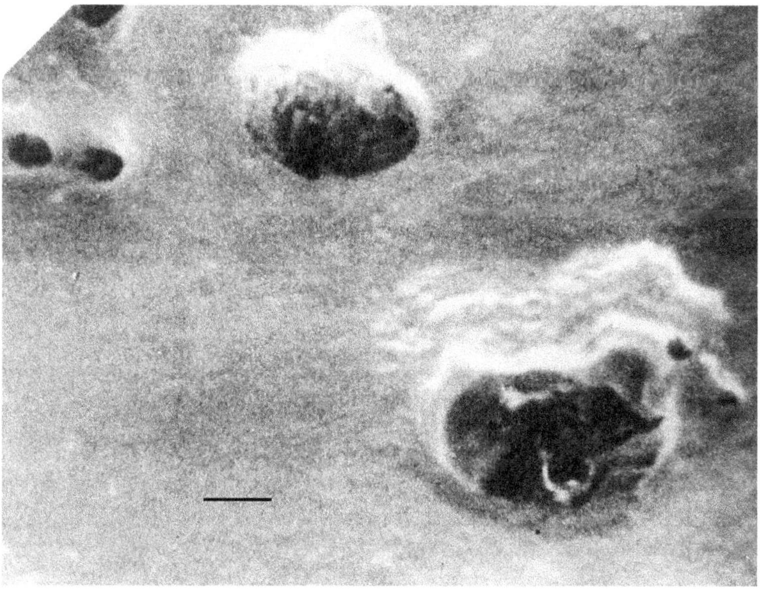

Fig. 45. *Bore hole in the extant foraminiferan* Rosalina globularis *d'Orbigny attributed to predaceous nematodes (from Sliter, 1971, Plate 1, Fig. 7). (Scale bar = 5 μm.)*

made by nematodes, then Cretaceous foraminifera with similar borings could be used indirectly to determine the distribution and abundance of select predatory nematodes. The malacologist Kabat (1990) concluded that the holes reported by Sliter were made by nematodes since the openings (less than 60 μm in diam.) were too small to be made by newly hatched predatory gastropods. Extant nematodes recovered from foraminifera tests include the leptosomatid, *Syringonomus typicus* Hope & Murphy, 1969, the leptolaimid, *Smithsoninema inaequale* Hope & Tchesunov, 1999 and reproducing populations of an undescribed species (Myers, 1943). No observations were made to determine if any of these reported nematodes bored through the test to enter the internal cavity of the protozoans, nor did the authors suggest any other method of entry.

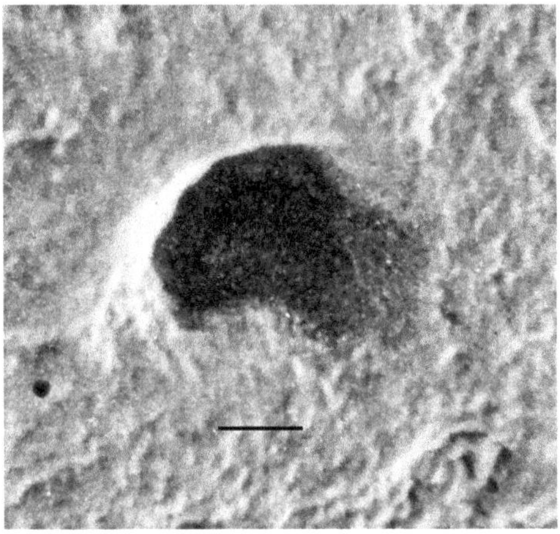

Fig. 46. *Bore hole in a Cretaceous foraminiferan attributed to predaceous nematodes (from Sliter, 1971, Plate 3, Fig. 12). (Scale bar = 50 μm.)*

Nematology Monographs & Perspectives, 2011, Vol. 9, 53-190

Nematodes from the Tertiary

Baltic amber nematodes

The Baltic amber forest, which flourished some 40-50 mya, contained a wide range of angiosperms and gymnosperms, including araucarians, which were probably the resin producers (Poinar, 1992). During the period that this forest existed, the climate changed from subtropical to warm temperate, thereby accounting for the diversity of life found in Baltic amber. During the cooling event there was apparently a southward migration, which explains why many of the invertebrate fossils in Baltic amber today have their closest descendants in Africa, Madagascar and Asia (Larsson, 1978).

A wide range of free-living and parasitic nematodes occurs in Baltic amber. Among them is *Oligoplectus succini* von Duisburg, 1862, representing the second description of a fossil nematode (Fig. 47). Plectids are common, widespread, microbotrophic nematodes that occur

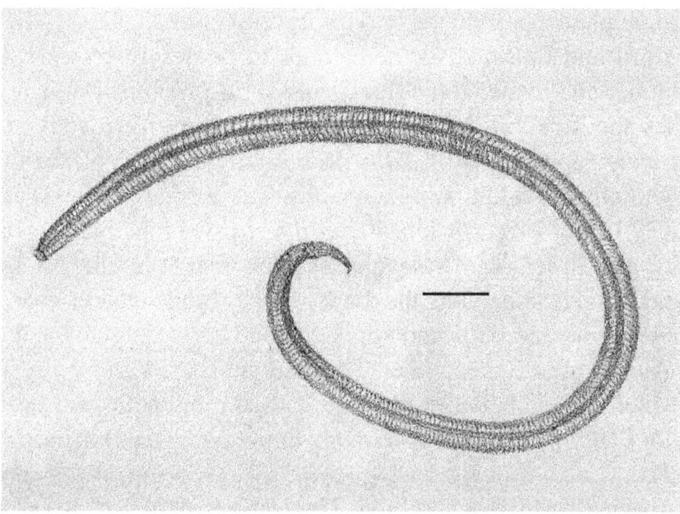

Fig. 47. *The Baltic amber* Oligoplectus succini *(von Duisburg, 1862) (modified from von Duisburg, 1862, Table 1, Fig. A2, opposite p. 35). (Scale bar = 428 μm.)*

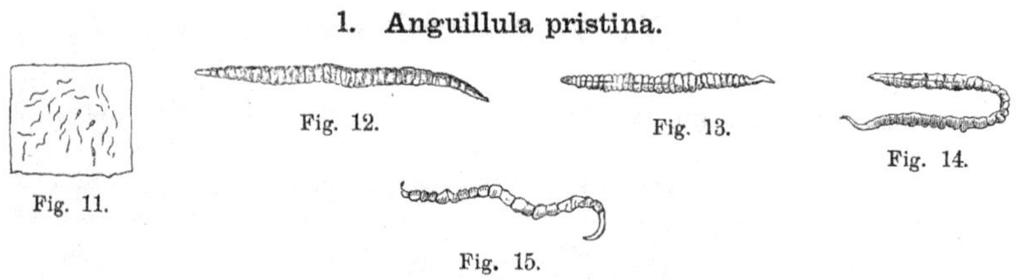

Fig. 48. *The Baltic amber* Vetus pristinus *(Menge, 1866) Taylor, 1935 (Menge, 1866, Figs 11-15, p. 7).*

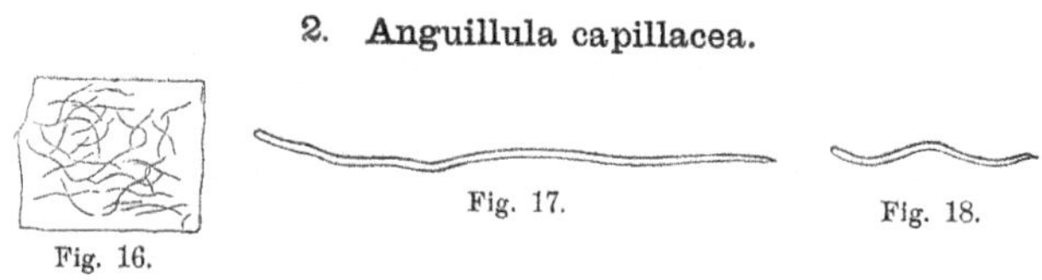

Fig. 49. *The Baltic amber* Vetus capillaceus *(Menge, 1866) Taylor, 1935 (Menge, 1866, Figs 16-18, p. 7).*

in freshwater, soil, moss and lichens (Goodey, 1963). They have the ability to survive periods of desiccation and *O. succini* probably lived under some plant cover on the bark of the amber tree.

The third and fourth fossil nematodes to be described were also in Baltic amber and included the free-living *Vetus pristinus* (Menge, 1866) (Fig. 48) and *V. capillaceus* (Menge, 1866) (Fig. 49). Both of these species were represented by 30 or more individuals in pieces of clear amber without any additional inclusions. The former species was quite small (139 μm in length) while *V. capillaceus* was much larger, ranging up to 2.0 mm in length. These species show that several types of free-living nematodes inhabited the bark of the Baltic amber tree. They probably fed on microbes and supplied food for springtails and mites, which also occur in Baltic amber. The medium sized *V. duisburgi* Taylor, 1935, which ranged between 650 and 850 μm in length, was described by Taylor (1935) from two specimens figured by von Duisburg (1862) (Fig. 50).

The species described herein as *Tipulacomis balticus** is associated with an adult cranefly (Diptera: Limoniidae) in Baltic amber. Several hundred dauer juveniles, as well as two females, are clustered around the abdominal segments of the cranefly (Figs 51-53; Plate 7). Based

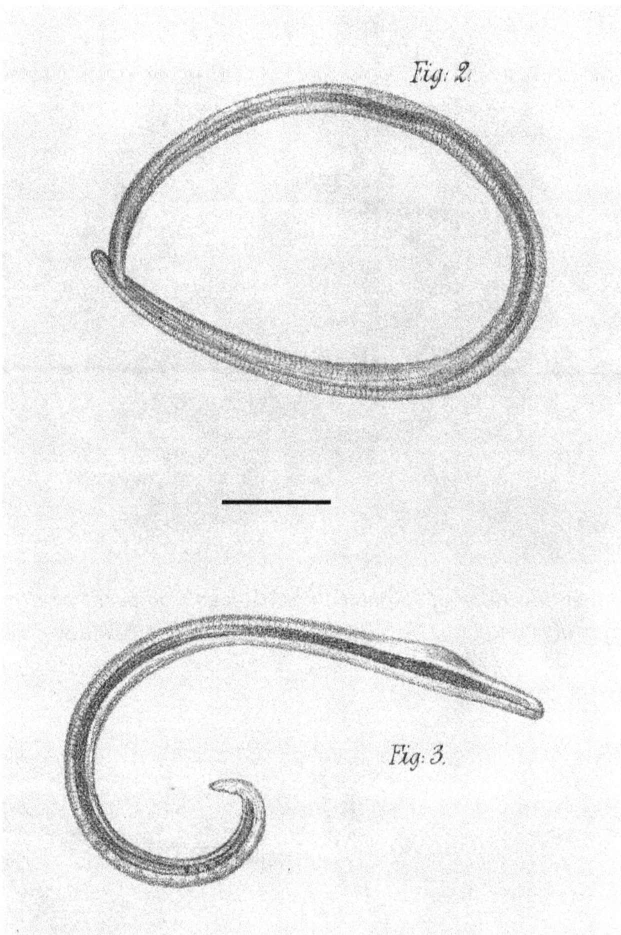

Fig. 50. *The Baltic amber* Vetus duisburgi *Taylor, 1935 (modified from von Duisburg, 1862, Table 1, Fig. B2, B3, opposite p. 35). (Scale bar = 68 μm.)*

on their size and shape, they were determined to belong to the family Rhabditidae. Members of this family are associated with extant craneflies in North America and Europe (Lam & Webster, 1971; Sudhaus, 1993). The nematodes were clearly developing in the same moist habitat as the cranefly larva, which was probably damp soil or rotting vegetation. As the habitat began to dry out the cranefly larva entered the pupal stage, and the nematodes formed dauer stages that collected around the pupa and then climbed onto the abdomen of the

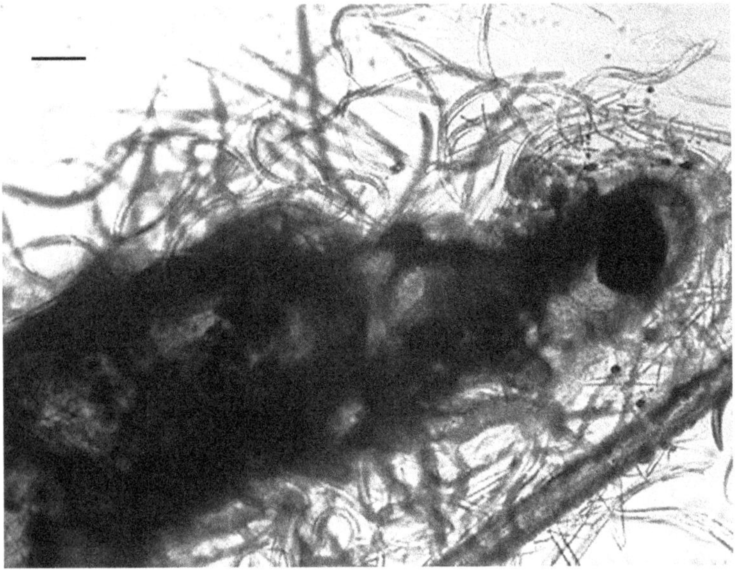

Fig. 51. *Dauer juveniles of* Tipulacomis balticus* *adjacent to a cranefly in Baltic amber (accession no. N-3-70) (see also Plate 7). (Scale bar = 77 μm.)*

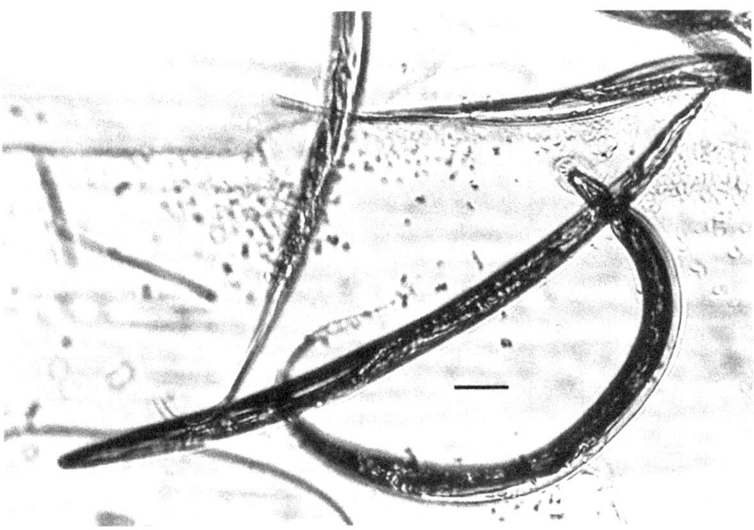

Fig. 52. *Detail of dauer juveniles of* Tipulacomis balticus* *in Baltic amber (accession no. N-3-70). (Scale bar = 21 μm.)*

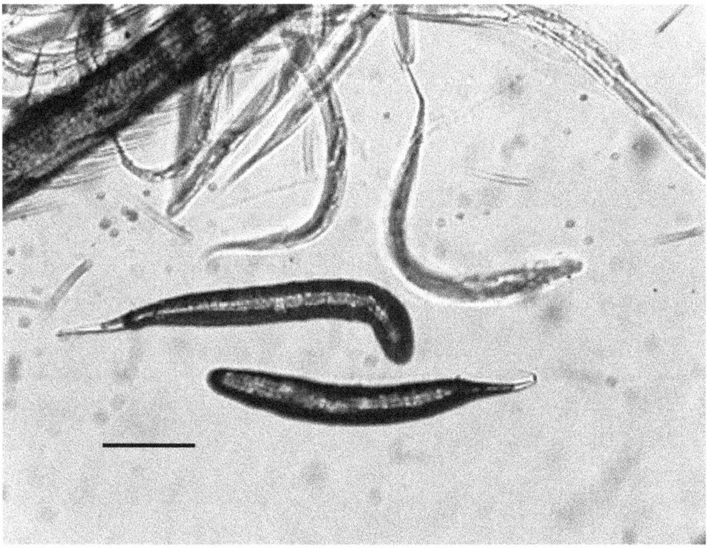

Fig. 53. *Two non-dauer stages of* Tipulacomis balticus* *adjacent to a cranefly in Baltic amber (accession no. N-3-70). (Scale bar = 94 μm.)*

adult fly as it emerged. This association appears to have been widespread in the Baltic amber forest, since a cranefly with a large population of dauer nematodes probably belonging to the same fossil species is illustrated by Weitschat and Wichard (2002; their Plate 7a, b).

Many free-living nematodes form dauer stages that attach themselves to insects. Dauer juveniles of the fossil rhabditid, *Psychodorhabditis balticus**, attached themselves to mothflies (Diptera: Psychodidae) in Baltic amber (Figs 54, 55). The two dauer stages adjacent to the mothfly are 1.1 mm and 1.2 mm in length, which is similar to the size of the extant rhabditid, *Rhabditis dubia* Bovien, 1937 (0.80-1.21 mm) that is phoretic on mothflies in Europe (Bovien, 1937). In the case of *R. dubia*, the dauer were long enough to wrap themselves tightly around the abdominal segments of the mothfly. It is not known exactly how the dauer stages of *P. balticus** were attached to the mothfly but it is possible that they also coiled around the abdomen. Mothflies breed in organic waste and *R. dubia* colonised cow dung. It is possible that the fossil mothfly was breeding in the dung of one of the mammals that left its hair in Baltic amber.

In another piece of Baltic amber were several hundred nematodes (Fig. 56) together with seven female braconid wasps, a bark beetle and

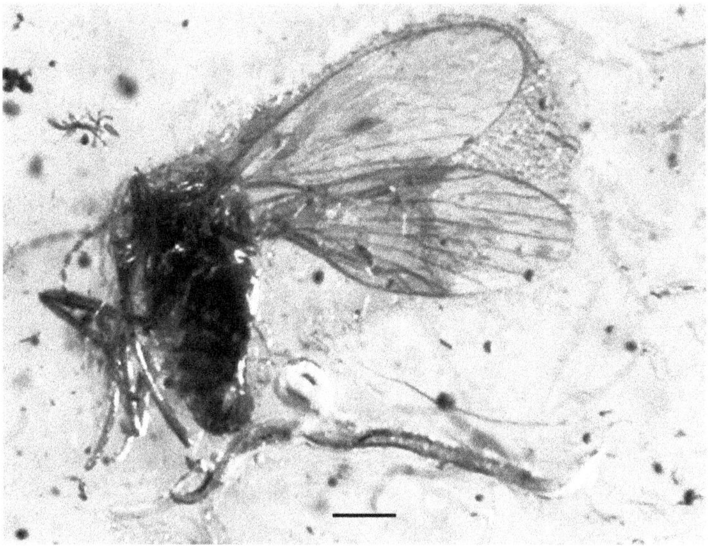

Fig. 54. *Two specimens of* Psychodorhabditis balticus* *adjacent to a mothfly in Baltic amber (accession no. N-3-74). (Scale bar = 220 μm.)*

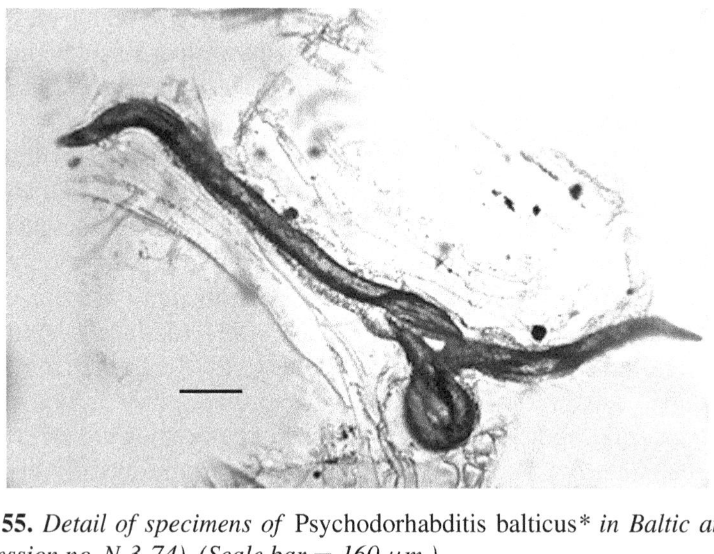

Fig. 55. *Detail of specimens of* Psychodorhabditis balticus* *in Baltic amber (accession no. N-3-74). (Scale bar = 160 μm.)*

a fungus gnat. It is interesting when so many different invertebrates occur together in a single piece of amber since it provides a glimpse

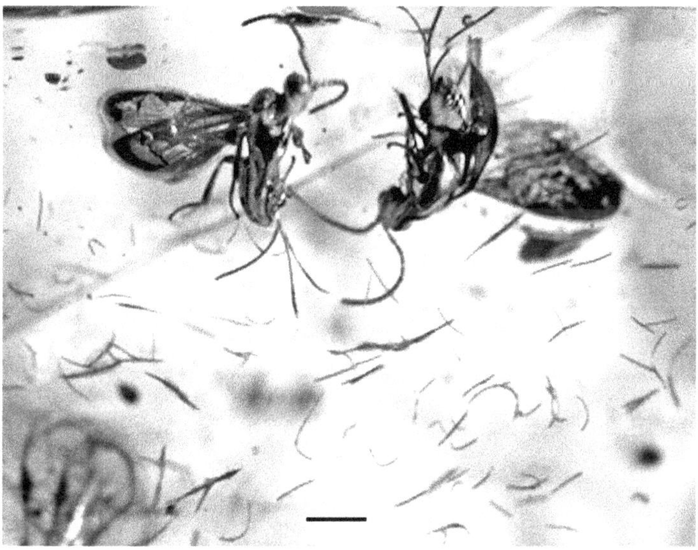

Fig. 56. *A population of the mycetophagous nematode* Palaeoaphelenchoides balticus*, *together with braconid wasps, in Baltic amber (accession no. N-3-80). (Scale bar = 800 μm.)*

of the micro-ecosystem of the amber forest. There is no evidence that the nematodes were attached to or parasitic on any of the insects.

The nematode *Palaeoaphelenchoides balticus** represented a reproducing population of stylet-bearing aphelenchoidids (Fig. 57) that were probably feeding on fungi growing in or around a damaged portion of the tree trunk. A possible scenario is that a bark beetle carrying the resistant stages of an aphelenchoidid made a gallery in the bark of the tree in which to deposit its eggs. Spores of the fungi that were the food source for the nematodes may have been introduced by the bark beetle. When the beetle's galleries reached the cambial layer, the tree responded to the injury by secreting resin, entrapping some of the associated life forms. The braconid wasps may have been bark beetle parasitoids since they are members of the subfamily Braconinae, which contains the extant bark beetle parasitoid genus, *Coeloides* Wesmael, 1838. Female *Coeloides* wasps congregate on trees and use their antennae to locate host larvae under the bark (Baker, 1972; Dajoz, 2000). Host related activities seem the most likely explanation for why so many female wasps were together in a single piece of amber.

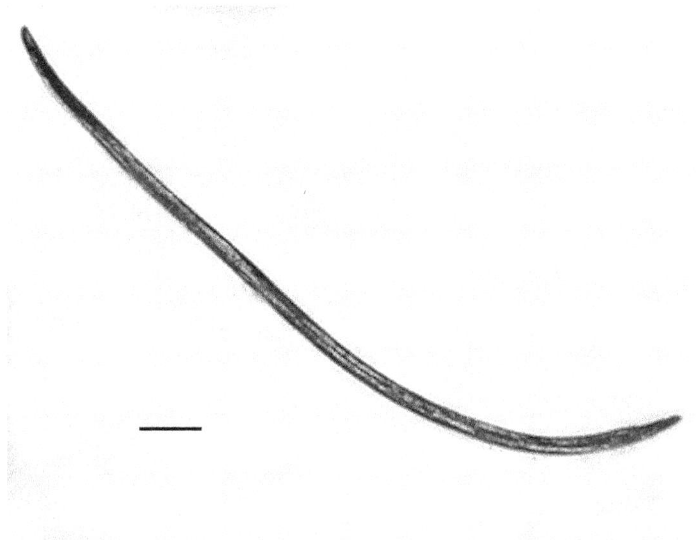

Fig. 57. *Detail of a specimen of* Palaeoaphelenchoides balticus* *in Baltic amber (accession no. N-3-80). (Scale bar = 74 μm.)*

There are many extant nematodes that live in the tunnels of bark beetles and develop on bacteria, protozoa and fungi in the insect galleries. Their presence in amber is usually only revealed by the resistant dauer stages associated with the insects. However, in one piece of Baltic amber an isolated pellet of beetle frass contained several hundred rhabditoid nematodes (Figs 58, 59). The frass probably represented the activities of a scolytid bark beetle, which are known from Baltic amber (Larsson, 1978).

These frass-associated nematodes, described as *Palaeorhabditis balticus**, consisted only of juveniles and females, which were probably hermaphrodites. Since no eggs were detected and developing juveniles occurred inside mature females, *P. balticus** can also be considered ovoviviparous. As dauer stages are absent, the nematodes were probably developing under optimum conditions in the beetle's gallery when they, along with the frass, were suddenly extruded by a resin flow.

A number of invertebrate-parasitic mermithids occur in Baltic amber. However, the earliest description of a fossil mermithid was from brown coal deposits in Germany (von Heyden, 1860) (Fig. 60), which are approximately the same age as Baltic amber. The fossil mermithid, *Heydenius antiquus*, was in the process of emerging from the body of a

Fig. 58. *Developing population of the rhabditid* Palaeorhabditis balticus* *in Baltic amber (accession no. N-3-81). (Scale bar = 1 mm.)*

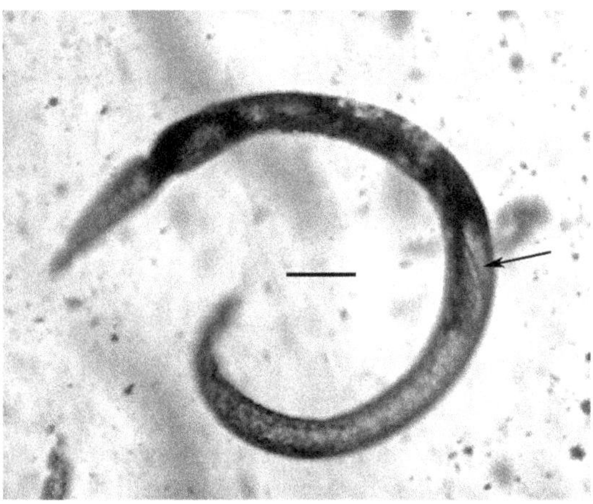

Fig. 59. *Detail of a specimen of* Palaeorhabditis balticus* *in Baltic amber. Arrow shows intrauterine juvenile (accession no. N-3-81). (Scale bar = 71 μm.)*

long-horned beetle (Coleoptera: Cerambycidae). There are several cases of mermithids infecting extant cerambycids (Poinar, 1975).

36.

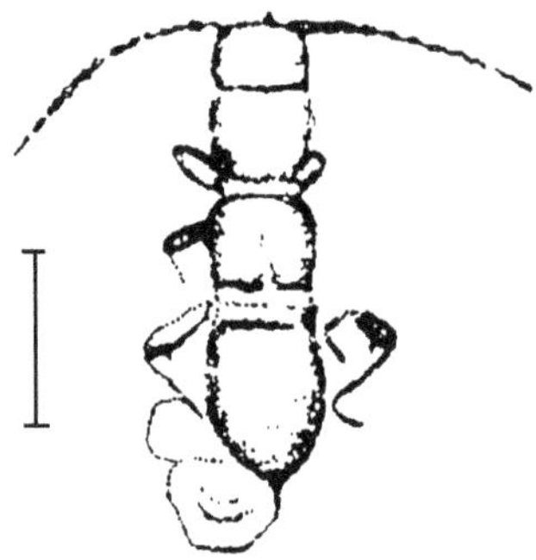

Fig. 60. Heydenius antiquus *(von Heyden, 1860) emerging from a beetle in Eocene brown coal deposits in Germany. This represents the earliest description of a fossil nematode. (Scale bar = 3 mm.)*

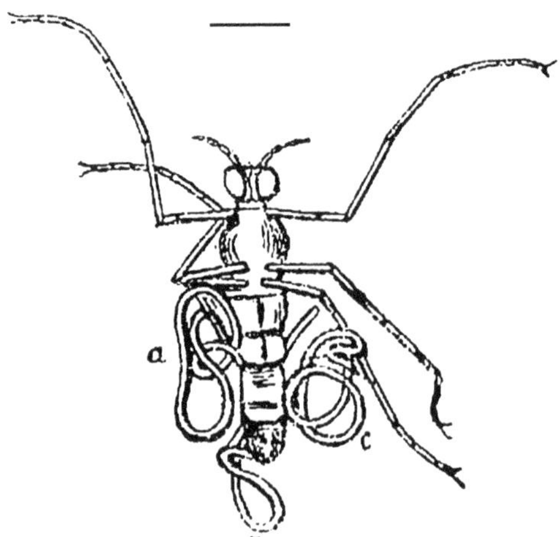

Fig. 7.

Fig. 61. *Drawing by Menge showing three specimens of* Heydenius matutinus *(Menge, 1866, Fig. 7, p. 5) parasitising a chironomid midge in Baltic amber. This is a collective species for mermithid parasites of chironomid midges in Baltic amber. (Scale bar = 823 μm.)*

In 1866, Menge described the first mermithid from amber, *Heydenius matutinus* (Menge, 1866). The host was a small adult non-biting midge (Diptera: Chironomidae) that was parasitised by three mermithids (Fig. 61). Two nematodes had completely emerged from and were adjacent to the midge, while the third had only partly emerged. The taxon *H. matutinus* is used here as a collective species for all mermithid nematodes parasitising midges of the family Chironomidae in Baltic amber. While several genera may actually be involved, as with present-day midge mermithids (Poinar, 1975) (Fig. 38), all of the amber forms appear to be either parasitic or post-parasitic stages and it is not possible to distinguish between them.

Certainly, *H. matutinus* is the most common fossil nematode parasite known and there are many examples in museums and private collections. A number of infected Baltic amber midges are shown in the present work, some with nematodes still within their body cavities, others with nematodes partially emerged and still others with mermithids that have completely emerged from their hosts (Figs 62-70; Plate 8). In some cases, the midges were parasitised by mites as well (Fig. 68; Plate 9). A rare case is shown in Figure 70 (Plate 10), where the mermithid had emerged but remained attached to the host by a strand of congealed haemolymph. While most fossil mermithids are pale-coloured as in

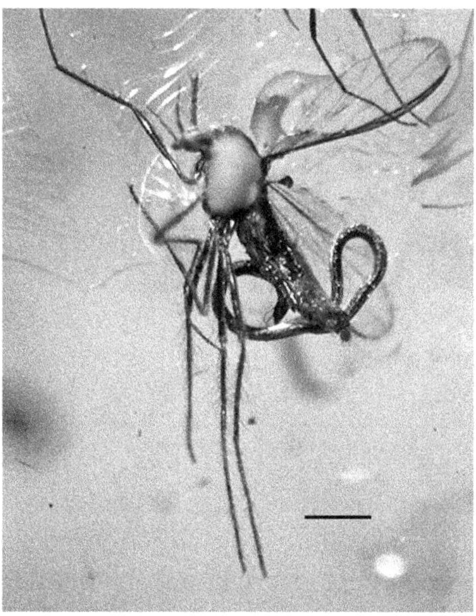

Fig. 62. *Baltic amber chironomid midge with two emerging specimens of* Heydenius matutinus *(Menge, 1866), similar to the specimens of Menge shown in Figure 61. (Scale bar = 630 μm.)*

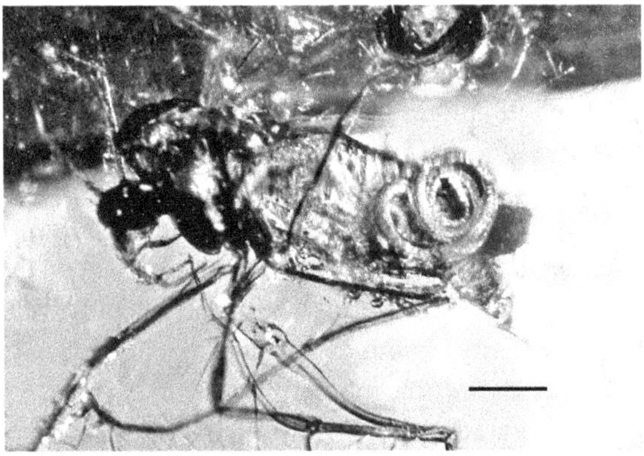

Fig. 63. *Specimen of* Heydenius matutinus *(Menge, 1866), coiled within the abdomen of a Baltic amber chironomid midge (specimen deposited in the Zoological Museum, Copenhagen). (Scale bar = 938 μm.)*

many extant forms, some are darkened by a superficial deposit formed during the fossilisation process (Fig. 69).

In 1872, Menge described the solitary mermithid *Heydenius quadristriatus* (Menge, 1872) (Fig. 71) in Baltic amber. It appeared to have no association with the fragments of ant and harvestman (phalangid) that were in the same piece of amber and, based on its large size (60 mm long), Menge felt that the host must have been a large orthopteran or caterpillar that had escaped capture by the resin or was later eaten by a predator. The species name *H. quadristriatus* is used here for mermithids in Baltic amber that have no obvious connection to any host. Two additional specimens were also assigned to this species (Fig. 72).

Ants are known to host a number of mermithid species and this parasitic association has often resulted in morphological modifications to all castes of ants, essentially producing morphotypes with a range of reduced or modified characters. Such modifications appear to have occurred in a winged male of *Prenolepis henschei* Mayr, 1868 parasitised by *Heydenius formicinus* Poinar, 2002 in Baltic amber (Fig. 73; Plate 11). More than half of the mermithid had emerged while the rest remained in the ant's abdomen. The mermithid had clearly completed the development of the parasitic phase and was ready to emerge when the ant flew into a pool of resin. Worker ants were also attacked by *H. formicinus* in the Baltic forest (Fig. 74). Mermithid parasitism of ants is common today (Figs 75, 76; see also Fig. 114).

Fig. 64. *A single specimen of* Heydenius matutinus *(Menge, 1866) that had emerged from its chironomid midge host in Baltic amber. Note an ectoparasitic mite (arrow) that released its hold on the fly (accession no. N-3-35A). (Scale bar = 864 μm.)*

Many aquatic insects mistake reflections from resin as a water source and the ant may have made the same mistake. While ants are not aquatic, some extant mermithids cause their ant hosts to seek water when the nematodes are ready to emerge (Kaiser, 1986). Such mermithids have an indirect life cycle where eggs deposited in a water source are ingested by an aquatic paratenic host. The nematodes encyst in the body cavity of the aquatic host and further development occurs in ants when the aquatic host is fed to the developing brood.

Further evidence of mermithid parasitism in Baltic amber was the first fossil record of a mermithid from a hemipteran. Planthoppers

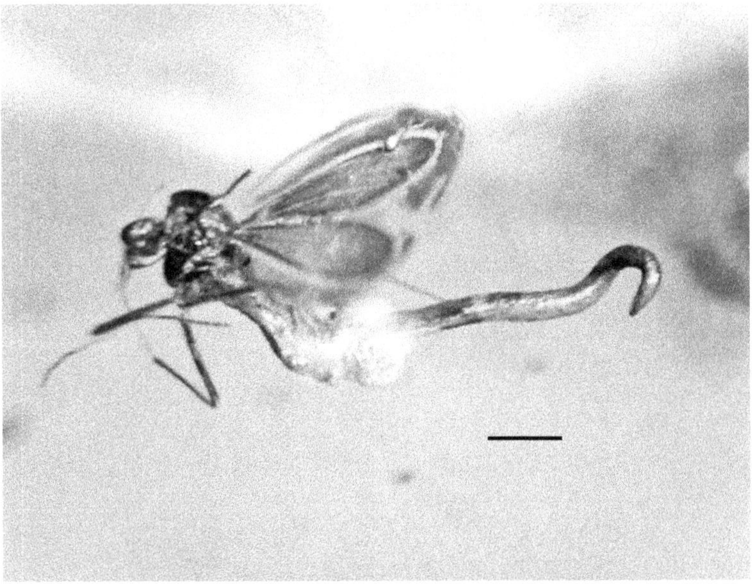

Fig. 65. *A single, stubby specimen of* Heydenius matutinus *(Menge, 1866) that was in the process of emerging from a Baltic amber chironomid midge (accession no. N-3-15C). (Scale bar = 544 μm.)*

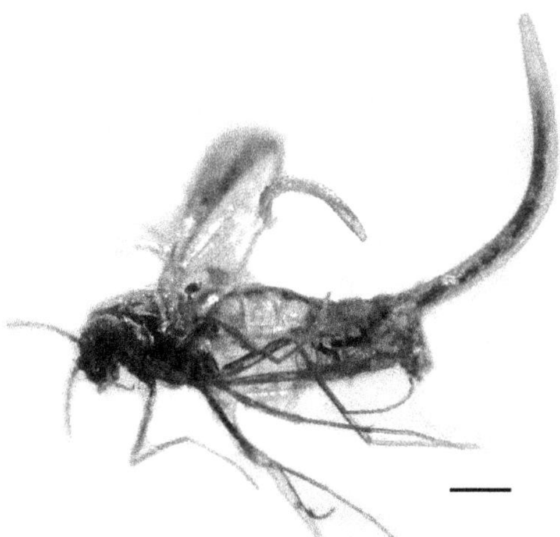

Fig. 66. *Large and smaller specimen of* Heydenius matutinus *(Menge, 1866) that had emerged from their chironomid midge host in Baltic amber (accession no. N-3-15E). (Scale bar = 226 μm.)*

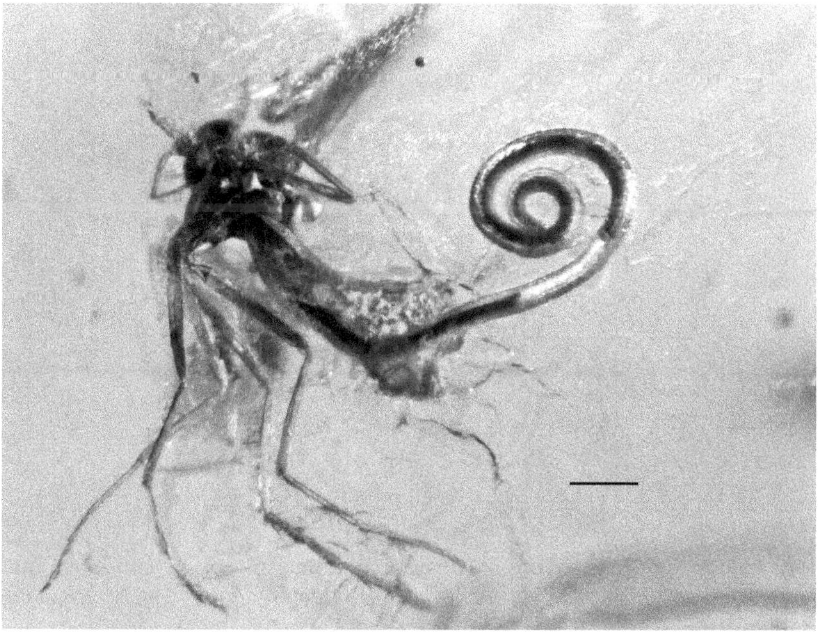

Fig. 67. Heydenius matutinus *(Menge, 1866), exiting from a chironomid midge in Baltic amber (accession no. N-3-54A). (Scale bar = 250 μm.)*

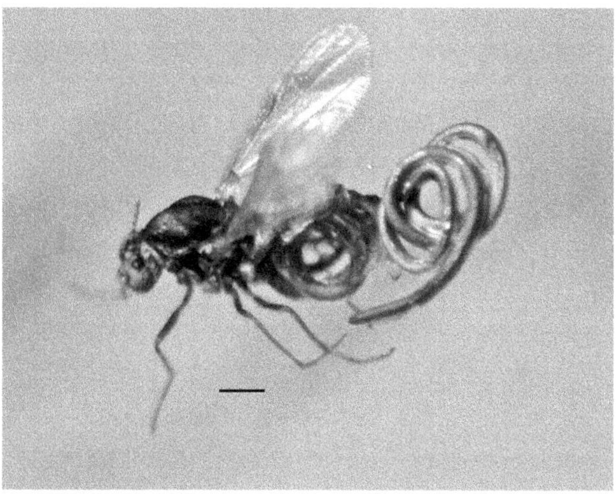

Fig. 68. *Two specimens of* Heydenius matutinus *(Menge, 1866), one completely emerged and the second still in the body cavity of its chironomid midge host in Baltic amber (accession no. N-3-56A) (see also Plate 8). (Scale bar = 53 μm.)*

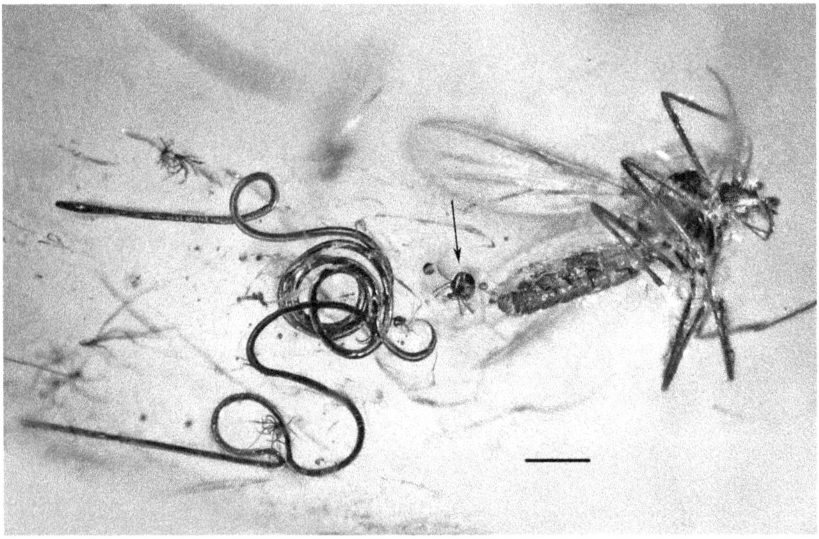

Fig. 69. *Two specimens of* Heydenius matutinus *(Menge, 1866) that had emerged from an adult female chironomid midge in Baltic amber. Note the ectoparasitic mite (arrow) adjacent to the midge (accession no. N-3-35B) (see also Plate 9). (Scale bar = 521 μm.)*

Fig. 70. *The mermithid* Heydenius matutinus *(Menge, 1866) still connected to its midge host by a thin strand of congealed blood in Baltic amber (accession no. N-3-76) (see also Plate 10). (Scale bar = 140 μm.)*

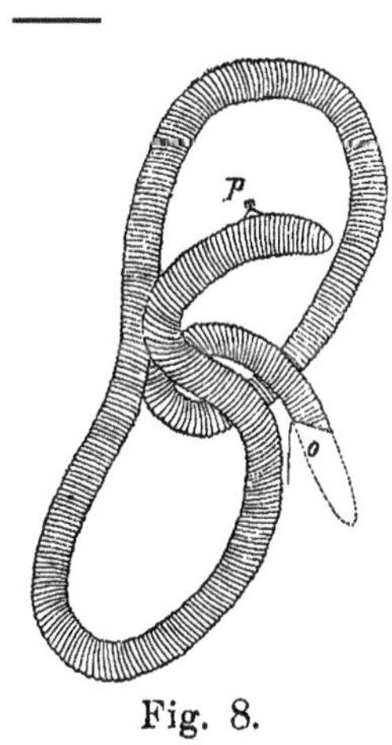

Fig. 71. *Menge's drawing of* Heydenius quadristriatus *(Menge, 1872) in Baltic amber. This specific name is used for mermithids in Baltic amber with no associated host. (Scale bar = 2.4 mm.)*

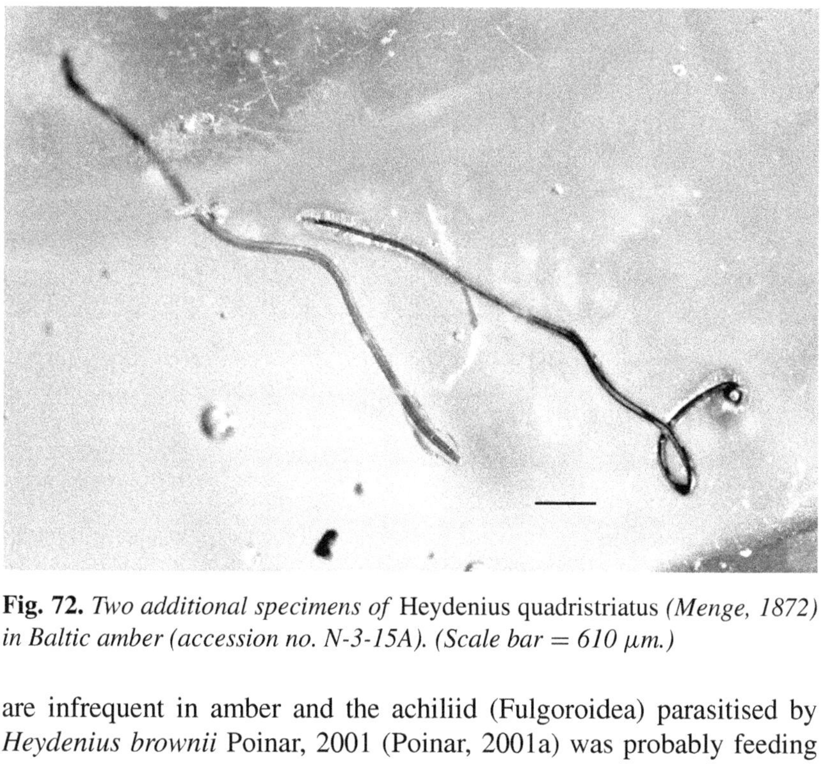

Fig. 72. *Two additional specimens of* Heydenius quadristriatus *(Menge, 1872) in Baltic amber (accession no. N-3-15A). (Scale bar = 610 μm.)*

are infrequent in amber and the achiliid (Fulgoroidea) parasitised by *Heydenius brownii* Poinar, 2001 (Poinar, 2001a) was probably feeding

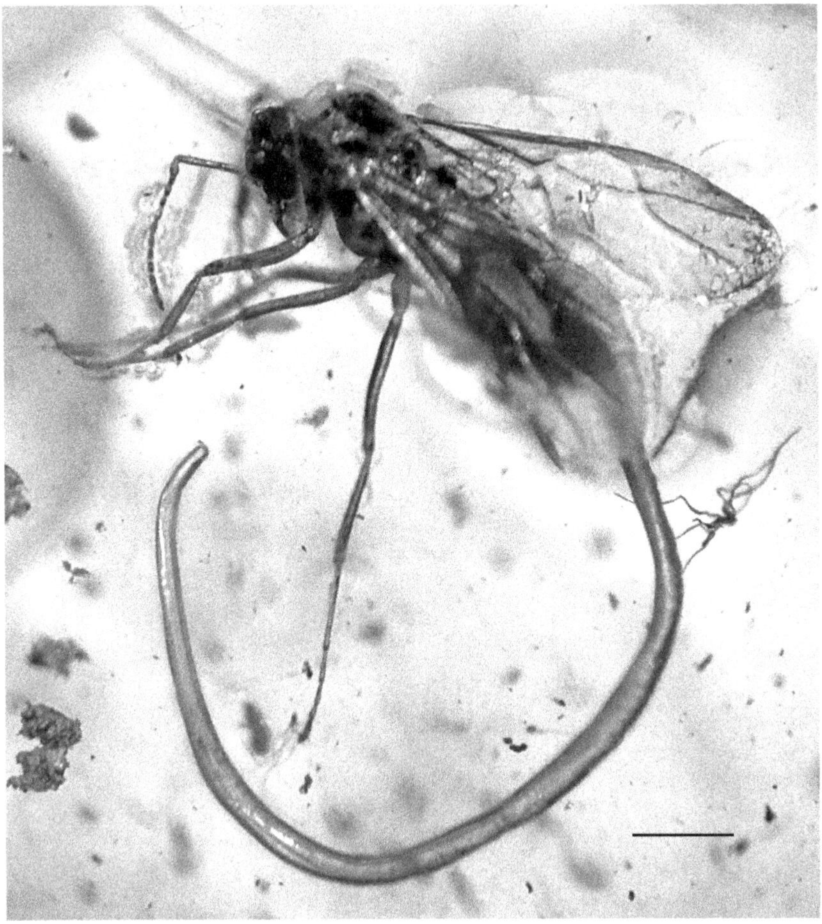

Fig. 73. *The mermithid* Heydenius formicinus *Poinar, 2002 emerging from a male winged ant (*Prenolepis henschei *Mayr) in Baltic amber (accession no. N-3-52) (see also Plate 11). (Scale bar = 640 μm.)*

on the leaves of the Baltic amber tree or a neighbouring plant (Fig. 77; Plate 12). This is one of several instances when a fossil association was discovered before a similar extant relationship. While *H. brownii* was the first mermithid parasite of a Holarctic planthopper, its discovery was soon followed by the report of extant mermithid-infected planthoppers in Ireland (Helden, 2008) (Fig. 78).

It is not known whether the planthopper was the sole host for *H. brownii*. While studying extant mermithids of the alfalfa weevil (*Hypera*

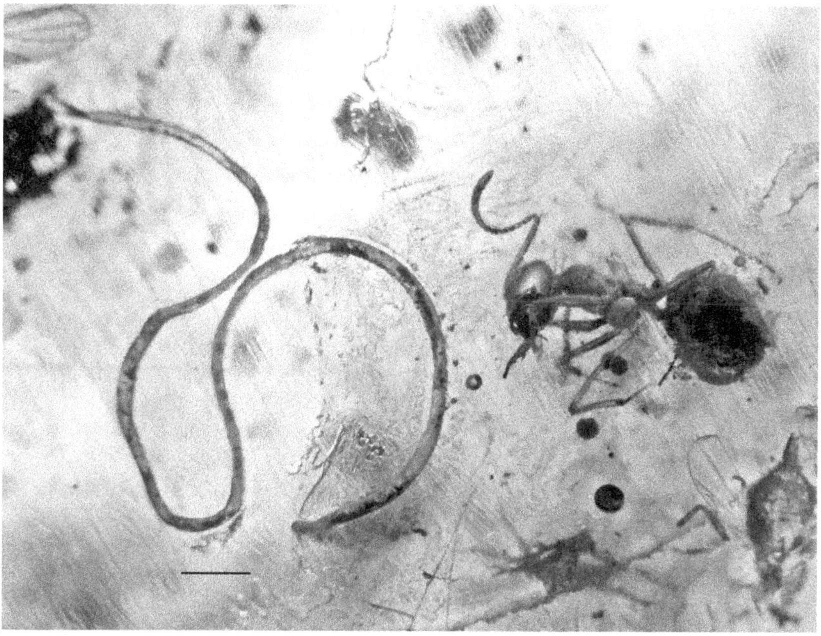

Fig. 74. *Specimen of* Heydenius formicinus *Poinar, 2002 that had emerged from a worker ant in Baltic amber (accession no. N-3-69). (Scale bar = 685 μm.)*

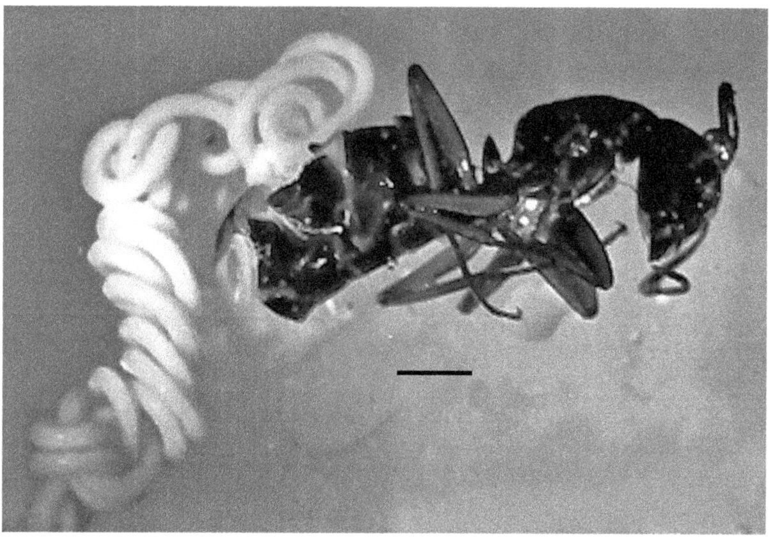

Fig. 75. *An extant worker carpenter ant* (Camponotus *sp.) with a mermithid emerging from its abdomen. (Scale bar = 900 μm.)*

Fig. 76. *An extant alate carpenter ant (*Camponotus *sp.) with a mermithid in its abdomen. (Scale bar = 930 μm.)*

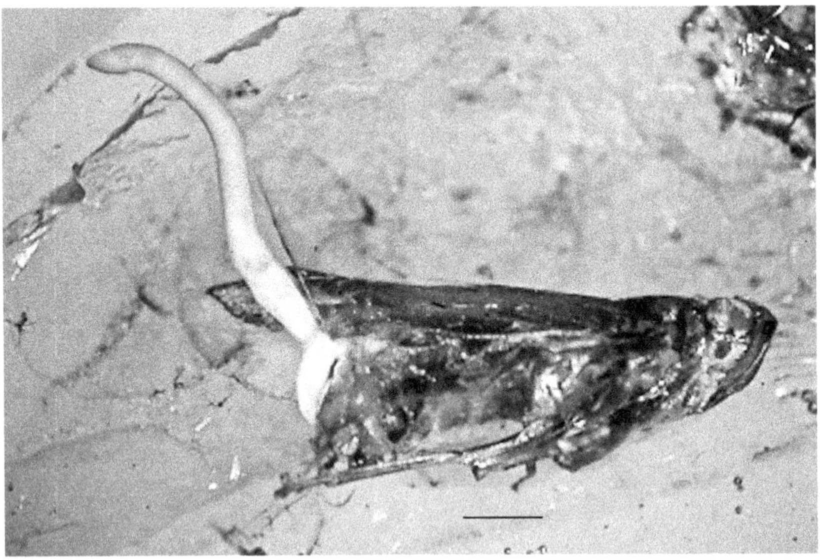

Fig. 77. *The mermithid* Heydenius brownii *Poinar, 2001 emerging from an achiliid planthopper in Baltic amber (accession no. N-3-51) (see also Plate 12). (Scale bar = 2.6 mm.)*

postica Gyll.) in eastern North America, the author discovered the same mermithid species developing in hemipterans (species of *Lygus* Hahn, 1833 and *Nabis* Latreille, 1802) as well as weevils. Since the alfalfa

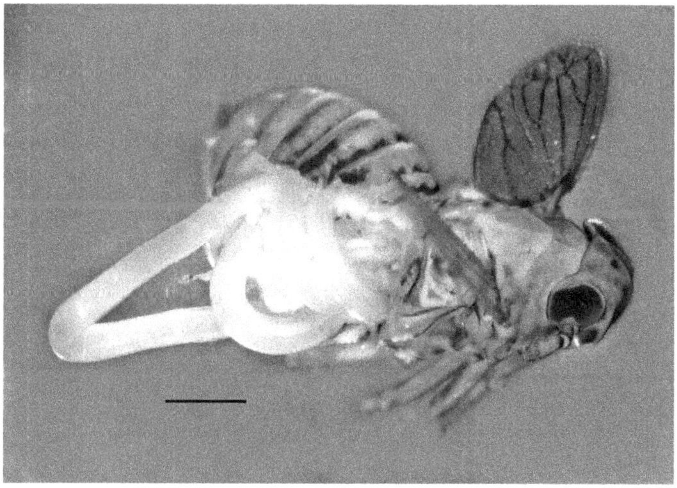

Fig. 78. *An extant mermithid emerging from the delphacid planthopper* Javesella dubia *in Ireland (photo courtesy of Alvin Helden). (Scale bar = 675 μm.)*

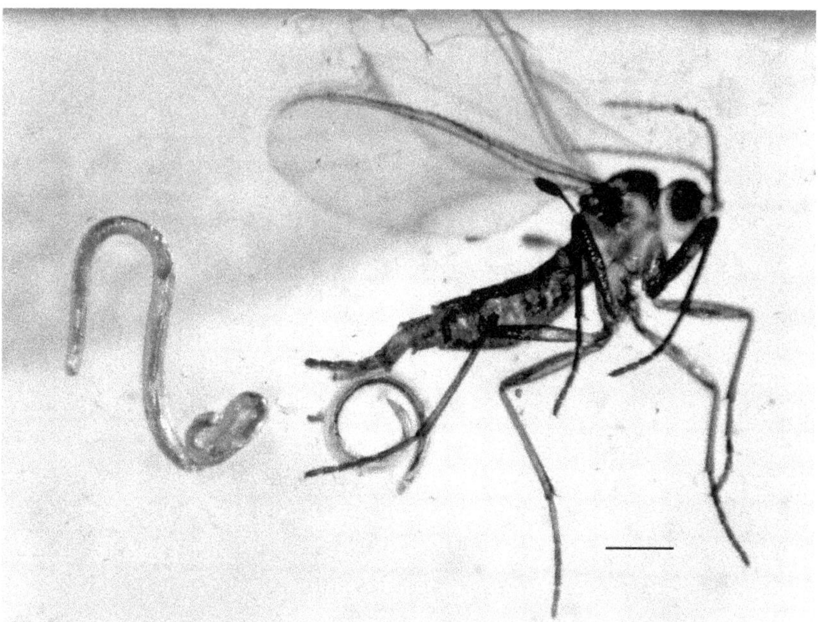

Fig. 79. *Two specimens of* Heydenius sciarophilus* *with their sciarid fungus gnat host in Baltic amber (accession no. N-3-46). (Scale bar = 295 μm.)*

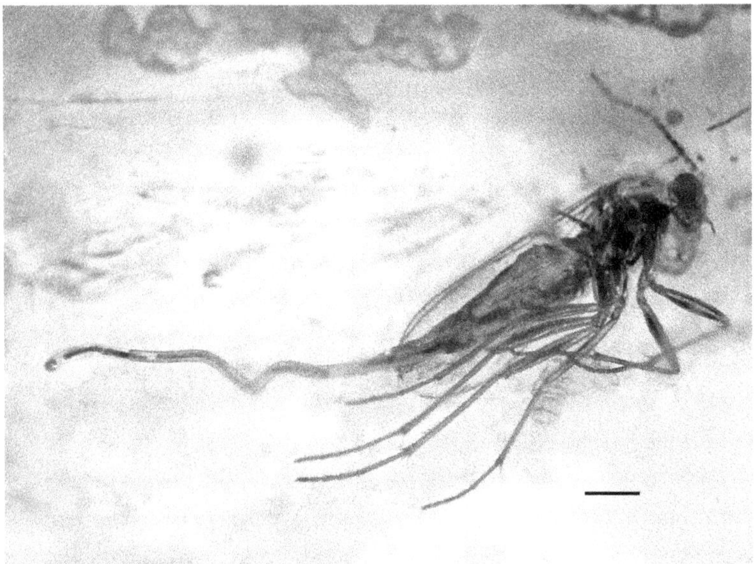

Fig. 80. *Specimen of* Heydenius sciarophilus* *emerging from a sciarid fungus gnat in Baltic amber (accession no. N-3-59). (Scale bar = 207 μm.)*

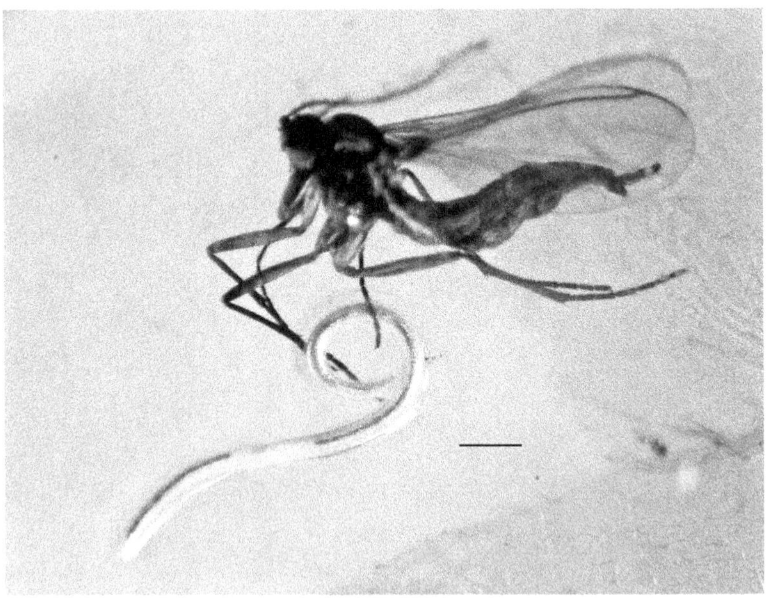

Fig. 81. *Specimen of* Heydenius sciarophilus* *that had emerged from a sciarid fungus gnat in Baltic amber (accession no. N-3-15B). (Scale bar = 282 μm.)*

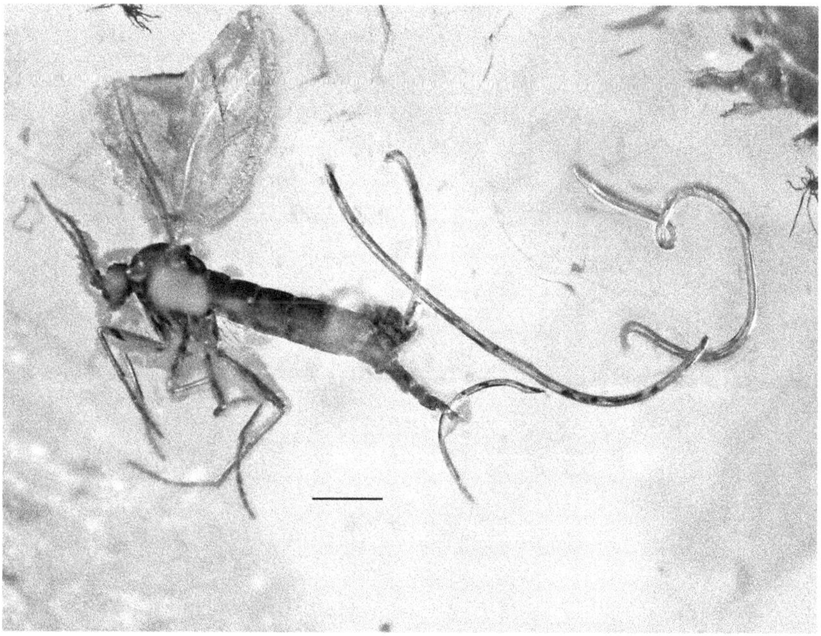

Fig. 82. *Three specimens of* Heydenius sciarophilus* *are adjacent to their Baltic amber fungus gnat host and a fourth is partly emerged (accession no. N-3-84) (see also Plate 13). (Scale bar = 429 μm.)*

weevil had recently been introduced into North America, it is possible that hemipterans were the original hosts. With a fossil record extending back to the Early Permian (Rasnitsyn & Quicke, 2002), hemipterans could have been ancient terrestrial hosts for mermithids.

Sciarid fungus gnats (Diptera: Sciaridae) feed on many types of fungi in many different habitats, including those growing in decaying plant matter. A variety of nematodes parasitised these gnats in the Baltic amber forest. One of the parasites was the mermithid *Heydenius sciarophilus**. Several pieces of Baltic amber with *H. sciarophilus** show that mermithid parasitism of sciarid flies was fairly common (Figs 79-83). Up to four individuals parasitised some flies (Fig. 82; Plate 13) and a few nematodes survived long enough to crawl into the resin. Some extant mermithids, such as *Orthomermis oedobranchus* Poinar, 1964, also parasitise sciarids (Poinar, 1964) (Fig. 84).

It is possible that a few specimens of *H. sciarophilus** are members of the family Tetradonematidae, which contains parasites of fungus gnats today (Keilin & Robinson, 1933; Ferris & Ferris, 1966). The Mermithi-

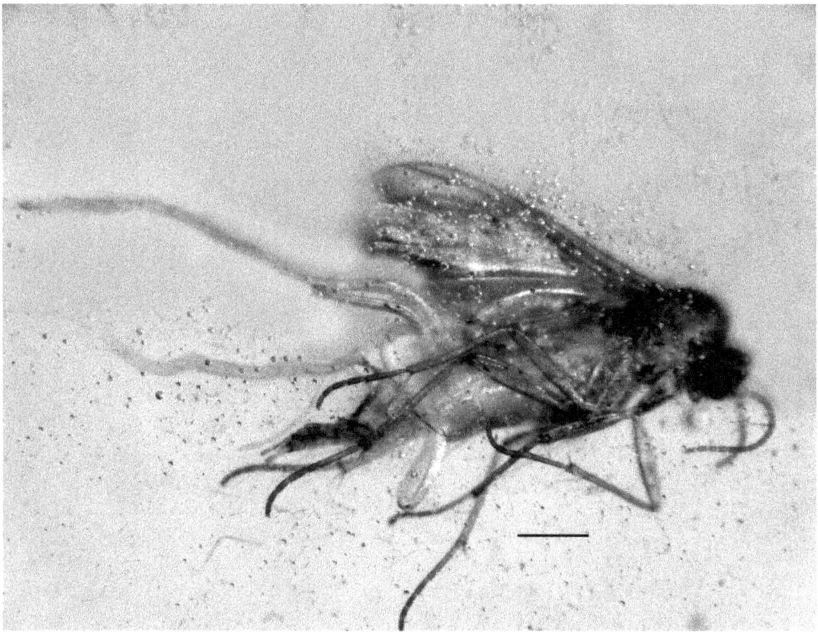

Fig. 83. *Three specimens of* Heydenius sciarophilus* *emerging from a sciarid fungus gnat in Baltic amber. Two additional mermithids are also present but outside the limit of the photo (accession no. N-3-85). (Scale bar = 400 μm.)*

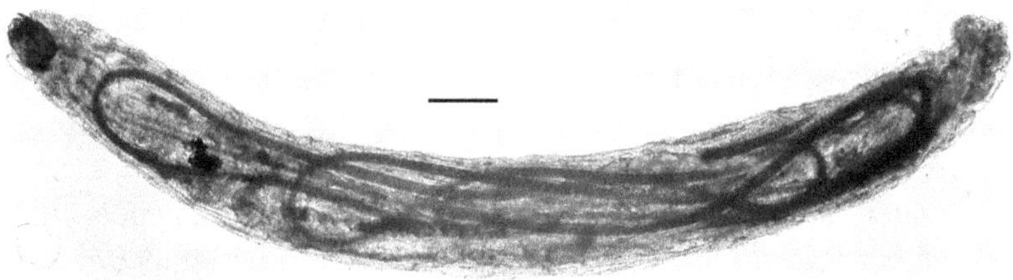

Fig. 84. *The extant mermithid* Orthomermis oedobranchus *Poinar, 1964 inside the larva of the sciarid,* Lycoriella solani *(Winn). (Scale bar = 387 μm.)*

dae and Tetradonematidae are closely related, but tetradonematids typically mate within the host and the eggs develop before the females enter the environment. One fossil female nematode from a sciarid did contain developing eggs and is described here as *Palaeotetradonema sciarae** (Fig. 85) in the Tetradonematidae. It may be a basal lineage to one of the

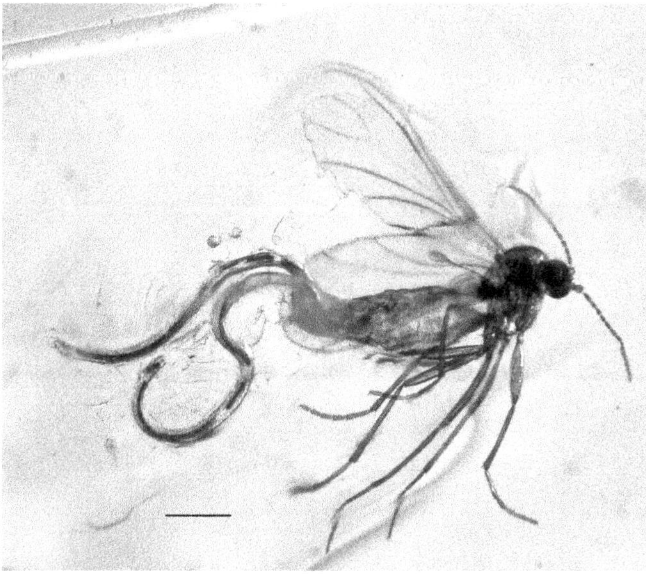

Fig. 85. *Two specimens of* Palaeotetradonema sciarae* *emerging from a sciarid fungus gnat in Baltic amber (accession no. N-3-62). (Scale bar = 511 μm.)*

extant genera (*Tetradonema* Cobb, 1919 or *Aproctonema* Keilin, 1917) that parasitise sciarids.

The preferred habitat of fungus gnats is moist decaying plant matter, and, since that is also an excellent site for nematodes, it is not surprising that other nematode groups attack these flies. Members of the genus *Tripius* Chitwood, 1935 parasitise fungus gnats worldwide today and *Tripius balticus** had similar habits in the Baltic amber forest (Figs 86-88). The morphological features of fourth-stage *T. balticus** that had emerged, or were in the process of emerging, from an adult sciarid closely resembled those of the extant *Tripius sciarae* Bovien, 1944 (Bovien, 1944; Poinar, 1965).

There is no doubt of their parasitic nature since several nematodes are still inside the body cavity of the fly host and others are emerging though openings made when the legs were torn away from the body (Fig. 87; Plate 14). In *T. sciarae*, the fourth-stage juveniles, which lack a stylet, leave the host and mature in the environment. Most leave the larval stage of the fly but some are carried into the adult stage where they either penetrate the intestine and leave through the anus or are expelled during oviposition (Fig. 89). There is no way of knowing whether the females of

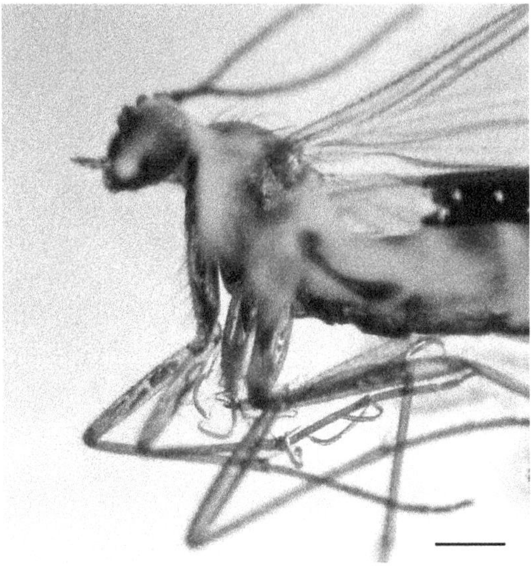

Fig. 86. *Parasitic juveniles of* Tripius balticus* *that had emerged from a sciarid fungus gnat in Baltic amber (accession no. N-3-67). (Scale bar = 322 μm.)*

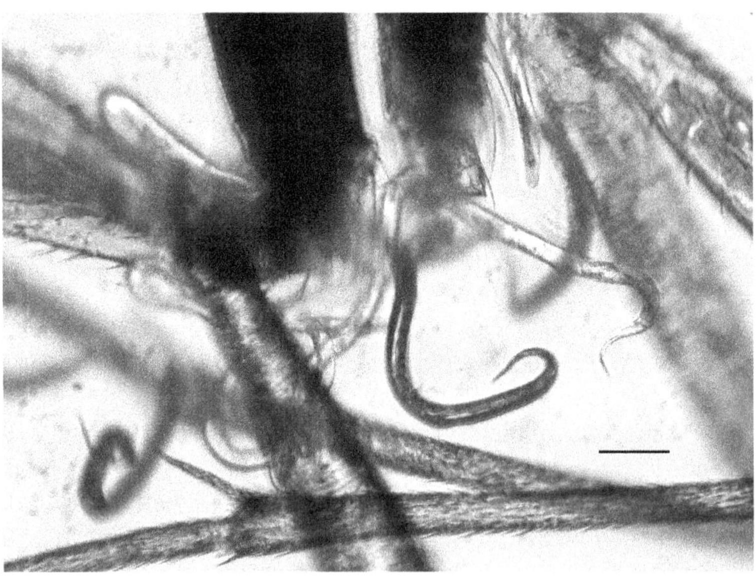

Fig. 87. *Parasitic juveniles of* Tripius balticus* *emerging from the tips of broken coxae of a sciarid fungus gnat in Baltic amber (accession no. N-3-67) (see also Plate 14). (Scale bar = 60 μm.)*

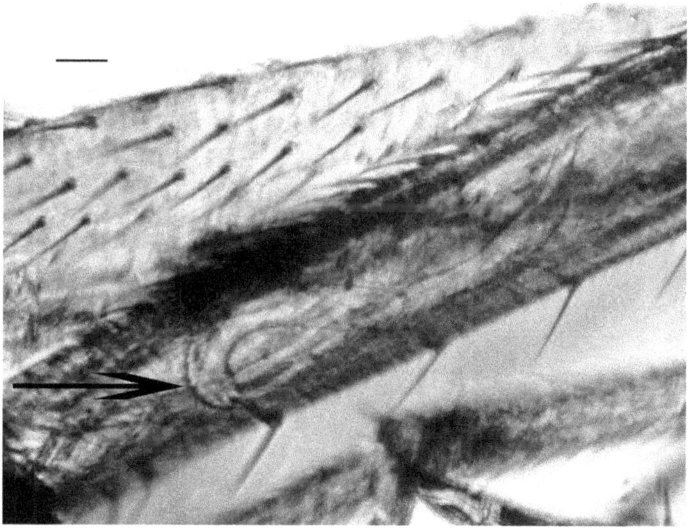

Fig. 88. *Parasitic juvenile (arrow) of* Tripius balticus* *inside a femur of its sciarid fungus gnat host in Baltic amber (accession no. N-3-67). (Scale bar = 23 μm.)*

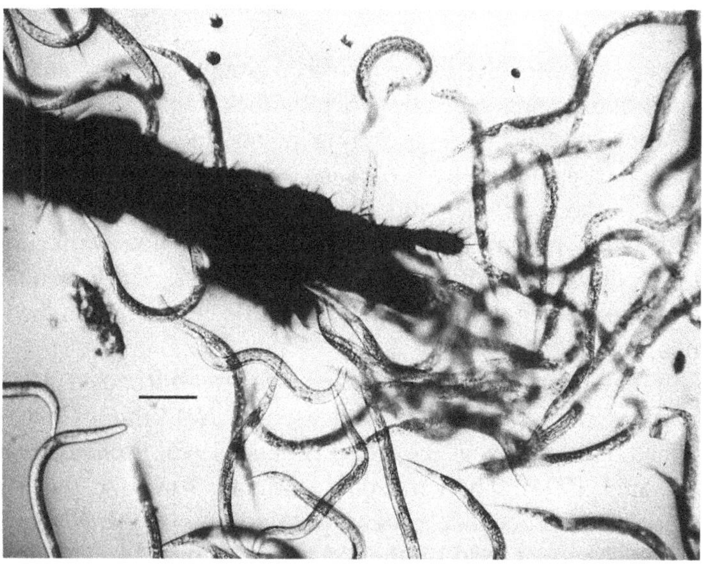

Fig. 89. *Parasitic juveniles of the extant* Tripius sciarae *(Bovien, 1944) being expelled through the ovipositor of the sciarid fungus gnat,* Bradysia paupera *Tuom. (Scale bar = 109 μm.)*

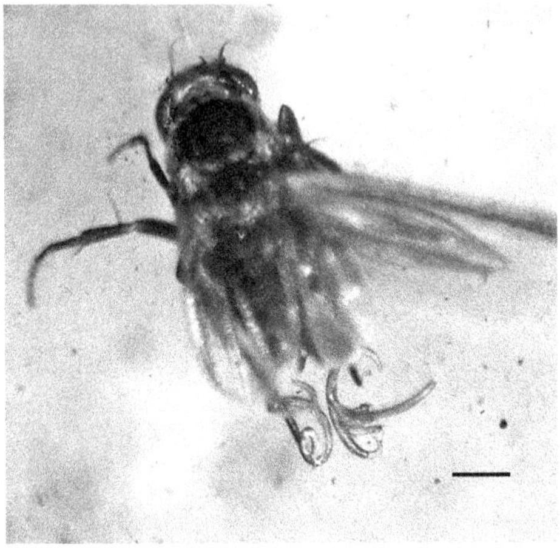

Fig. 90. *Specimens of the allantonematid* Howardula helenoschini *Poinar, 2003 emerging from a phorid fly,* Triphleba *sp., in Baltic amber (accession no. N-3-64) (see also Plate 15). (Scale bar = 347 μm.)*

*T. balticus** had developed the unique method of entering host larvae by auto-digesting the anterior portion of their ensheathing cuticle, an action that stabilises the nematode during the penetration process (Poinar & Doncaster, 1965).

Fungus gnats were not the only insect groups in Baltic amber that scavenged in decaying vegetation. Scuttle flies (Diptera: Phoridae) can tolerate foul habitats, such as rotting plants and animal waste, and one necrophagous phorid fly (*Triphleba* spp.) was discovered with five specimens of the allantonematid *Howardula helenoschini* Poinar, 2003 clustered at the posterior tip of its abdomen, obviously having just exited from the host's body cavity (Fig. 90; Plate 15). The lemon-shaped parasitic female of *H. helenoschini* could be detected through the host's abdomen.

Present-day phorid flies of the genera *Megaselia* Rondani, 1856 and *Woodiphora* Schmidt, 1927 are parasitised by *Howardula* Cobb, 1921 spp. nematodes throughout the world (Poinar, 1975; Richardson *et al.*, 1977; Disney, 1994). The nematodes normally leave the fly host *via* the intestine or reproductive system (Richardson *et al.*, 1977), and that is probably the route used by the five fossil nematodes. The presence of ensheathing cuticles indicates a rapid development to the free-living adult stages. While extant *Howardula* females normally produce up to 50 juveniles in phorids, the small number of fossil progeny could

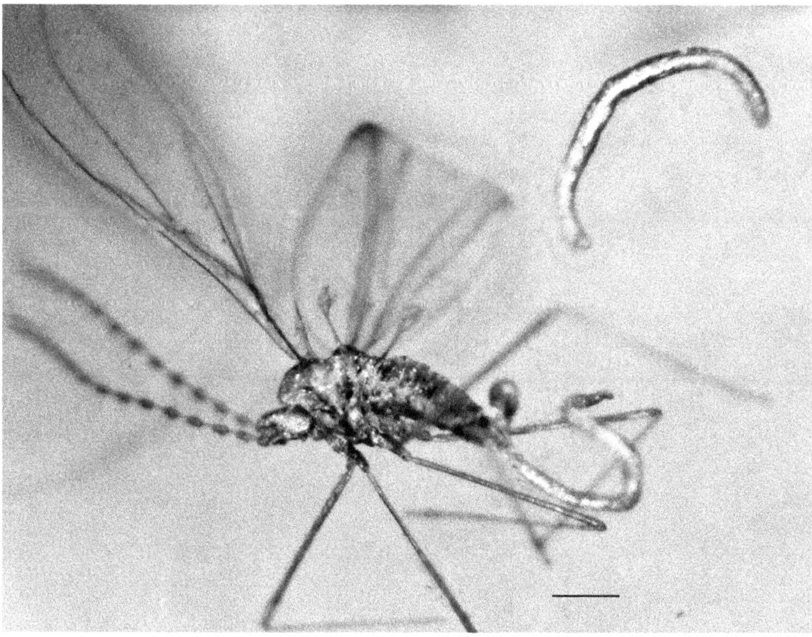

Fig. 91. *Two specimens of* Heydenius cecidomyae*, *a parasite of a gall gnat in Baltic amber (accession no. N-3-53). (Scale bar = 236 μm.)*

either have been due to a reduced food supply in a small host (1.2 mm long), represents a small sample of the total progeny, or indicates an early evolutionary stage of parasitism in this nematode clade. Since the fossil nematodes are larger than those of extant species of *Howardula* (Richardson *et al.*, 1977; Poinar *et al.*, 1998a), an evolutionary trend might be increased progeny correlated with a smaller body size.

After leaving the host under natural conditions, the five specimens of *H. helenoschini* would have matured and mated and the fertilised females would have searched for *Triphleba* larvae. The infective stage female of *H. helenoschini* probably entered a late instar host larva, which resulted in the parasites passing through the pupa and into the adult fly. Since the host's abdomen appeared empty aside from the parasitic female, it is possible that it was at least partially sterilised from the infection.

A group of small, delicate flies that breed in habitats containing fungi are gall gnats or cecidomyids. Not all feed on fungi since many form galls on higher plants. An adult female in Baltic amber was parasitised by *Heydenius cecidomyae**. One of the nematodes had completely

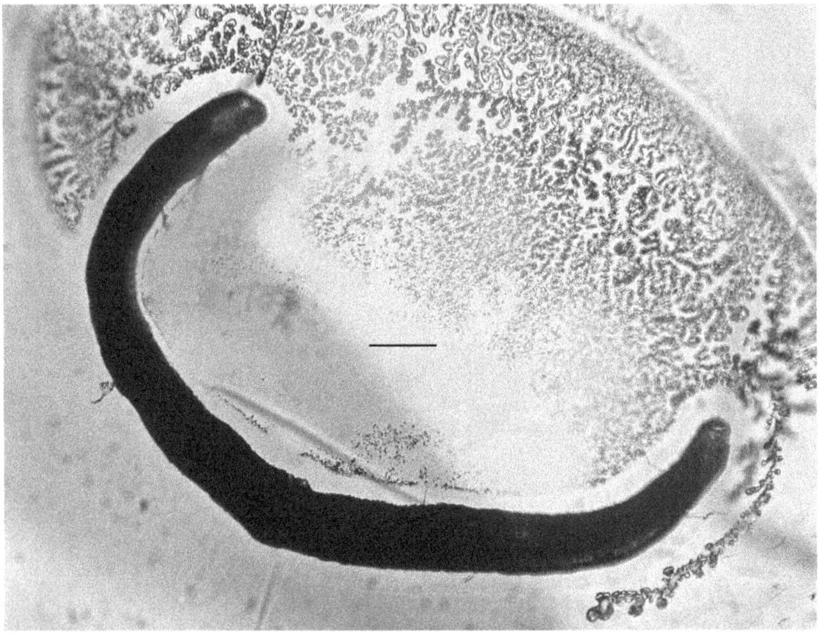

Fig. 92. *Detail of a specimen of* Heydenius cecidomyae* *adjacent to its gall gnat host in Baltic amber (accession no. N-3-53). (Scale bar = 87 μm.)*

emerged from the fly while the second still had a small portion of its body in the gall gnat (Figs 91, 92). Parasitism would have been initiated in the fly larva and then passed though the pupal stage and into the adult. Apparently there are no reports of extant gall gnats parasitised by mermithids.

Flies are not the only insects that frequent decomposing matter. Rove beetles of the family Staphylinidae also come to these sites, not just to feed on refuse but to dine on other scavengers. Rove beetles are common in all amber deposits and, while most are predators, that does not prevent nematodes from parasitising them. An adult rove beetle in Baltic amber was surrounded by over 90 specimens of the allantonematid *Palaeoallantonema baltica** (Figs 93-95). The nematodes that had exited the beetle consisted of two morphotypes, one short and stout, which may represent females from the first parasitic generation, and a second short and slender group, which probably represents pre-adults of the second parasitic generation.

Haematophagous flies are rare in amber and the entrapment of many may be associated with their need to rest after taking a blood meal. This

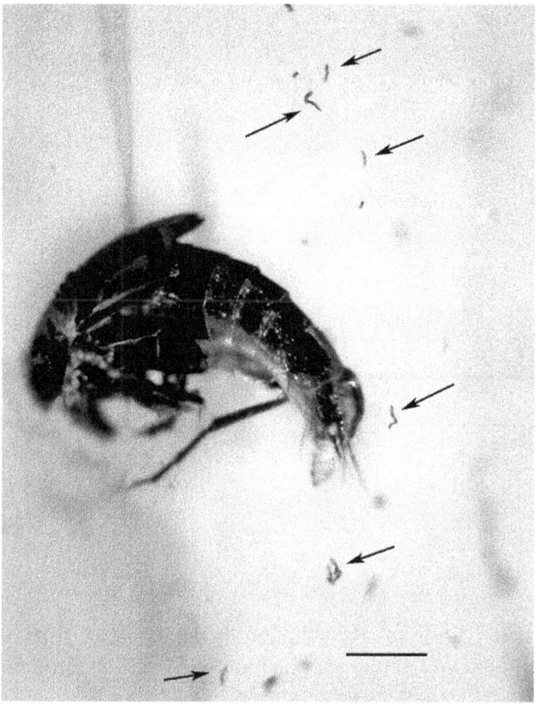

Fig. 93. *Parasitic juveniles (arrows) of the allantonematid* Palaeoallantonema baltica* *from a rove beetle (Staphylinidae) in Baltic amber (accession no. N-3-37). (Scale bar = 843 μm.)*

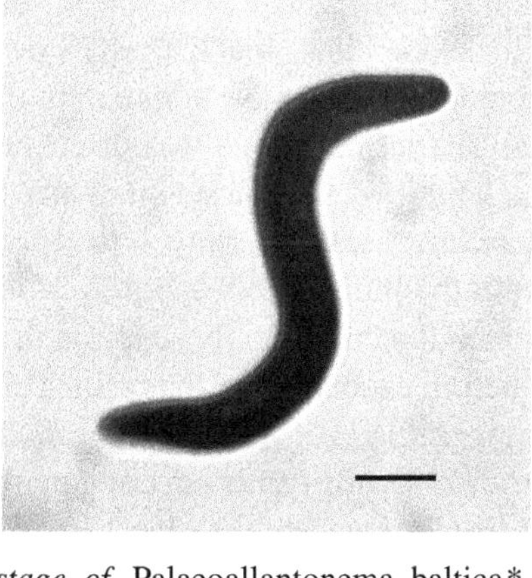

Fig. 94. *Parasitic stage of* Palaeoallantonema baltica* *from a rove beetle (Staphylinidae) in Baltic amber (accession no. N-3-37). (Scale bar = 337 μm.)*

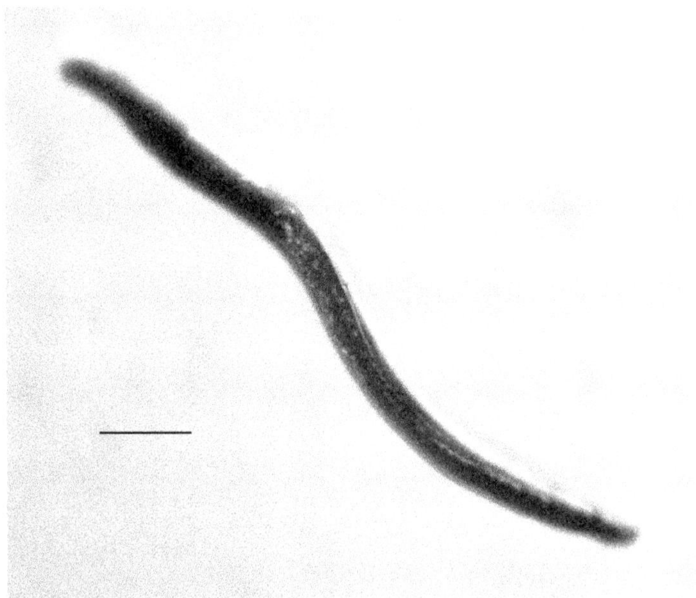

Fig. 95. *Parasitic adult of* Palaeoallantonema baltica* *from a rove beetle (Staphylinidae) in Baltic amber (accession no. N-3-37). (Scale bar = 30 μm.)*

behaviour occurs even if the insect has not completely finished its meal, as when disturbed by movements of the host. In a wooded area, the most natural place to rest is on the bark of a tree, which is probably how most bloodsucking flies ended up in amber.

Of all the blood-sucking flies, black flies (Simuliidae) and biting midges (Ceratopogonidae) are the most common in Baltic amber. In one amber piece, a small filarial nematode was discovered adjacent to the head of a female black fly (Fig. 96). Based on its size (294 μm long) and shape, *Cascofilaria baltica** was determined to represent a microfilaria that had escaped from the proboscis of the fly just after it had finished feeding. The swollen abdomen of the fly suggests that it had acquired a blood meal shortly before encountering the resin. The fossil falls within the size range of the microfilaria of species of *Onchocerca* Diesing, 1841 (190-395 μm long), a worldwide genus infecting mammals and vectored by black flies. This would represent the oldest fossil record of a filarial nematode, although this group may extend back at least to the Jurassic (see Table 5).

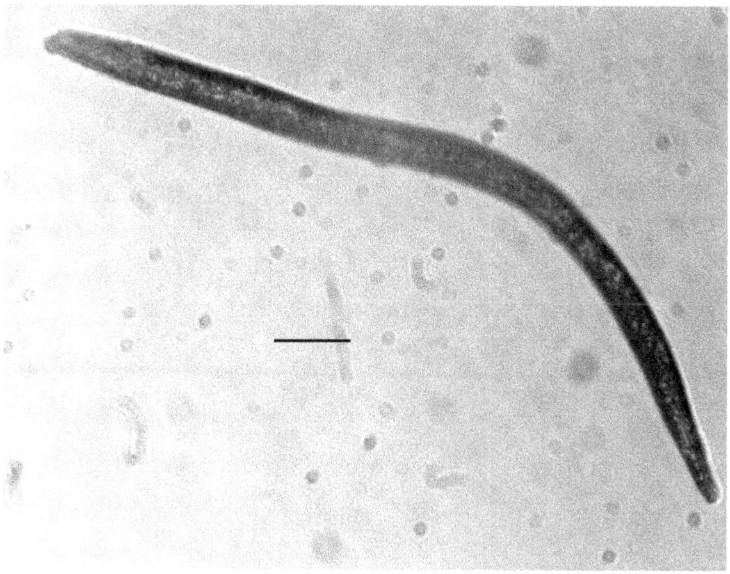

Fig. 96. *The filariid* Cascofilaria baltica* *(arrow) adjacent to a female blackfly (Simuliidae) in Baltic amber (accession no. SY-1-172). (Scale bar = 425 μm.)*

Spider biodiversity was especially high in the Baltic amber forest. There are more spider families in Baltic amber than there are in all of Europe today and, thus far, some 500 species have been described from this amber source (Wunderlich, 2004). With such diversity, it is not surprising that at least some were parasitised by mermithid nematodes. One of the hosts was a crab spider (Thomisidae), and the mermithid *Heydenius araneus* Poinar, 2000 is adjacent to the arachnid (Fig. 97). Its large size (24 mm in length) explains why the spider's abdomen is so flaccid and transparent.

The extant, worldwide, mermithid genus *Aranimermis* Poinar & Benton, 1986 appears to specialise on spiders. These nematodes have an indirect life cycle which involves an aquatic paratenic host. In the case of *Aranimermis giganteus* Poinar & Early, 1990 (Fig. 98), a parasite of New Zealand trapdoor spiders, the paratenic hosts are larval caddis flies, mayflies and stoneflies. The infective stage mermithids enter the body cavity of these paratenic hosts and are carried though the pupal and into the adult stage. The spiders are infected when they feed on adult paratenic hosts. It is possible that *H. araneus* had a similar indirect life cycle.

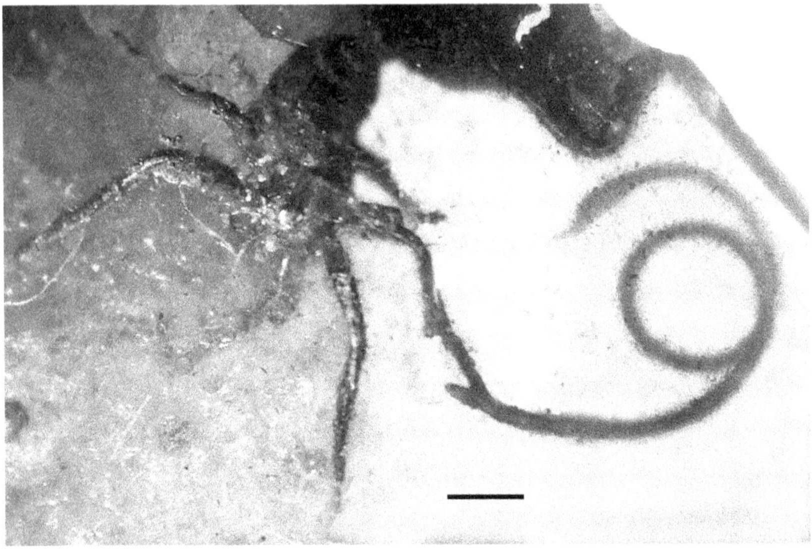

Fig. 97. *The spider mermithid,* Heydenius araneus *Poinar, 2000, adjacent to its clubionid spider host in Baltic amber (Andrew Cholewinski amber collection). (Scale bar = 2 mm.)*

Wunderlich (2004) reported some phoretic nematodes on several spiders in Baltic amber. Two rhabditids that were adjacent to a male *Eotrechalea annulata* Wunderlich, 2004 are described here as *Palaeoarachnus balticus**. Phoretic nematodes on extant spiders have been little studied, but rhabditids have been noted on the Dutch dune spider, *Centromerus sylvaticus* (Blackwall, 1841) (Noordam *et al.*, 1998).

The fossil record of spiders extends back to the Carboniferous and they probably represent another ancient host group for mermithids. If mermithids also attacked scorpions in the Silurian as they do today (Poinar & Stockwell, 1988) (Fig. 99), these arachnids could have been even earlier terrestrial hosts. And if mermithids attacked eurypteroids, especially the freshwater and terrestrial forms, these parasites could extend back to the Ordovician (Lehmann & Hillmer, 1983).

Eocene trace fossils

Before leaving the Eocene, attention should be given to some trace fossils from this period attributed to nematodes. Some sinusoidal trails in sedimentary deposits from the Eocene Green River Formation in Utah,

Fig. 98. *The extant mermithid* Aranimermis gigantea *Poinar & Early, 1990 in the body cavity of the New Zealand mygalomorph spider* Cantuaria borealis *Forster. (Scale bar = 3 mm.)*

USA, are thought to have been made by free-living microbotrophic nematodes (Moussa, 1969) (Fig. 100). Wave tracks of nematodes vary greatly depending on their size, shape and speed of travel (Croll, 1976). While these trails could have been made by aquatic or semi-aquatic nematodes crawling through sediment, the wave pattern is similar to that produced by some other limbless invertebrates. Another likely contender for the Green River trails is the elongate, legless larva of biting midges, which make similar sinusoidal trails (Fig. 101). The size range of larval biting midges overlaps with that of aquatic nematodes. These insect larvae occur in mud along the edges of streams and ponds and were abundant in the Eocene. A sinusoidal trail in Baltic amber may also have been made by a nematode (Fig. 102) but it is not certain.

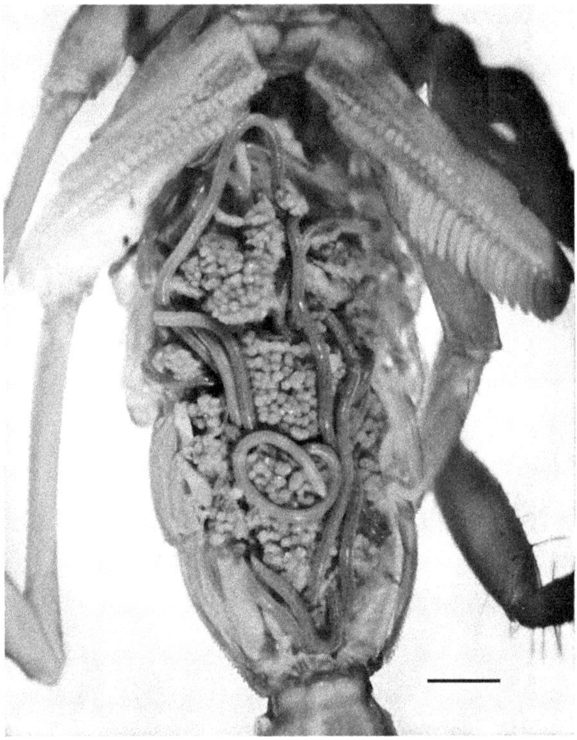

Fig. 99. *Extant mermithid parasite of a scorpion (Poinar & Stockwell, 1988). (Scale bar = 1.2 mm.)*

Dominican amber nematodes

A number of fossil nematodes occur in mid-Tertiary amber from the Dominican Republic. A study aimed at reconstructing the original Dominican amber forest from fossils in the amber showed that the ancient forest was very similar to that of a present-day tropical moist forest (Poinar & Poinar, 1999). These types of forests teem with plant and animal life and, while a wide range of biota occurs in Dominican amber (Perez-Gelabert, 2008), it only represents a fraction of what existed millions of years ago.

Eight long, slender nematodes ranging from 945 to 1260 μm in length were found free in a piece of amber and may have been living in the bark of a resin-producing tree (Figs 103, 104). Described as *Vetus elongatus**, their narrow body, rounded tail tip and stylet are characteristic of entomogenous females of some members of the family

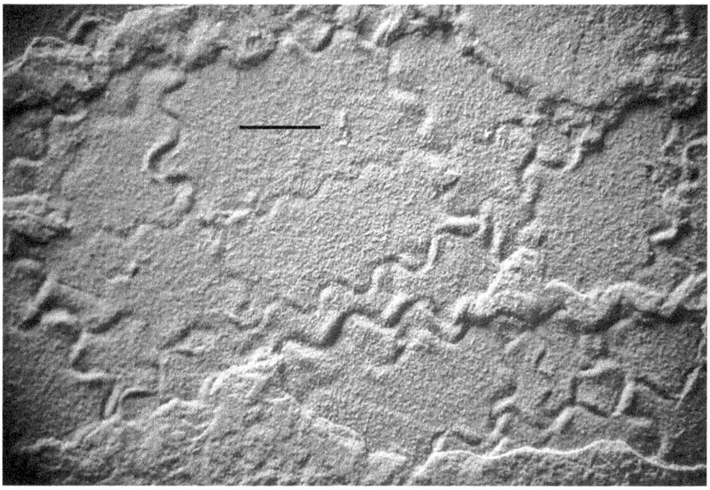

Fig. 100. *Tracks attributed to aquatic nematodes in the Eocene Green River Formation in Utah (Poinar collection). (Scale bar = 3 mm.)*

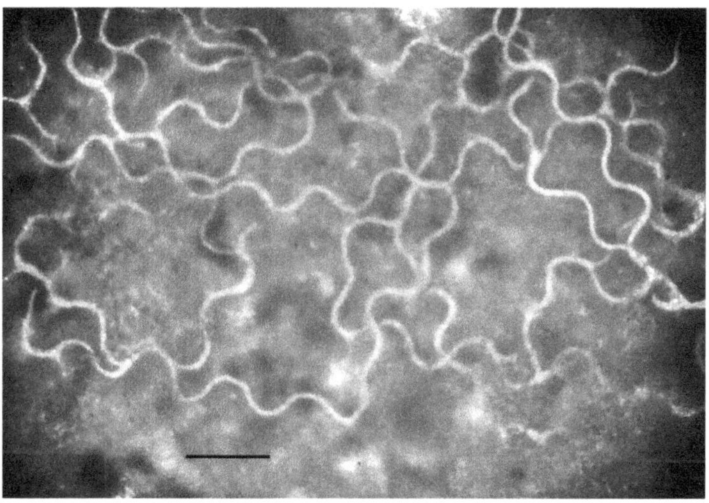

Fig. 101. *Tracks similar to those in the Green River Formation made by extant biting midge larvae (Ceratopogonidae) on an agar plate. (Scale bar = 3 mm.)*

Neotylenchidae, such as those of the genus *Deladenus* Thorne, 1941. The extant *D. siricidicola* Bedding, 1968 has two separate life cycles, one on fungi associated with *Sirex* wood wasps and a parasitic one in wood wasps. This species has been used successfully for the biological

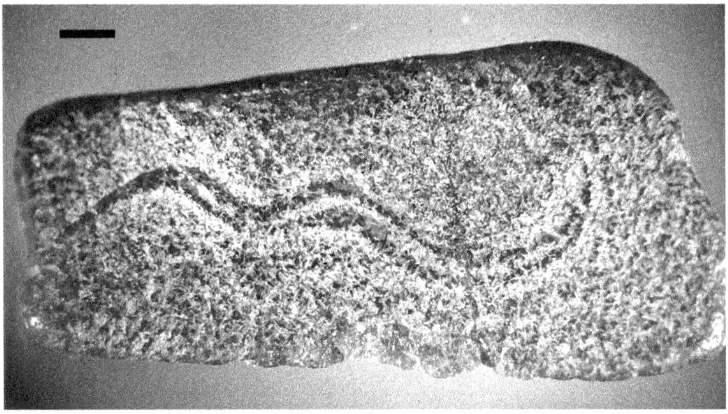

Fig. 102. *Impression on the surface of a piece of Baltic amber possibly made by a nematode. (Scale bar = 640 μm.)*

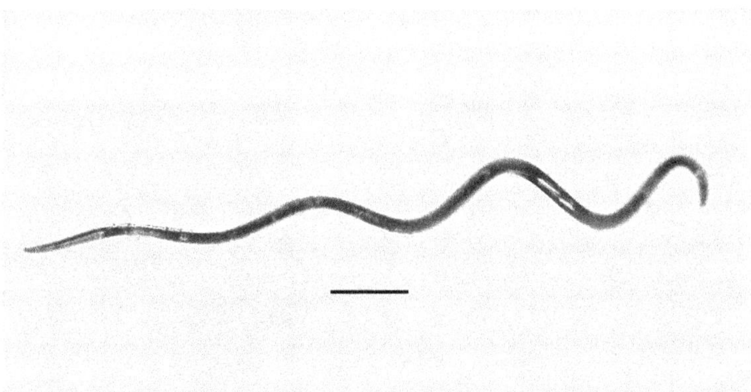

Fig. 103. Vetus elongatus*, *a possible infective stage entomogenous neotylenchid, in Dominican amber (accession no. N-3-33). (Scale bar = 113 μm.)*

control of wood wasps (Bedding, 1972a, 1993). It is possible that *V. elongatus** had a similar life cycle in the Dominican amber forest. Since there were no arthropods present, the insect hosts remain unknown.

Ants and termites are fairly well represented in Dominican amber since representatives of both groups occur on the bark or in wood of the amber-producing tree, *Hymenaea protera* Poinar, 1991, a member of the legume family. Both of these insect groups have nematode associates or parasites today. One of the most common genera of ants in Dominican amber is *Azteca* Forel, 1878, and it is obvious by their presence that the workers spent much time patrolling the bark. Some pieces of Dominican

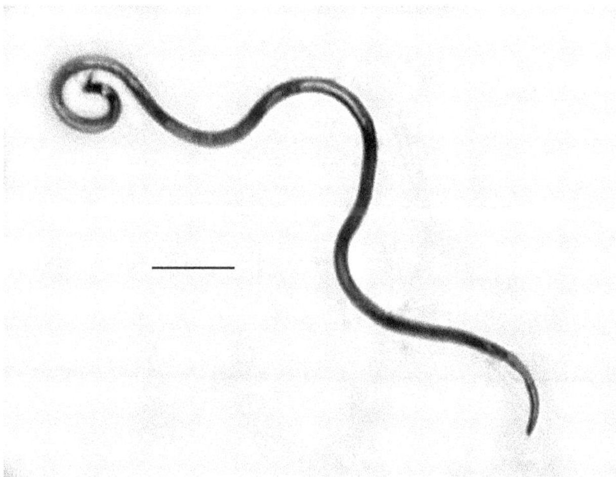

Fig. 104. *A second specimen of* Vetus elongatus* *in Dominican amber (accession no. N-3-33). (Scale bar = 113 μm.)*

Fig. 105. *Three dauer juveniles of* Formicodiplogaster myrmenema* *associated with a worker* Azteca alpha *Wilson ant in Dominican amber (accession no. N-3-18G). (Scale bar = 150 μm.)*

amber contain *Azteca* ants fighting with pseudoscorpions or carrying insects back to their nests, while other pieces contain portions of *Azteca* carton nests. Extant *Azteca* ants construct their nests from fibrous plant material and it is obvious that their ancestors did the same.

Some fossil *Azteca* ants in Dominican amber are associated with nematodes (Figs 105-109). Fossil phoretic nematodes associated with

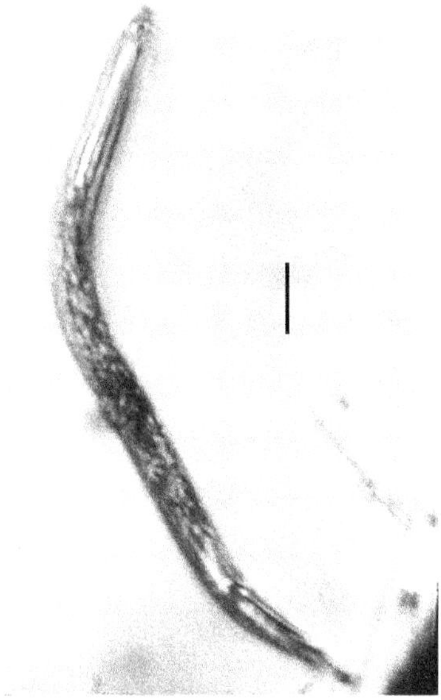

Fig. 106. *Detail of one of the specimens of* Formicodiplogaster myrmenema* *shown in Figure 105 (accession no. N-3-18G). (Scale bar = 16 μm.)*

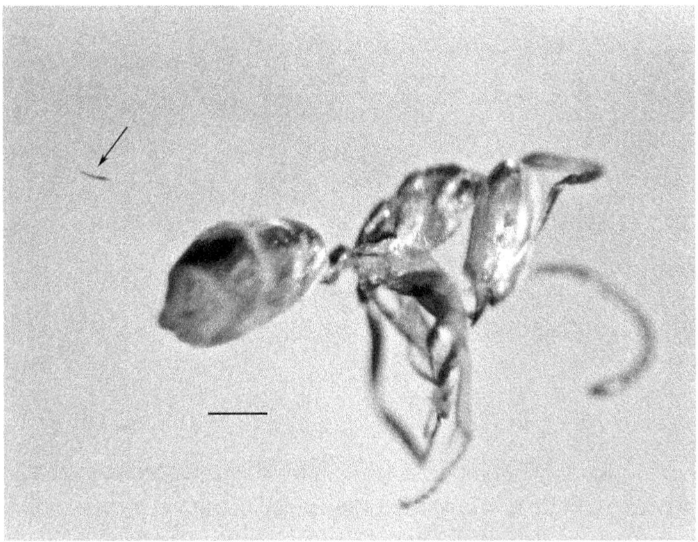

Fig. 107. *Specimen (arrow) of* Formicodiplogaster myrmenema* *associated with a worker* Azteca alpha *ant in Dominican amber (accession no. N-3-18A). (Scale bar = 491 μm.)*

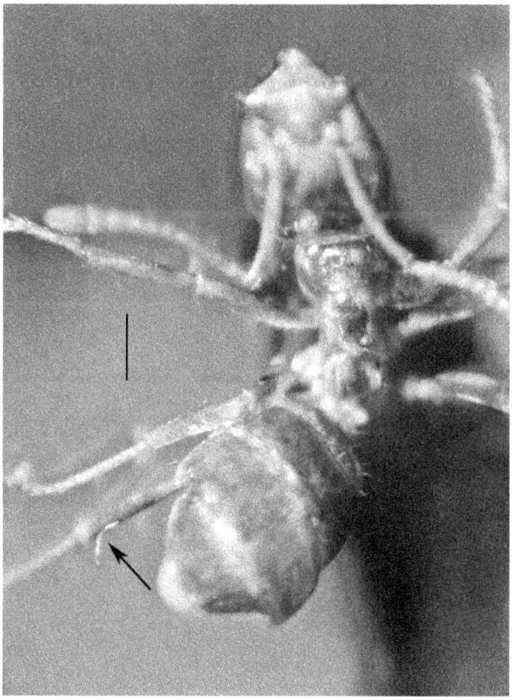

Fig. 108. *Another example (arrow) of* Formicodiplogaster myrmenema* *associated with a worker* Azteca alpha *ant in Dominican amber (accession no. N-3-18F). (Scale bar = 215 μm.)*

worker *Azteca* ants are in the dauer stage, while those associated with *Azteca* nests represent actively reproducing colonies of adults and juveniles. Based on the size and shape of the fossil dauer nematodes, they are considered members of the family Diplogastridae and described as *Formicodiplogaster myrmenema**. Diplogastrids are also associated with extant *Azteca* ants in Costa Rica (Paul E. Hanson, pers. comm.).

Studies conducted on phoretic associations between extant nematodes and ants (Wahab, 1962; Nickle & Ayre, 1966; Markin & McCoy, 1968) indicate that the dauer juveniles are carried in the salivary glands of worker ants. It is possible that the nematodes cause lesions that lead to the ant's death, but this is unclear. What is interesting about the fossil dauer juveniles of *F. myrmenema** is that they are always located near the ant's abdomen, suggesting that they were lodged in the intersegmental membranes. While it is possible that some juveniles also occurred in the salivary glands, they were not detected. The presence of reproducing

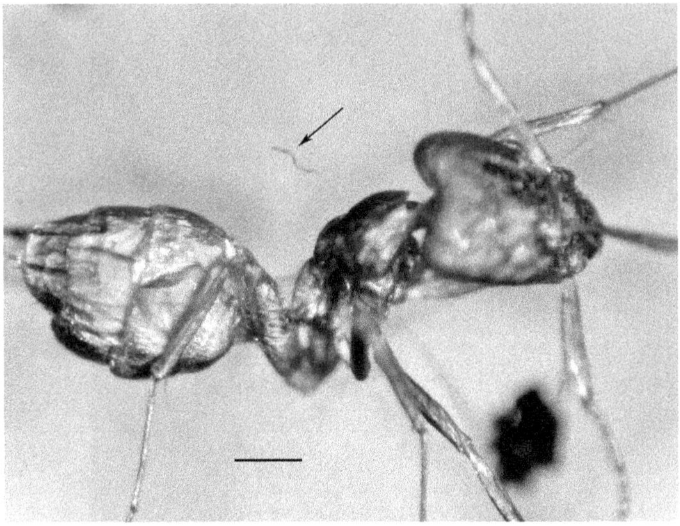

Fig. 109. *Another specimen (arrow) of* Formicodiplogaster myrmenema* *associated with a worker* Azteca alpha *ant in Dominican amber (accession no. N-3-18B). (Scale bar = 191 μm.)*

colonies of *F. myrmenema** in pieces of amber with portions of *Azteca* nests show that the nematodes were probably feeding on microorganisms in the ant nests.

As noted earlier, extant ants are also hosts of mermithid nematodes (Poinar, 1975; Poinar *et al.*, 2006b, 2007b). A species of the ant genus *Linepithema* Mayr, 1866, which belongs to the same subfamily as *Azteca* (the Dolichoderinae), served as host to the Dominican amber mermithid *Heydenius myrmecophila* Poinar, Lachaud, Castillo & Infante, 2006. The specimen (Figs 110, 111) was a parasitic juvenile that had emerged from the ant host. It is unusual for a parasitic juvenile to emerge before maturity but, when the body wall of the host has been damaged, extant developing mermithids often leave prematurely and the abdomen of the fossil ant had ruptured. There is no way of knowing if *H. myrmecophila* had an indirect life cycle that involved an aquatic paratenic host. Mermithid parasites of ants were previously noted in Baltic amber and this parasite-host association was probably widespread throughout the Tertiary, just as it is today. Mermithid parasites of extant neotropical ants have been little studied and are being discovered continuously, as shown by the recent report of a mermithid parasitising fire ants in Argentina (Poinar *et al.*, 2007b) (Fig. 112).

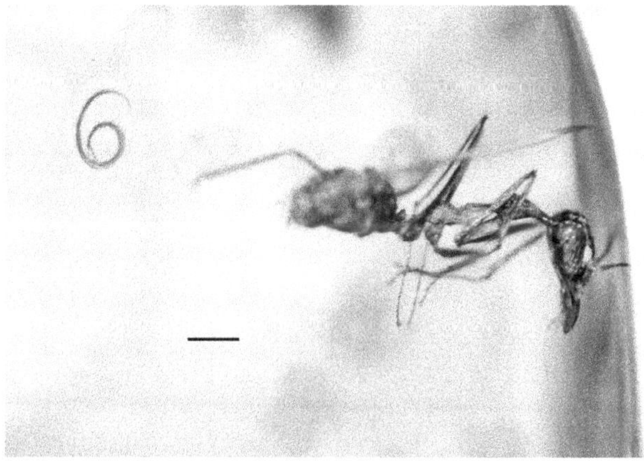

Fig. 110. *A mermithid,* Heydenius myrmecophila *Poinar, Lochaud, Castillo & Infante, 2006, that had emerged from a worker* Linepithema *sp. ant in Dominican amber (accession no. N-3-65). (Scale bar = 543 μm.)*

Another genus of ants in Dominican amber is *Cephalotes* L., 1758, the flat-headed ants. Extant species colonise hollow stems and cavities in plants and the workers and soldiers use their large, flat heads to block the nest entrances. They have the peculiar habit of collecting, among other items, bird droppings to feed to their developing brood, a feature which has been used to good advantage by some nematodes.

A *Cephalotes* worker ant in Dominican amber with a hole in its abdomen was surrounded by hundreds of small nematode eggs, some with emerging juveniles (Figs 113-115). These eggs are very similar in size and shape to those of the extant *Cephalotes* tetradonematid nematode, *Myrmeconema neotropicum* Poinar & Yanoviak, 2008 (Figs 116, 117), and the fossil species is described as *Myrmeconema antiqua**. It is likely that *M. antiqua** had a life cycle similar to the bizarre one of *M. neotropicum*. Females of *M. neotropicum* mature in worker ants and release infective eggs into the insect's body cavity. At this time, the colour of the ant's gaster (abdomen) changes from its normal black hue to a shiny red. The infected ants also become more lethargic and suddenly develop the habit of raising their abdomen. The bright abdomen resembles red berries that are relished by birds. When a bird ingests an infected ant, the eggs of *M. neotropicum* pass through the digestive tract largely unharmed and are deposited in the droppings. Worker ants then collect and feed the infested bird droppings to their

Fig. 111. *Detail of* Heydenius myrmecophila *Poinar, Lochaud, Castillo & Infante, 2006 in Dominican amber (accession no. N-3-65). (Scale bar = 2 mm.)*

brood. The ingested nematode eggs hatch in the gut and the juveniles penetrate into the body cavity of the ant larvae.

This fossil shows that the unique *Myrmeconema* life cycle has been in existence for some 20-40 million years. The hole in the abdomen of the parasitised Dominican amber *Cephalotes* could have been made by the bill of a bird that mistook the gaster for a berry and was attempting to eat it. Unfortunately, since colours fade in amber, it is not possible to determine if the gaster of the fossil ant was red.

Tetradonematids were not the only nematode parasite that *Cephalotes* ants had to contend with in the Dominican amber forest. A small allantonematid also selected this genus of ants as a host and seven preadults of *Palaeoallantonema cephalotae** had emerged from a worker of *Cephalotes serratus* (Vierbergen & Scheven, 1995) in a piece of Dominican amber (Figs 118, 119). A faint stylet as well as a vulval opening could be detected in a fourth-stage female nematode. It is curious that several years ago, when this specimen was discovered, the author had just finished examining some extant allantonematids from workers of

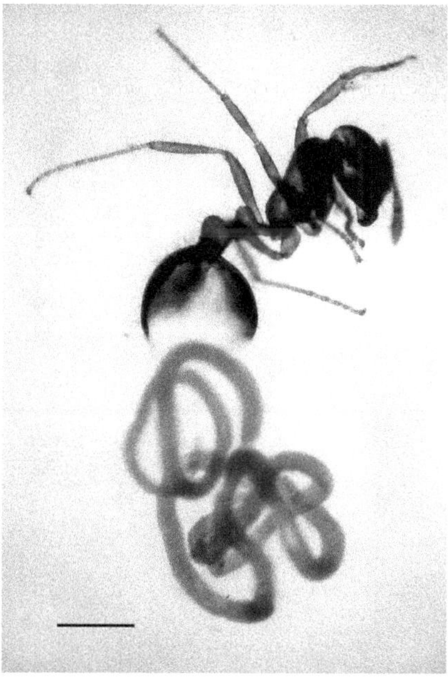

Fig. 112. *The extant mermithid* Allomermis solenopsi *Poinar, Porter, Tang & Hyman, 2007 from the fire ant* Solenopsis invicta *Buren. (Scale bar = 800 µm.)*

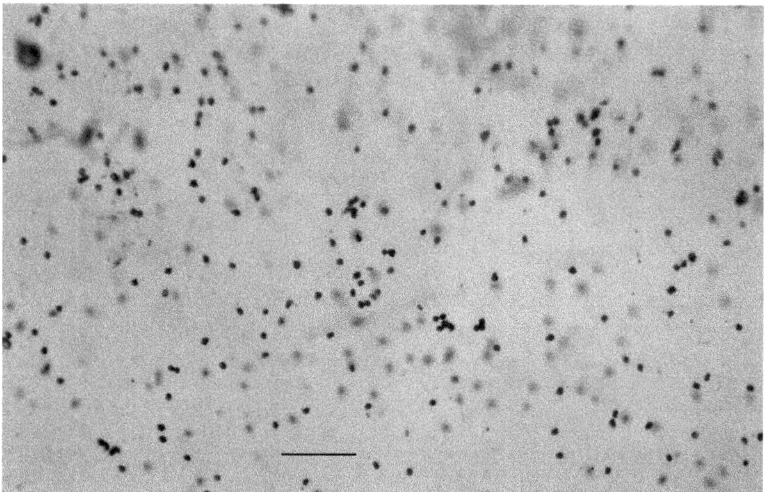

Fig. 113. *Eggs of* Myrmeconema antiqua* *from an ant of the genus* Cephalotes *Latreille in Dominican amber (accession no. N-3-75). (Scale bar = 360 µm.)*

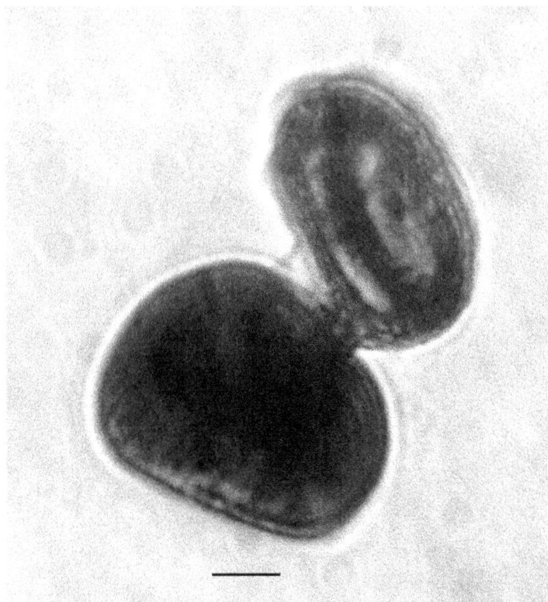

Fig. 114. *Detail of two eggs of* Myrmeconema antiqua* *adjacent to an ant of the genus* Cephalotes *Latreille in Dominican amber. Note outline of infective stage nematode in upper egg (accession no. N-3-75). (Scale bar = 8 μm.)*

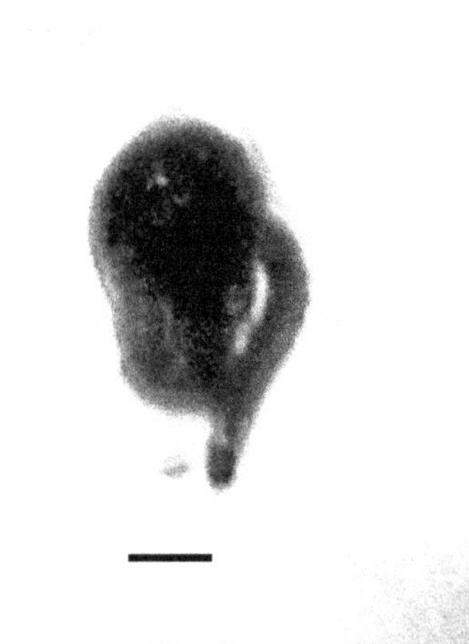

Fig. 115. *Infective juvenile of* Myrmeconema antiqua* *emerging from an egg in Dominican amber (accession no. N-3-75). (Scale bar = 11 μm.)*

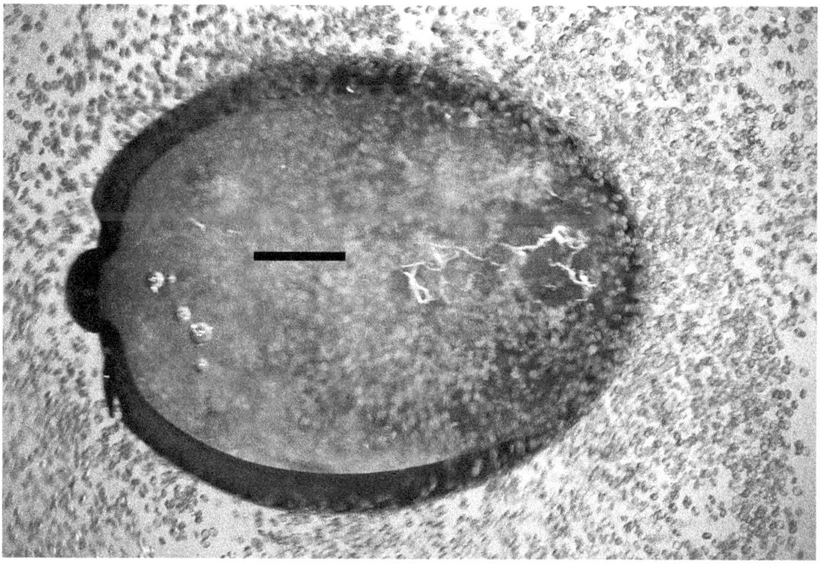

Fig. 116. *Eggs of the extant tetradonematid* Myrmeconema neotropicum *Poinar & Yanoviak, 2008 surrounding the detached abdomen of its neotropical ant host* Cephalotes atratus *(Linnaeus). (Scale bar = 855 μm.)*

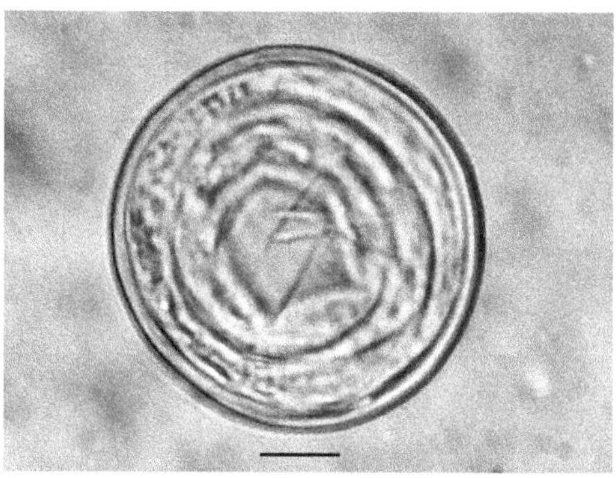

Fig. 117. *Egg of the extant* Myrmeconema neotropicum *Poinar & Yanoviak, 2008 containing an infective stage juvenile. (Scale bar = 12 μm.)*

Cephalotes christopherseni (Forel, 1912) collected in Peru. This species has not yet been described, but the emerging juveniles are very similar to *P. cephalotae**, although the pre-adults are somewhat larger than the fossil species (Fig. 120). Previous to these two cases only a single extant allantonematid was known from ants, namely *Formicitylenchus ore-*

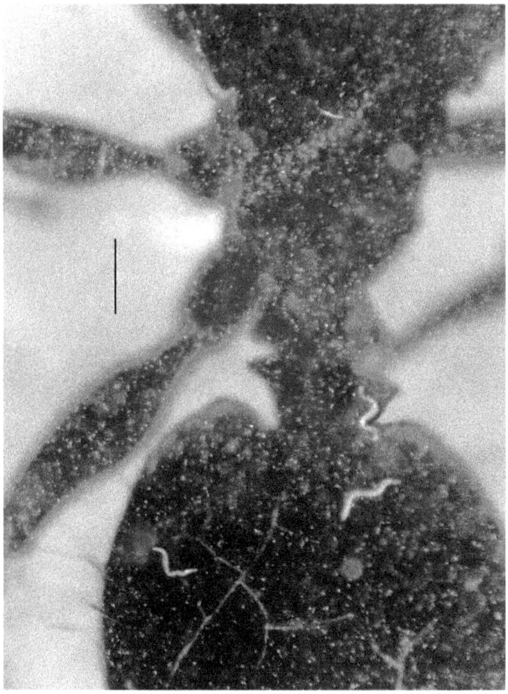

Fig. 118. *Parasitic juveniles of the allantonematid* Palaeoallantonema cephalotae* *that emerged from the ant* Cephalotes serratus *Vierbergen & Scheve in Dominican amber (accession no. N-3-77). (Scale bar = 298 μm.)*

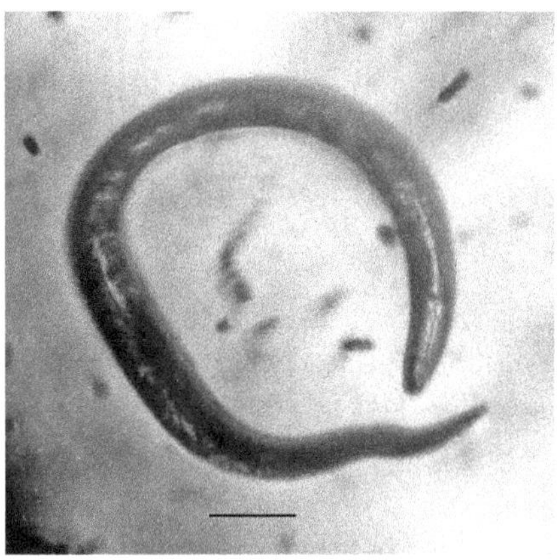

Fig. 119. *Detail of a specimen of* Palaeoallantonema cephalotae* *from the ant* Cephalotes serratus *Vierbergen & Scheve, in Dominican amber (accession no. N-3-77). (Scale bar = 24 μm.)*

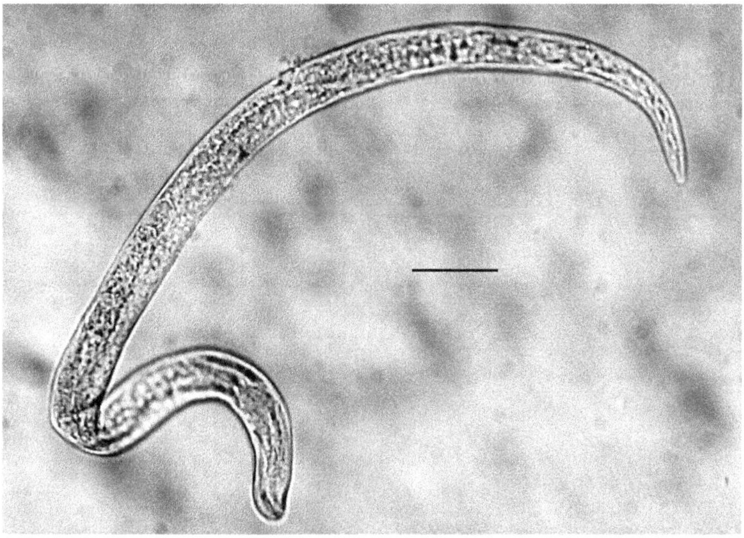

Fig. 120. *An extant allantonematid juvenile from the ant* Cephalotes christopherseni *(Forel) from Panama. (Scale bar = 23 μm.)*

gonensis Poinar, 2003 (Poinar, 2003a) parasitising *Camponotus* spp. in western North America. These three records suggest that allantonematid parasitism of ants may be a New World phenomena, since extensive investigations of European ants have not revealed their presence.

After ants, the most common social insects in Dominican amber are bees, especially small stingless bees of the genus *Proplebeia* (Wille & Chandler, 1964). While stingless, they actively defend their colonies by biting their attackers. Bees, especially social species, have fewer nematode associates than ants. Perhaps it is because they are more stringent in cleaning their hives, are more careful about bringing in foreign objects, or their nests are not conducive to nematode habitation. However, in the Dominican amber forest some diplogastrid nematodes lived in the arboreal hives of the stingless bee, *Proplebeia dominicana* (Wille & Chandler, 1964) (Hymenoptera: Apidae). Specimens of *Apifilaria proplebiae** are represented by dauer stages that were carried by worker bees (Figs 121-123). Their presence on two workers in separate pieces of Dominican amber would have been insignificant in extending their range since only queens establish new hives. Most stingless bees caught in amber are older workers carrying out the final task in their lives. Newly emerged adult bees care for the

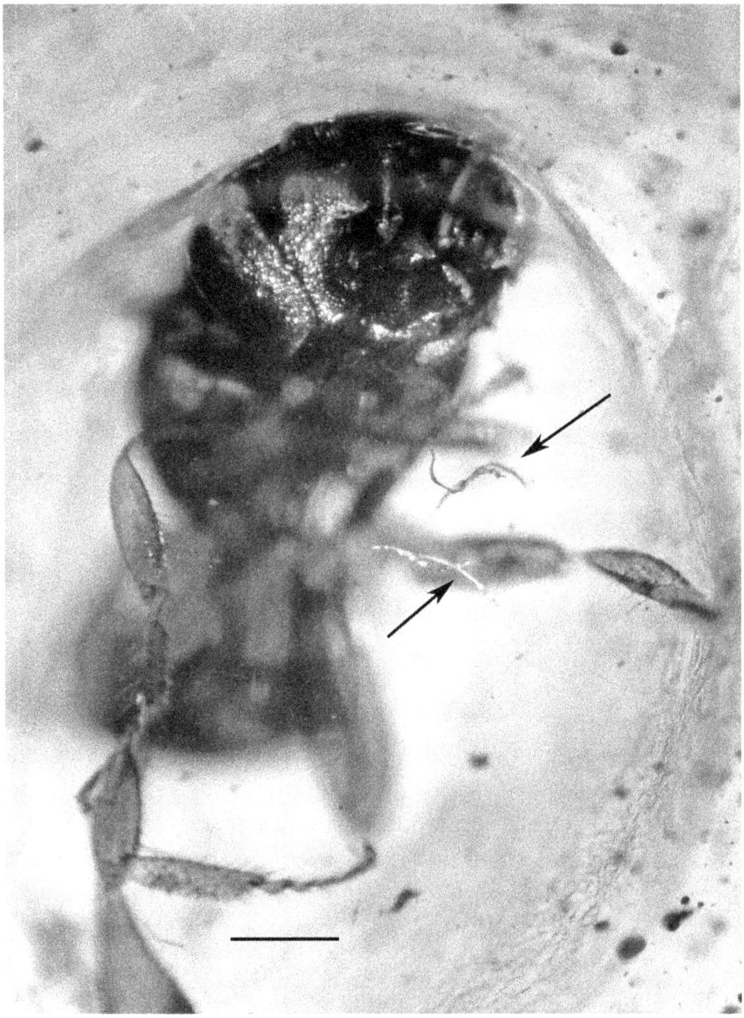

Fig. 121. *Several dauer juveniles (arrows) of* Apifilaria proplebiae*, *adjacent to their stingless bee host,* Proplebeia dominicana *Wille & Chandler, in Dominican amber (accession no. N-3-25). (Scale bar = 523 μm.)*

brood, older ones forage for pollen, and in the final days of their lives they collect resin. Some never returned from this task, not because they became entrapped in resin, but because they were targeted by ambush bugs that waited by the resin deposits (Poinar & Poinar, 1999).

The shape and size of the nematodes and a comparison with the dauer stages of *Acrostichus* (= *Aduncospiculum*) *halicti* (Giblin & Kaya, 1984)

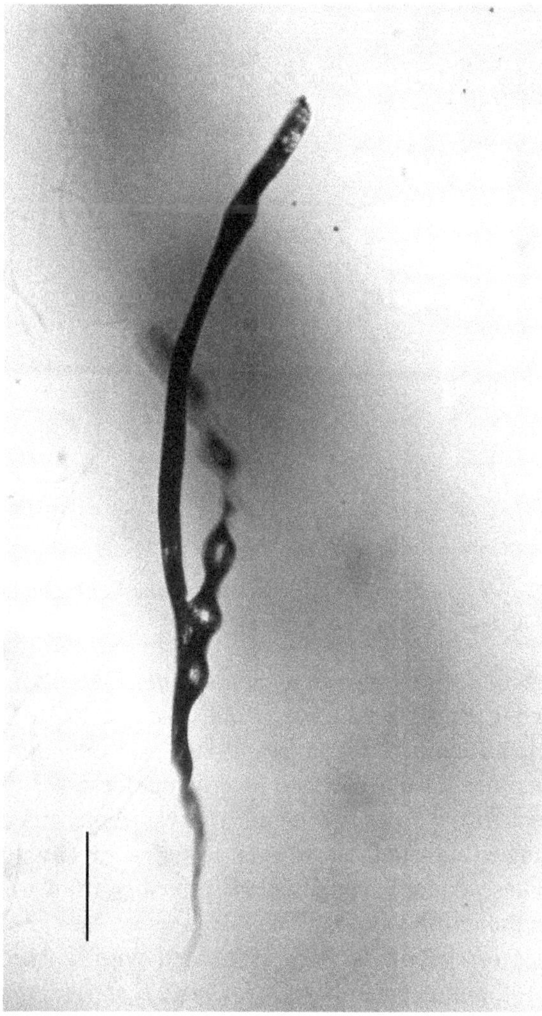

Fig. 122. *Two dauer juveniles of* Apifilaria proplebiae**. One shows distortions resulting from the fossilisation process (accession no. N-3-25). (Scale bar = 71 μm.)*

that are associated with solitary bees today suggest that *A. proplebiae** is a diplogastrid that was colonising the hives of stingless bees. Another diplogastrid is associated with solitary bees of the genus *Andrena* L., 1758 (Hymenoptera: Andrenidae) (Erteld, 1994).

Termites are another group of eusocial insects found in amber. Nematodes of the families Rhabditidae and Diplogastridae have interesting

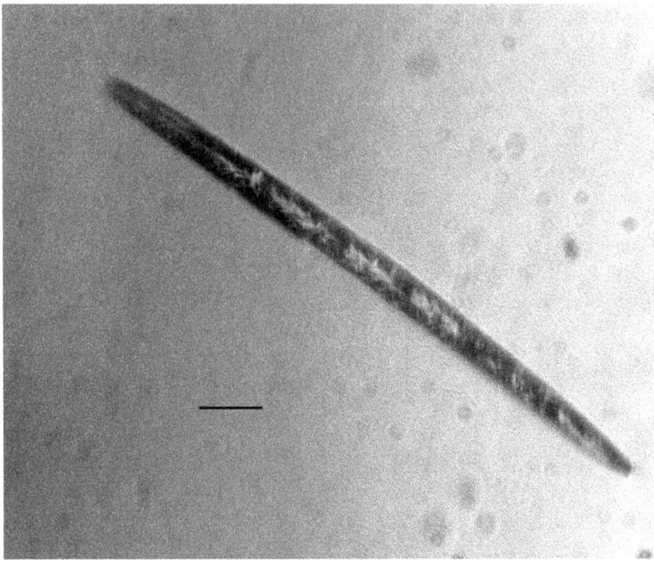

Fig. 123. *Dauer juvenile of* Apifilaria proplebiae* *in Dominican amber (accession no. N-3-47). (Scale bar = 38 μm.)*

associations with a wide range of extant termites (Merrill & Ford, 1916; Banks & Snyder, 1920; Pemberton, 1928; Massey, 1971; Poinar, 1972, 1975; Coppel & Liang, 1987; Miller, 1989), and some termites in Dominican amber were also associated with nematodes.

One diplogastrid, *Filotermes minutus**, was represented by numerous dauer stages surrounding worker termites in Dominican amber (Fig. 124). These dauer juveniles were probably from a large colony developing in the termite's nest.

Two species of rhabditids were associated with worker termites in other pieces of Dominican amber. One, *Prescafilotermes dominicanus**, occurred with representatives of the termite family Rhinotermitidae. Five pieces of Dominican amber contained not only dauer stages but also reproductive stages of *P. dominicanus** (Figs 125-128). It is obvious from some amber pieces that this nematode develops on the faecal pellets of termites, possibly feeding on symbiotic bacteria and protozoa that are voided with faecal material (Fig. 127).

A dauer of a second rhabditid species, *Prescafilotermes elongatus**, was associated with termites of the family Termitidae (Figs 129, 130). The dauer of this species is fairly large (654 μm long) and similar to some extant termite-associated rhabditids, such as *Rhabditis ossicula*

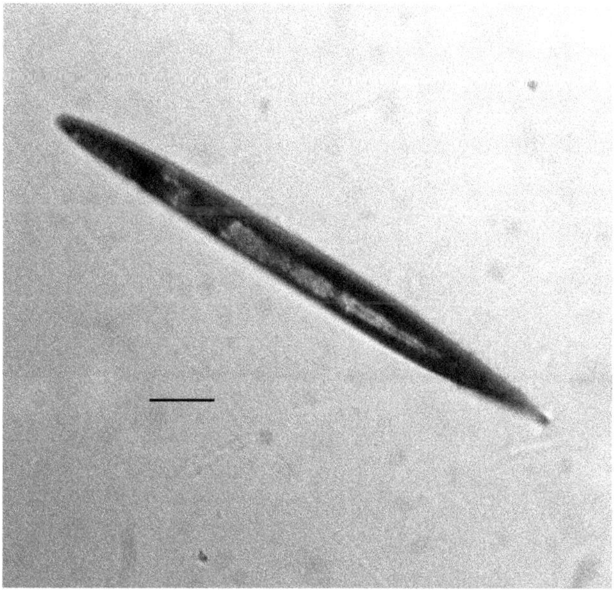

Fig. 124. *Dauer juvenile of* Filotermes minutus* *that was adjacent to a termite in Dominican amber (accession no. N-3-26F). (Scale bar = 17 μm.)*

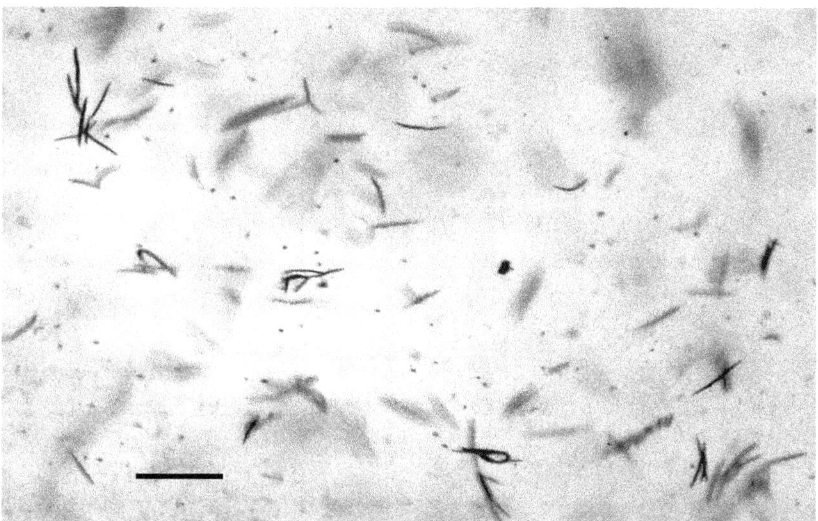

Fig. 125. *Developing stages of* Prescafilotermes dominicanus* *together with worker termites in Dominican amber (accession no. N-3-26A). (Scale bar = 571 μm.)*

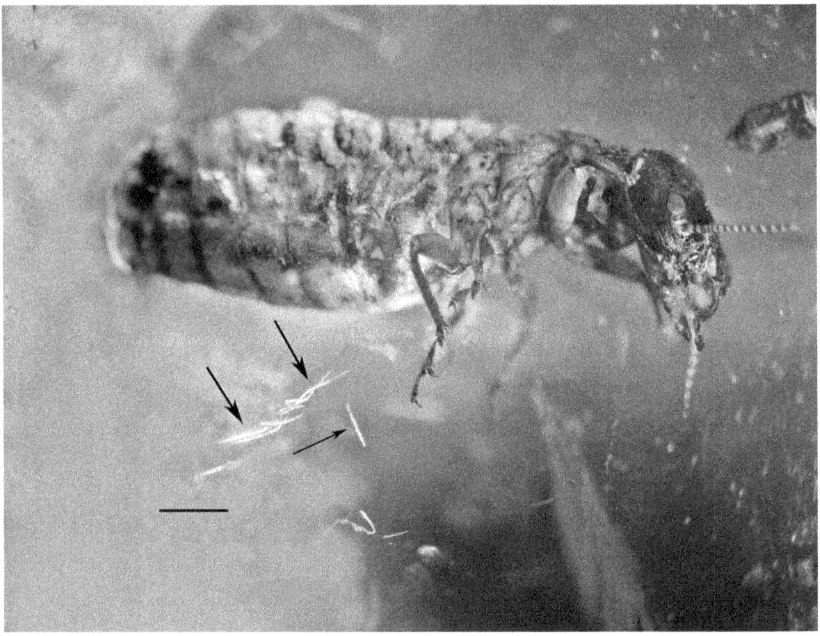

Fig. 126. *Dauer juveniles of* Prescafilotermes dominicanus* *adjacent to a worker termite in Dominican amber (accession no. N-3-26B). (Scale bar = 501 μm.)*

Massey, 1971 with a dauer length of 425 μm, and *Rhabditonema fastidiosum* Massey, 1971 with a dauer length of 538 μm.

The dauer stages of some extant nematodes enter the head glands of healthy termites and remain there without further development. Sometimes they pass into the termite's gut (Fig. 131). The association can be considered internal phoresis, but it is not known if the dauer juveniles can break through the head glands or gut wall of the termite and cause a lethal septicaemia. However, when the termite dies, the dauer juveniles exsheath and develop on the carcass. If the termite dies in the nest or termitarium, the nematodes, having finished with the cadaver, may continue their development on faecal matter or surrounding debris. Some extant termite dauer stages are quite active and will enter the host's trachea (see Fig. 27) as well as the maxillary glands (Miller, 1989).

Determining the exact location of the fossil dauer stages on their termite carriers is difficult. The dauer of extant nematodes occur in the head glands, trachea, gut and intersegmental membranes of the abdomen. While some of the fossil forms described here could be

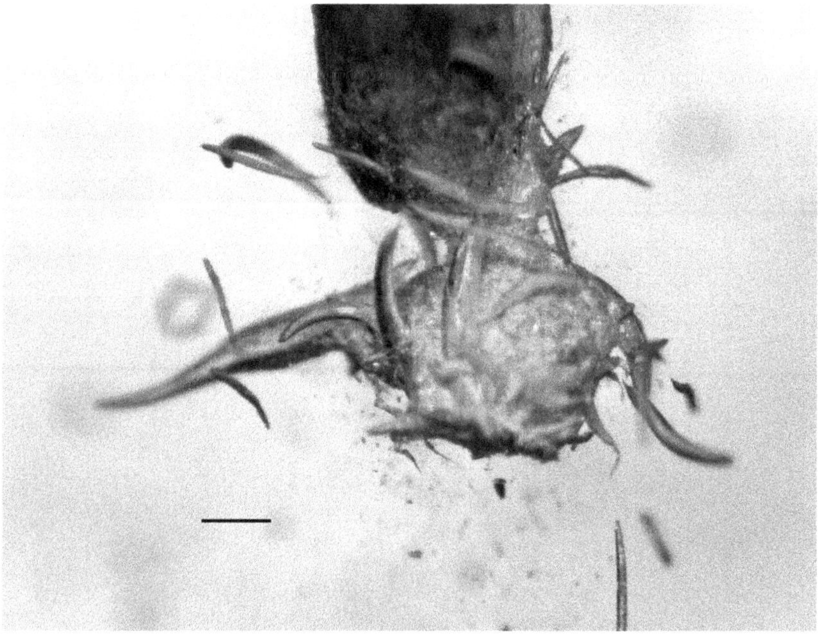

Fig. 127. *Small group of* Prescafilotermes dominicanus* *developing on termite faeces in Dominican amber (accession no. N-3-26C). (Scale bar = 406 μm.)*

concealed in the head glands, all were adjacent to the abdomen, suggesting that they were carried in the intersegmental folds of the termite's abdomen.

Some nematodes appear to have been feeding on microorganisms developing in rotting portions of the bark of the Dominican amber tree. While the 89 specimens of *Vetus maximus** represented all developmental stages, including eggs and adults (Figs 132, 133), it was not possible to place them in a definite family. However, they are probably either rhabditids or diplogastrids.

Bark lice are common in amber since, as their name implies, they spend most of their lives on tree bark and have a good chance of contacting resin deposits. One adult bark louse in Dominican amber was associated with a nematode described here as *Psocidonema dominicana** (Figs 134, 135). One specimen had left the insect and was free in the amber while another was still attached to the thorax. They appear to be dauer juveniles of a rhabditid and exactly where they were developing in the environment is unknown, possibly in a damaged area on a branch

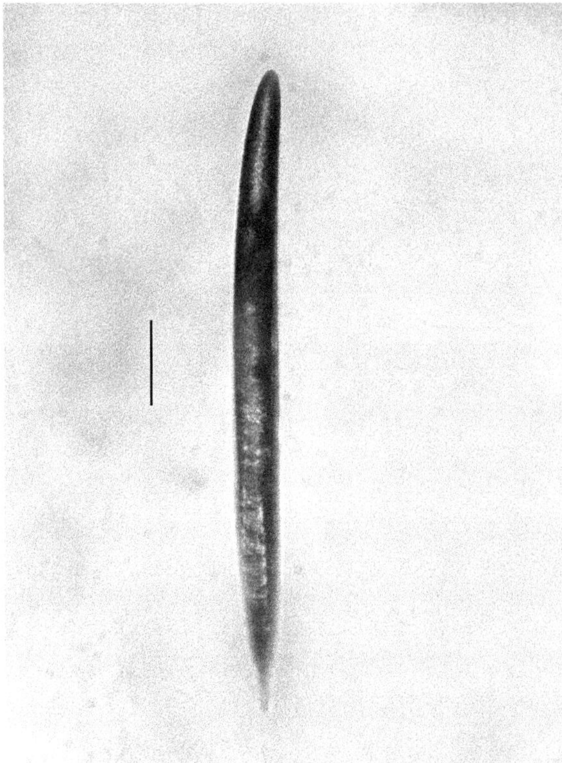

Fig. 128. *Dauer juvenile of* Prescafilotermes dominicanus* *in Dominican amber (accession no. N-3-26D). (Scale bar = 41 µm.)*

that was visited by the bark louse. Apparently no parasitic or phoretic nematodes have been described from extant members of this common insect group.

Both mycetophagous and microbotrophic nematodes are associated with bark beetles (Scolytidae) and pinhole beetles (Platypodidae). Because beetle galleries are ideal habitats for nematode development, there is often more than one species living in tunnels made by a particular beetle. Since most of these nematodes are carried from tree to tree by the insects, it is not unusual to find the dauer stages of several species associated with a single beetle.

Two aphelenchoidid nematodes feeding on fungi in beetle galleries provide a glimpse of mycetophagous nematodes in the amber forest. One, *Bursaphelenchus similus**, is represented by a large population

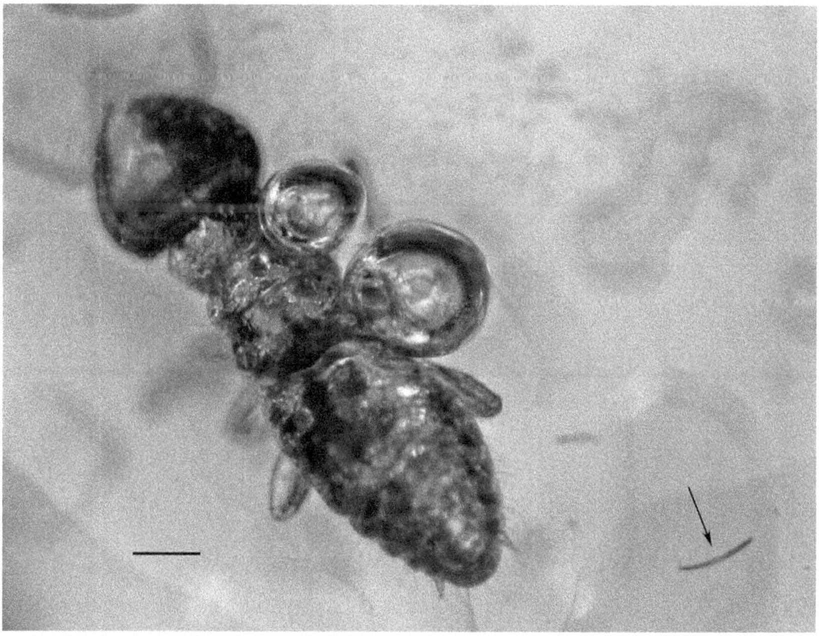

Fig. 129. *Dauer juvenile (arrow) of* Prescafilotermes elongatus* *adjacent to a nasute termite soldier in Dominican amber (accession no. N-3-26E). (Scale bar = 568 μm.)*

of aphelenchoidid nematodes adjacent to a platypodid beetle and some frass that probably originated from the beetle's gallery (Figs 136-138). These large, stylet-bearing forms, with a curved male tail and digitate female tail, align with members of the genus *Bursaphelenchus* Fuchs, 1937. It is likely that *B. similus** was developing in the gallery of a platypodid when the beetle, frass and nematodes were caught in a resin flow. Apparently the developmental conditions were optimum since there is no evidence of resistant juveniles. Some extant species of mycetophagous *Bursaphelenchus* species can cause serious damage to pine trees (Fig. 139).

The second species of aphelenchoidid associated with an adult platypodid beetle was *Cryptaphelenchus dominicus**. It would appear from the nearly 60 specimens of dauer and third-stage females in the piece of amber that developmental conditions in the beetle gallery were suboptimal, forcing most of the population to form resistant stages. Most nematodes inhabiting beetle galleries have the capability of forming dauer (resistant) juveniles under suboptimum conditions and the great

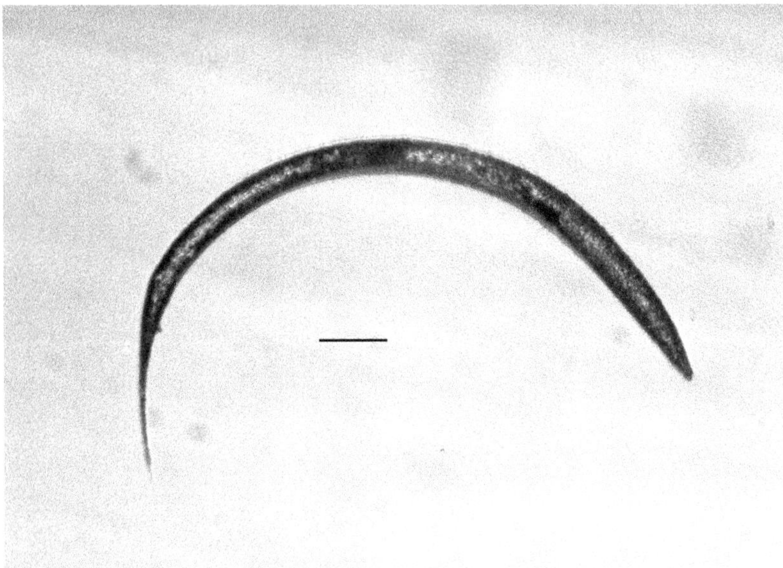

Fig. 130. *Dauer juvenile of* Prescafilotermes elongatus* *in Dominican amber (accession no. N-3-26E). (Scale bar = 52 μm.)*

majority of these form phoretic associations with their beetle carriers. This ensures that they are transported from one food source to another.

Many specimens of *C. dominicus** were located near the head of the beetle with some emerging between the head and thorax (Figs 140-142; Plate 16). Their location and morphological features, including a stylet (Fig. 143), are typical for species of *Cryptaphelenchus* Fuchs, 1937 (Rühm, 1956; Massey, 1974).

The dauer stages of microbotrophic nematodes were also associated with platypodid beetles in Dominican amber. Twelve dauer juveniles of *Panagrolaimus neotropicus** were positioned near the posterior end of an adult *Cenocephalus* platypodid beetle (Figs 144, 145). The partially opened elytra and extended membranous wings exposed the position of some of the nematodes. In addition to being under the elytra, dauer of *P. neotropicus** were in the intersegmental membranes of the abdomen. The dauer of the extant *Panagrolaimus verrucosus* Fuchs, 1930 occur under the elytra and in dorsal abdominal folds of its bark beetle host (Rühm, 1956).

Dauer juveniles of *Scolytonema dominicana** were carried by both bark (Figs 146, 147) and pinhole beetles (Figs 148-150) in separate

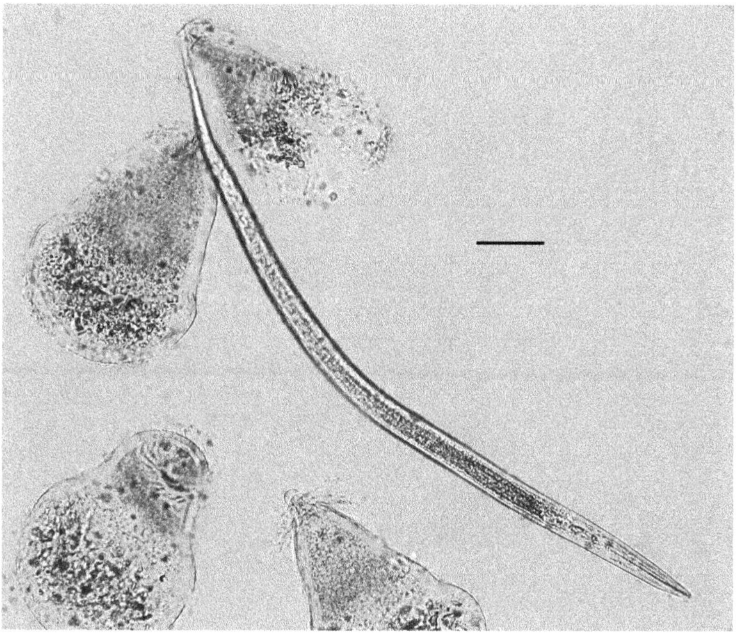

Fig. 131. *A rhabditid dauer juvenile associated with the extant termite* Zootermopsis angusticollis *(Hagen). Note associated mutualist flagellates. (Scale bar = 38 μm.)*

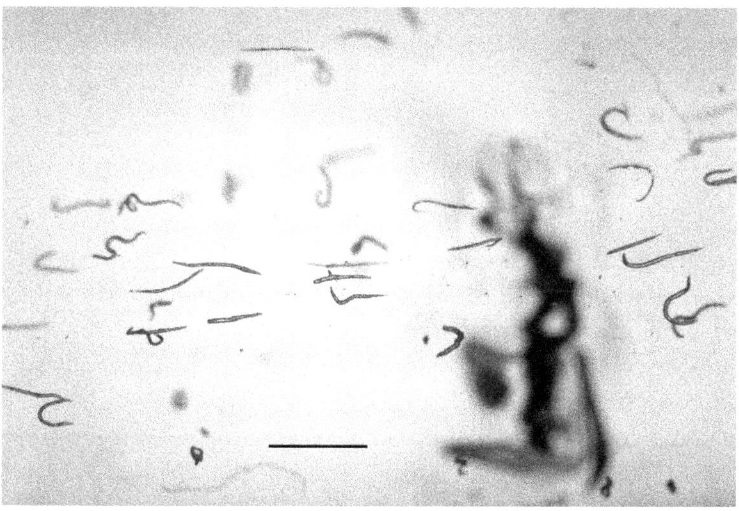

Fig. 132. *Reproducing population of* Vetus maximus* *in Dominican amber (accession no. N-3-4). (Scale bar = 106 μm.)*

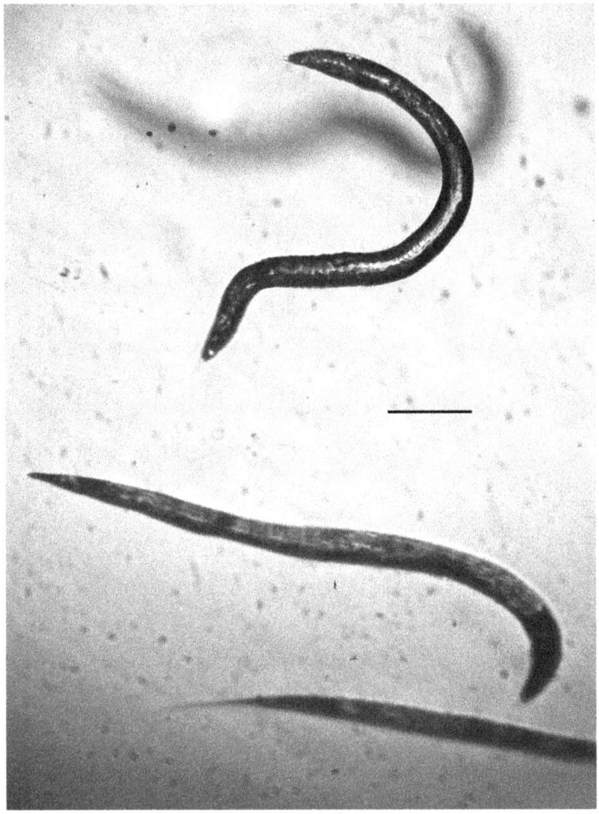

Fig. 133. *Detail of specimens of* Vetus maximus* *in Dominican amber (accession no. N-3-4). (Scale bar = 133 μm.)*

pieces of Dominican amber. This species has an attenuated head that is rounded at the tip, and a slender pointed tail. They are fairly short (192-275 μm in length) but not enough details are visible to place them in an extant family. There were some smaller dauer juveniles with different features associated with two pinhole beetles and these were described as *Scolytonema hispaniolae** (Figs 151, 152). One rare association was a male platypodid in Dominican amber that was not only carrying both species of *Scolytonema** but also a pseudoscorpion (Fig. 153; Plate 17). The pseudoscorpion undoubtedly lived in the beetle's tunnel or near the entrance of the tunnel where it probably fed on mites and nematodes. As with nematodes that inhabit the beetle galleries, these arachnids are

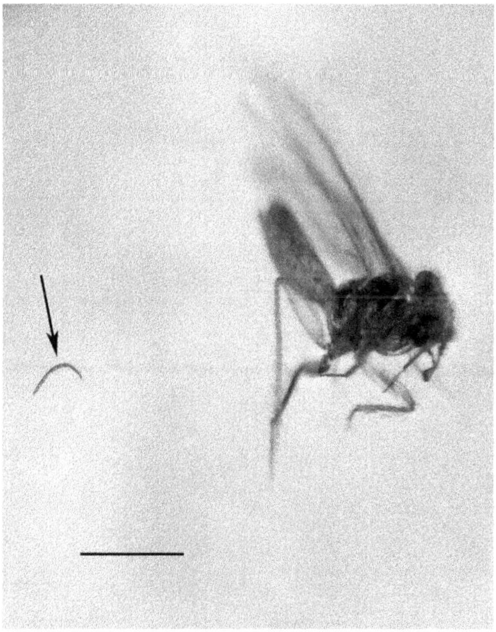

Fig. 134. *Dauer juvenile of* Psocidonema dominicana* *with its bark louse host in Dominican amber (accession no. N-3-92). (Scale bar = 568 μm.)*

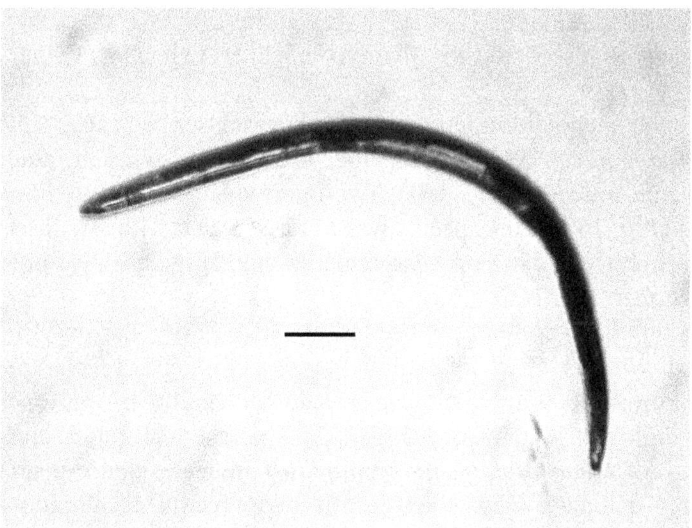

Fig. 135. *Detail of dauer juvenile of* Psocidonema dominicana* *in Dominican amber (accession no. N-3-92). (Scale bar = 40 μm.)*

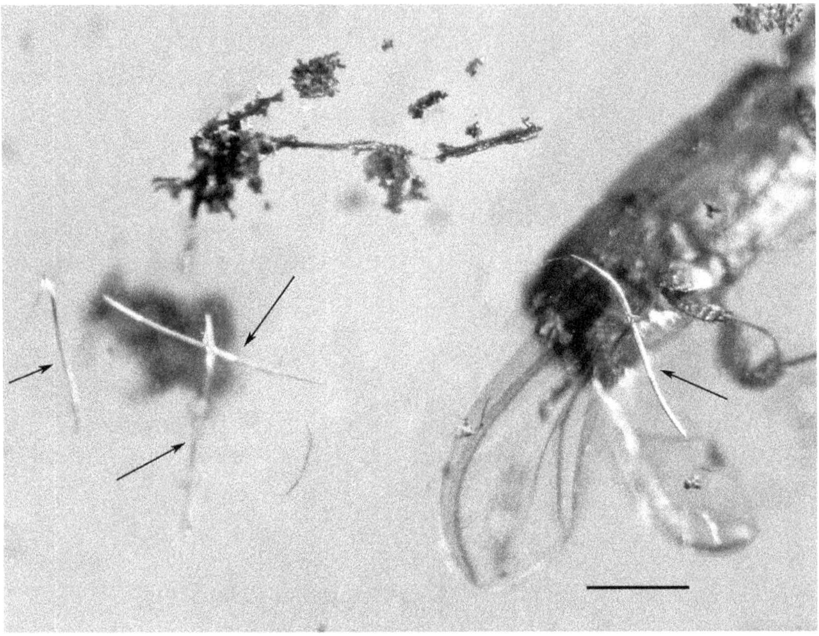

Fig. 136. *Developing stages (arrows) of* Bursaphelenchus similus* *adjacent to their bark beetle host in Dominican amber (accession N-3-88). (Scale bar = 115 μm.)*

dependent on insects to carry them from habitat to habitat (Poinar *et al.*, 1998b).

An aphelenchoidid in Dominican amber is described as *Oligaphelenchoides dominicanus**. This nematode is quite large, with the adults just over 1 mm in length (Fig. 154). Two intertwined adults may have been mating (Fig. 155). This species was not associated with any insects but was apparently feeding on a large mat of fungal hyphae also preserved in the amber (Fig. 156).

Among the wood-boring beetles are some very narrow 'straight-snouted' weevils. They are extremely rare as fossils and *Dominibrentus leptus* Poinar, 2009 in Dominican amber (Poinar, 2009b) was associated with a diplogastrid nematode (Fig. 157). The length (1.2 mm) and diam. (58 μm) of *Palaeodiplogaster brentiphila** indicates that it is an adult, probably a female (Fig. 158). It was undoubtedly developing in the galleries of the weevil, similar to extant diplogastrids that live in the galleries of bark beetles (Rühm, 1956). No dauer juveniles were present in the amber. Some extant diplogastrids, such as species of *Mikoletzkya*

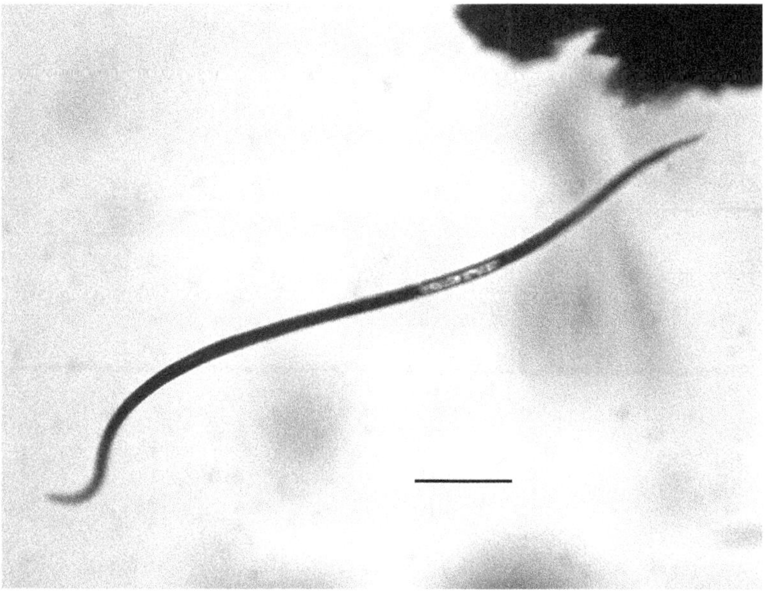

Fig. 137. *Female of* Bursaphelenchus similus* *in Dominican amber (accession no. N-3-88). (Scale bar = 100 μm.)*

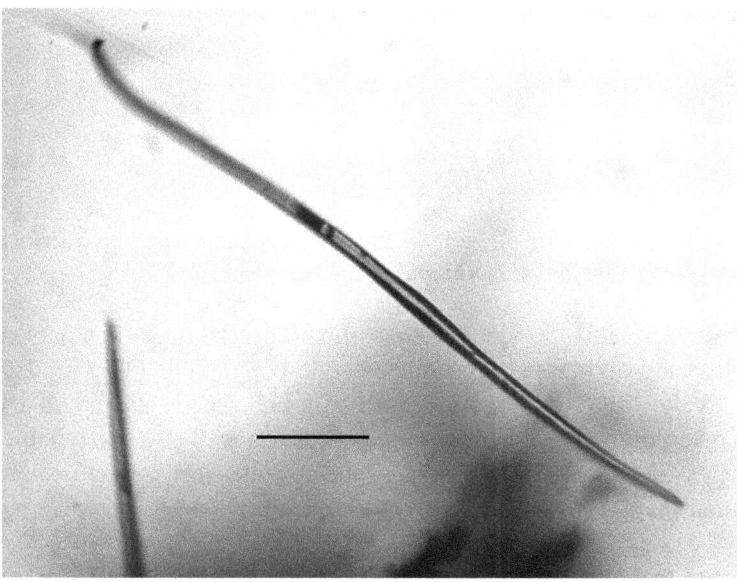

Fig. 138. *Male of* Bursaphelenchus similus* *in Dominican amber (accession no. N-3-88). (Scale bar = 100 μm.)*

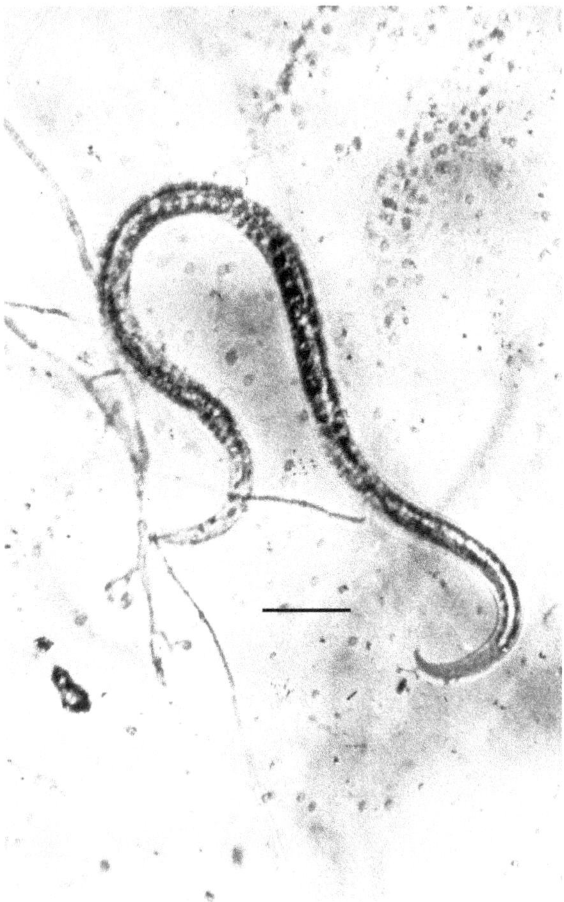

Fig. 139. *Specimen of the extant* Bursaphelenchus xylophilus *(Steiner & Buhrer) feeding on fungal mycelium. (Scale bar = 43 μm.)*

Weingärtner, 1955, prey on the eggs of bark beetles and when nematode numbers are high appear to reduce beetle populations (Thorne, 1935). It is likely that *P. brentiphila** lived in a bark beetle gallery that was then taken over by the brentid weevil. Brentids are known to inhabit the galleries of bark and platypodid beetles, often ejecting the occupants (Zimmerman, 1994). It is also possible *P. brentiphila** fed on the weevil's eggs.

Some fruit flies breed in decaying bark and populations of these drosophilids in the Dominican amber forest were parasitised by an allantonematid described as *Palaeoparasitylenchus dominicanus** (Fig. 159;

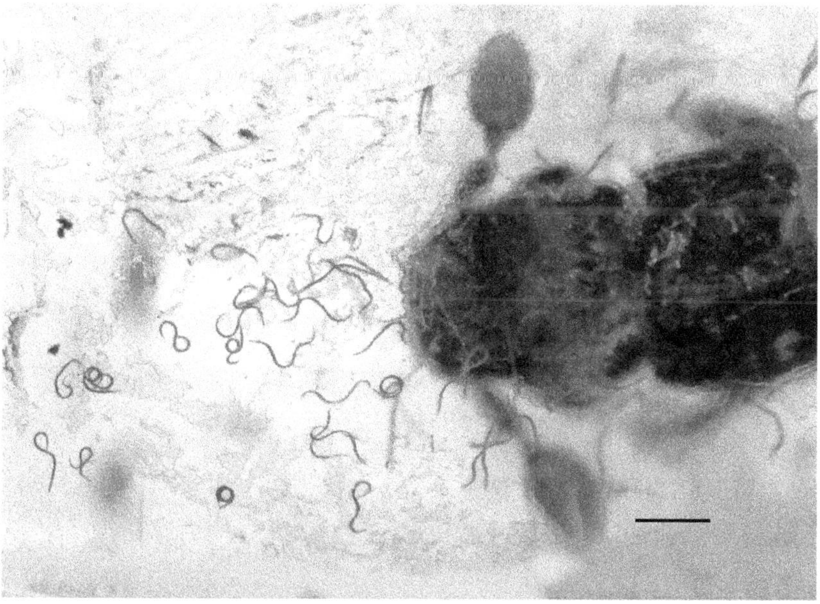

Fig. 140. *Cluster of dauer juveniles of* Cryptaphelenchus dominicus* *adjacent to their platypodid beetle carrier in Dominican amber (accession no. N-3-19) (see also Plate 16). (Scale bar = 260 μm.)*

Plate 18). The fossil female fly was heavily infected and large numbers of nematodes were adjacent to it, including second generation females (Figs 160-165). Other parasites were in the process of exiting the host through an injury, probably made as the fly attempted to escape from the resin. Some nematodes were still inside the fly's body (Fig. 160), leaving no doubt that the various stages found in the amber originated from the drosophilid. Many of the nematodes possessed stylets and pharyngeal glands and some had ensheathing cuticles.

Extant fruit flies are parasitised by two genera of allantonematids, *Howardula* Cobb, 1921 and *Parasitylenchus* Wachek, 1955 (Welch, 1959; Poinar *et al.*, 1997, 1998a), and since there is evidence that the fossil species underwent two amphimictic cycles in the host, it was assigned to the *Parasitylenchus* clade. In this lineage, juveniles produced by second generation females can mature to adult free-living stages inside the host. The life cycle of *P. dominicanus** was probably similar to that of the extant species *Parasitylenchus nearcticus* Poinar, Jaenike & Dombeck, 1997 from eastern North America (Fig. 166). The free-living

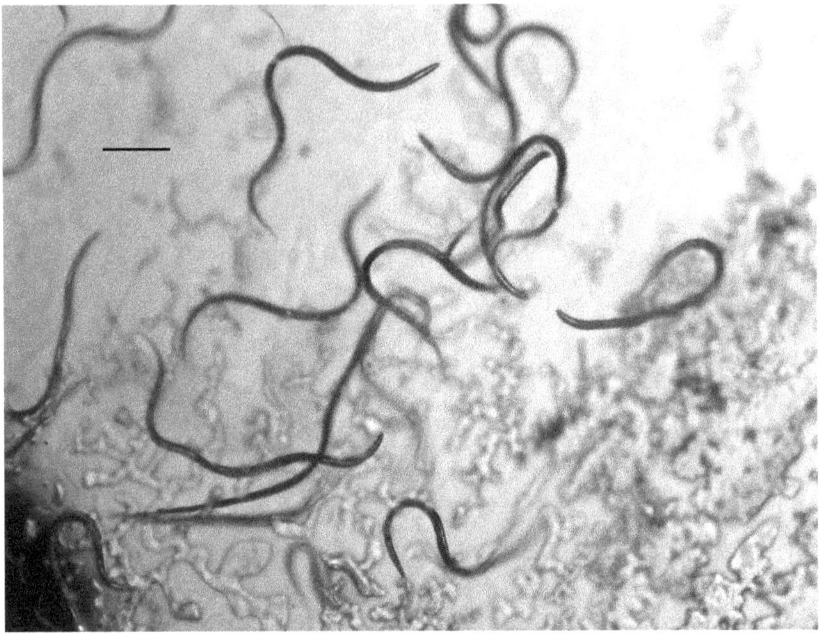

Fig. 141. *Small group of dauer juveniles of* Cryptaphelenchus dominicus* *in Dominican amber (accession no. N-3-19). (Scale bar = 60 μm.)*

females that exited the fly would have mated in the environment then searched for and penetrated the bodies of drosophilid larvae. The first generation of nematodes would have occurred in the fly larva and the second in the pupa and adult.

Mothflies of the family Psychodidae breed in rotting organic matter and this probably included decaying bark of the Dominican amber tree. Two dauer nematodes of *Psychodorhabditis attenuatus** associated with an adult female mothfly in Dominican amber show that the wrap-around method of nematode attachment is an ancient one (Figs 167-169). One dauer was still wrapped around the abdomen of the mothfly while the other was free in the amber. The wrap-around method of attachment to mothflies was first reported for the extant nematode, *Rhabditis dubia* Bovien, 1937, and subsequently confirmed by Sudhaus and Kühne (1989). Another piece of Dominican amber contained a single dauer juvenile, described here as *Psychodorhabditis dominicanus**, adjacent to an adult mothfly (Figs 170, 171). Since this dauer was almost twice as long as *P. attenuatus** it was considered a separate species. Its size is similar to the dauer juveniles of *R. dubia*.

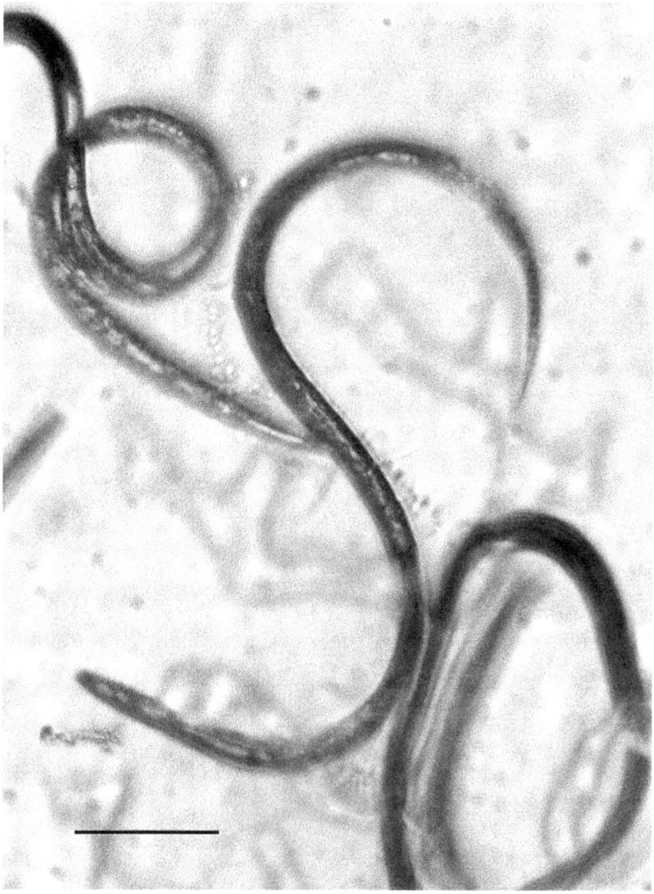

Fig. 142. *Detail of dauer juveniles of* Cryptaphelenchus dominicus* *in Dominican amber (accession no. N-3-19). (Scale bar = 21 μm.)*

Mothflies had to contend with at least one parasitic nematode in the Dominican amber forest. In one piece of amber, two juvenile mermithids, *Heydenius psychodae**, had completely emerged from an adult mothfly (Figs 172, 173; Plate 19). Both nematodes appear to have nearly completed their development and so, under normal circumstances, would have moulted to the adult stage, mated and oviposited in the habitat of the larval mothflies. Development of *H. psychodae** would have been direct (lacking a paratenic host) and infective juveniles hatching from the eggs would have penetrated the fly larvae. If the nematodes had completed their development before the host pupated,

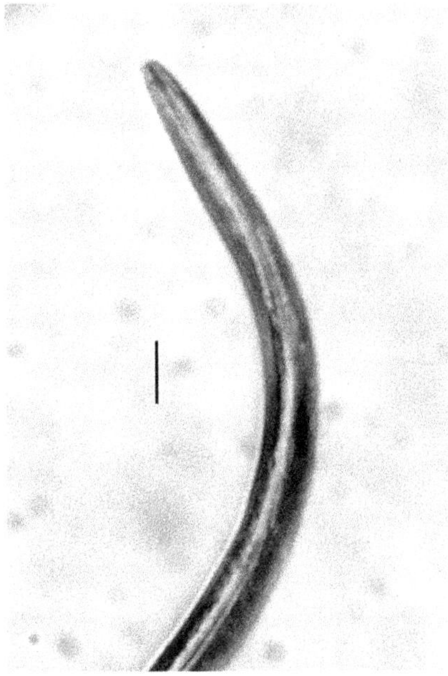

Fig. 143. *Stylet in a dauer juvenile of* Cryptaphelenchus dominicus* *in Dominican amber (accession no. N-3-19). (Scale bar = 10 μm.)*

Fig. 144. *Dauer juveniles of* Panagrolaimus neotropicus* *associated with a platypodid beetle in Dominican amber (accession no. N-3-21). Arrow shows dauer juvenile on wing membrane. (Scale bar = 654 μm.)*

their exit would have killed the fly larva. Obviously some infective juveniles of *H. psychodae** entered more mature larvae and were carried through the pupal and into the adult stage of the mothfly. Adult

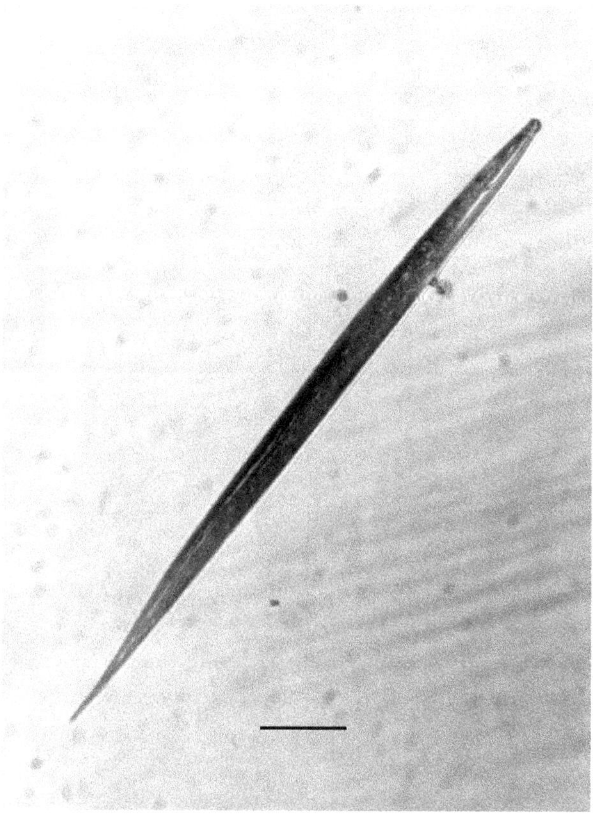

Fig. 145. *Detail of dauer juvenile of* Panagrolaimus neotropicus* *in Dominican amber (accession no. N-3-21). (Scale bar = 27 μm.)*

parasitism would have helped distribute *H. psychodae**. There appears to be no record of mermithids parasitising extant mothflies.

Other detritivores in the Dominican amber forest were scavenger flies or scatopsids (Diptera: Scatopsidae). These flies breed in decaying organic matter and, from their frequent occurrence in amber, there was obviously an adequate supply of that resource. Some of the amber scavenger flies were parasitised by *Heydenius scatophilus**. Infection rates in some habitats must have been high since five specimens of this mermithid were adjacent to a large scatopsid fly in one piece of amber (Figs 174, 175). The mode of parasitism was probably similar to that of *H. psychodae**, with a direct cycle where the infective stages hatch in the host's environment and invade scavenger fly larva. When

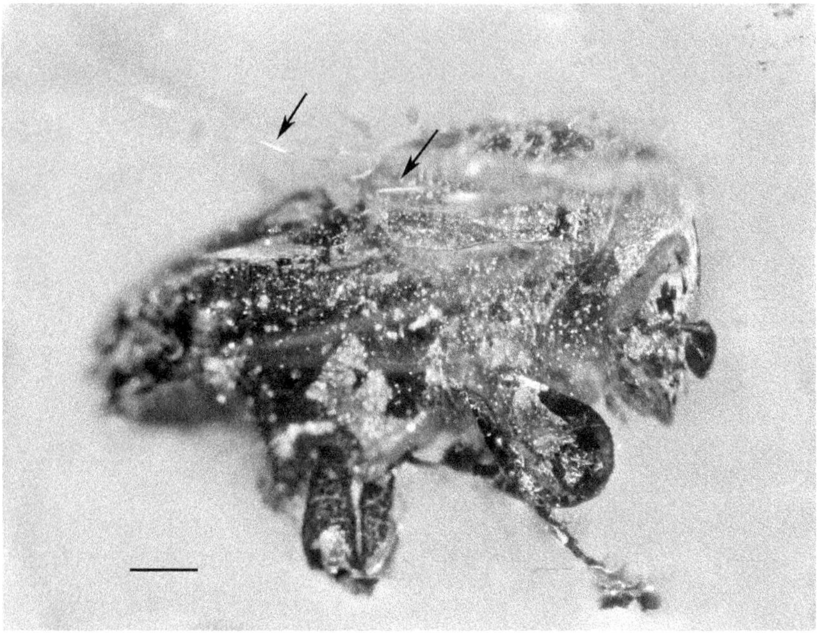

Fig. 146. *Dauer juveniles (arrows) of* Scolytonema dominicana* *on the scolytid bark beetle,* Scolytus poinari *Bright, in Dominican amber (accession no. N-3-60). (Scale bar = 458 μm.)*

Fig. 147. *Dauer juvenile of* Scolytonema dominicana* *on the bark beetle,* Scolytus poinari *Bright, in Dominican amber (accession no. N-3-60). (Scale bar = 22 μm.)*

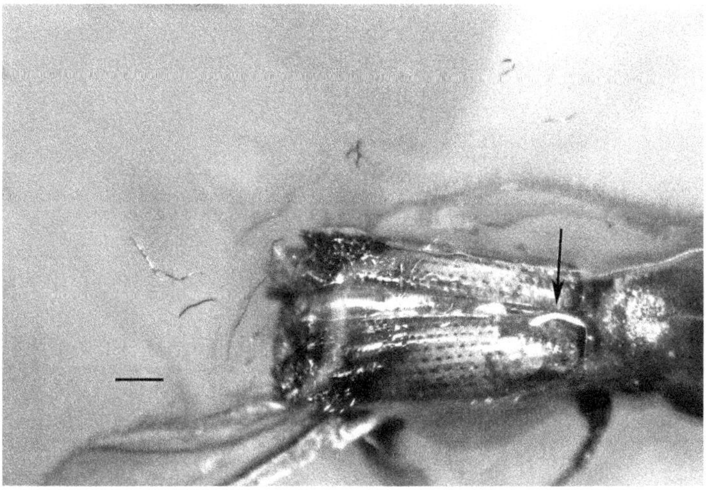

Fig. 148. *Dauer juveniles and adult (arrow) of* Scolytonema dominicana* *associated with a platypodid beetle in Dominican amber (Ru Smith collection). (Scale bar = 229 μm.)*

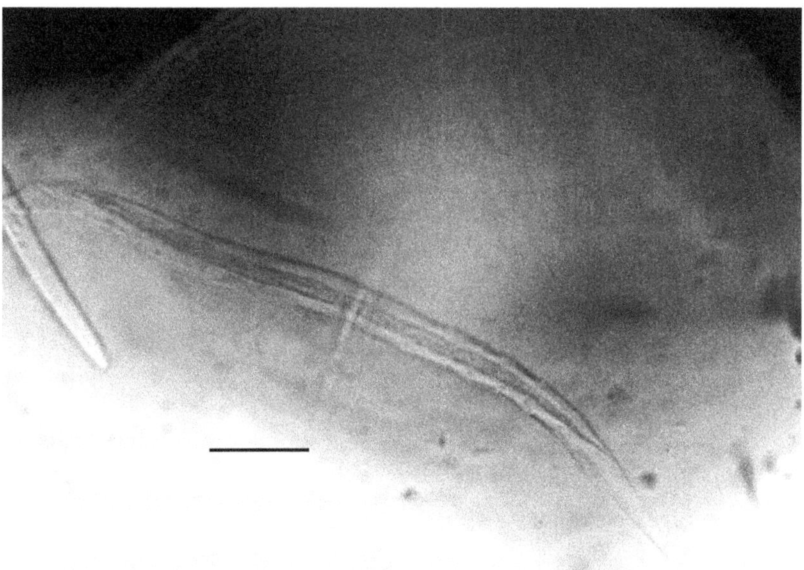

Fig. 149. *Detail of a dauer of* Scolytonema dominicana* *associated with a platypodid beetle in Dominican amber (Ru Smith collection). (Scale bar = 31 μm.)*

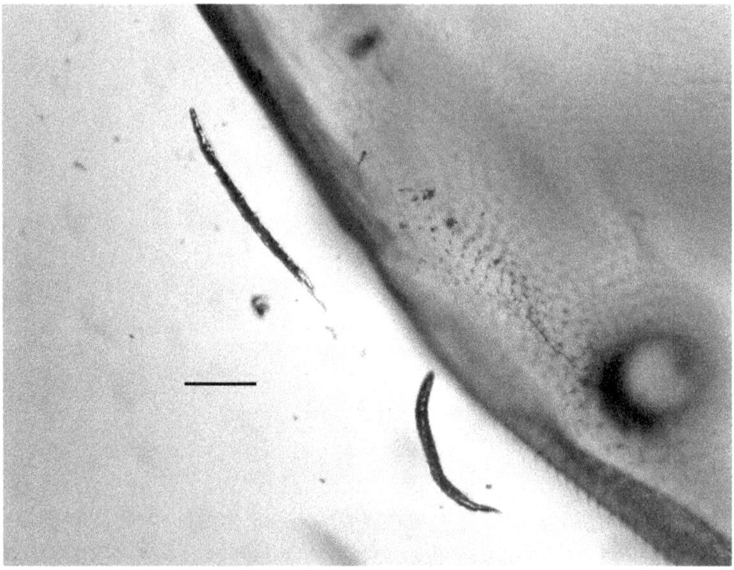

Fig. 150. *Two dauer juveniles of* Scolytonema dominicana* *adjacent to a platypodid beetle in Dominican amber (accession no. N-3-90). (Scale bar = 71 μm.)*

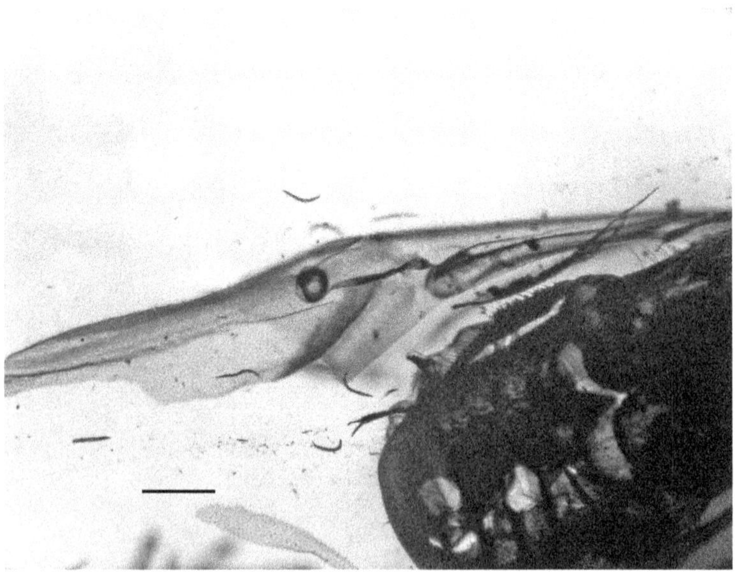

Fig. 151. *Dauer juveniles of* Scolytonema hispaniolae* *adjacent to a platypodid beetle in Dominican amber (accession no. N-3-89). (Scale bar = 250 μm.)*

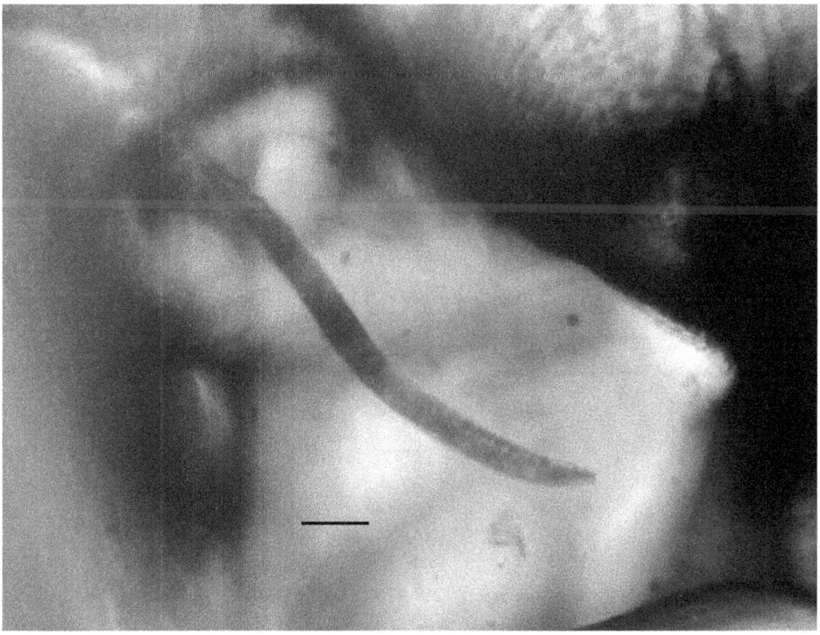

Fig. 152. *Detail of a dauer juvenile of* Scolytonema hispaniolae* *adjacent to a platypodid beetle in Dominican amber (accession no. N-3-89). (Scale bar = 20 μm.)*

multiple parasitism occurs with mermithids, both sexes are usually present and the ratio of males to females often depends on the amount of nourishment available. Three of the mermithids are over 4 mm in length suggesting that they were females, while the smallest one at under 3 mm in length was likely a male. There appears to be no report of mermithids infecting extant scavenger flies.

Insects that visit waste material, such as dung beetles, often carry nematodes from one food source to another (Sachs, 1950). Adjacent to a small dung beetle (Fig. 176) in Dominican amber are two dauer juveniles of *Scarabinema minuta** (Fig. 177). The nematodes were probably under the elytra or in the intersegmental membranes, which are the two most common locations of dauer nematodes carried by dung beetles (Sachs, 1950). It is unlikely that *S. minuta** belongs to those coprophilous lineages such as *Pelodera coarctata* (Leuckart) that form thickened cysts on the exoskeleton of dung beetles (Triffet & Oldham, 1927) (Figs 178, 179). No such cysts were noted on the fossil beetle and the juvenile nematodes are much larger (500-800 μm) than those of *P. coarctata*. The

Fig. 153. *A platypodid beetle carrying dauer juveniles of* Scolytonema dominicana* *and* S. hispaniolae* *as well as a pseudoscorpion in Dominican amber (accession no. N-3-95) (see also Plate 17). (Scale bar = 685 μm.)*

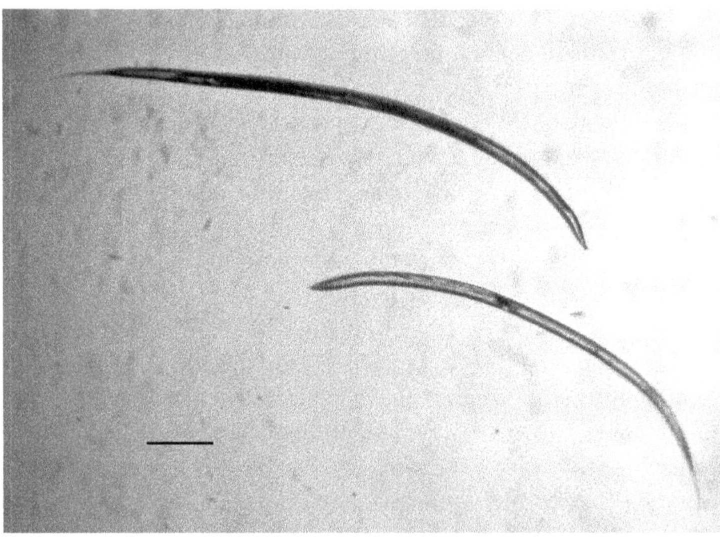

Fig. 154. *Two juveniles of* Oligaphelenchoides dominicanus* *in Dominican amber (accession no. N-3-91). (Scale bar = 80 μm.)*

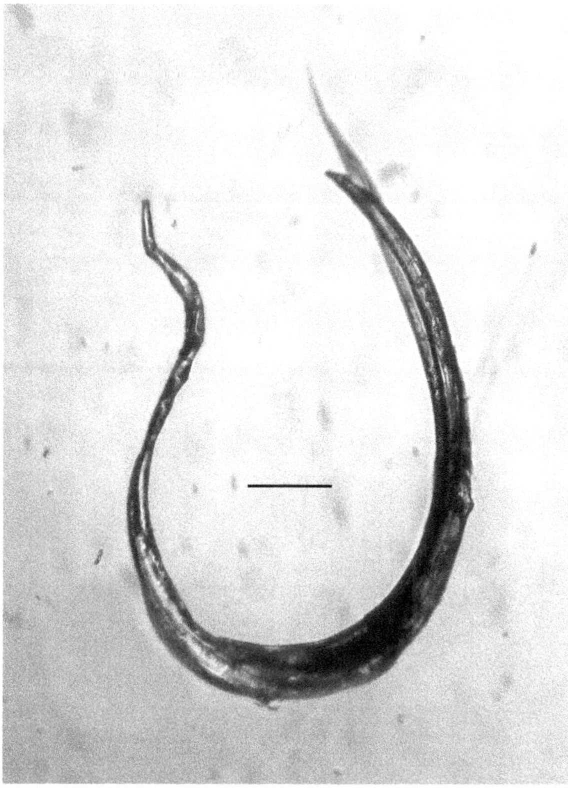

Fig. 155. *Two possible mating specimens of* Oligaphelenchoides dominicanus* *in Dominican amber (accession no. N-3-91). (Scale bar = 100 μm.)*

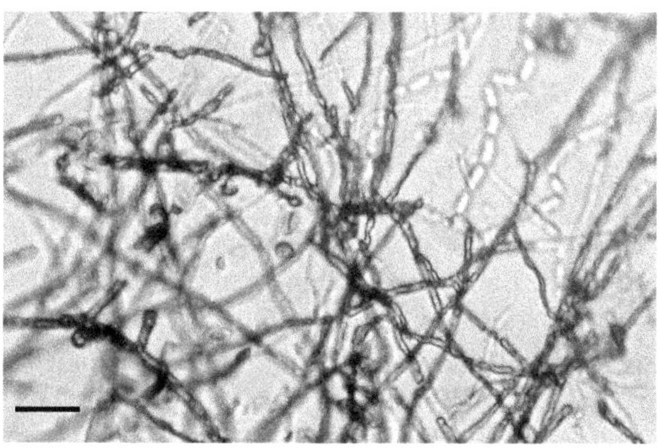

Fig. 156. *Fungal mat associated with* Oligaphelenchoides dominicanus* *in Dominican amber (accession no. N-3-91). (Scale bar = 24 μm.)*

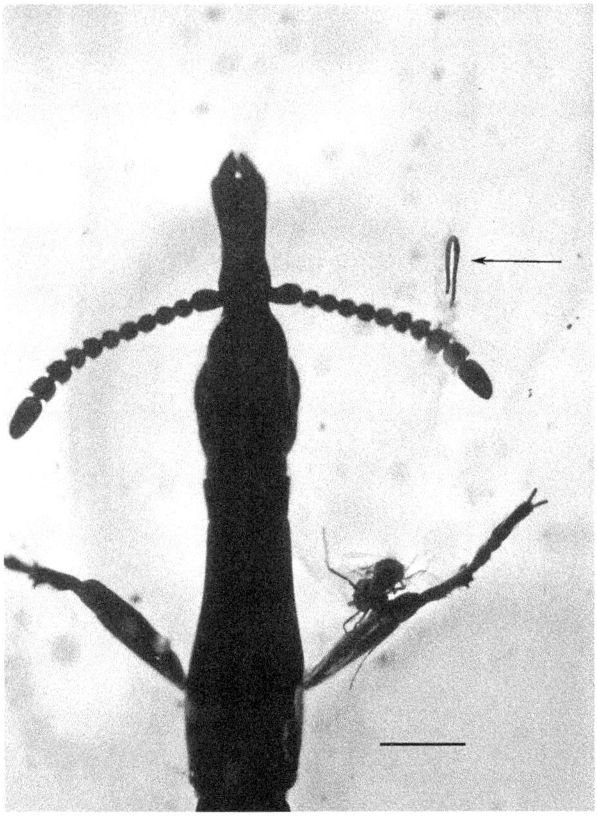

Fig. 157. *Adult (arrow) of* Palaeodiplogaster brentiphila* *adjacent to its brentid weevil host,* Dominibrentis leptus *Poinar 2009 in Dominican amber (Ron Cauble collection). (Scale bar = 960 μm.)*

third-stage juveniles of some diplogastrids, typified by *Eudiplogaster aphodii* Bovien, 1937, develop in the haemocoel of dung beetles (Bovien, 1937; Poinar *et al.*, 1976). However, when these nematodes eventually emerge from the adult beetles they are larger than *S. minuta**.

Another group of flies in the Dominican amber forest that bred in decaying organic matter were the milichids (Diptera: Milichidae). The biology of these small flies is largely unknown but it is obvious that some in the Dominican amber forest served as hosts for mermithid nematodes. The mermithid *Heydenius dipterophilus** is adjacent to an apparent milichid in Dominican amber (Fig. 180). The nematode had completely emerged from the host, and it would probably have had

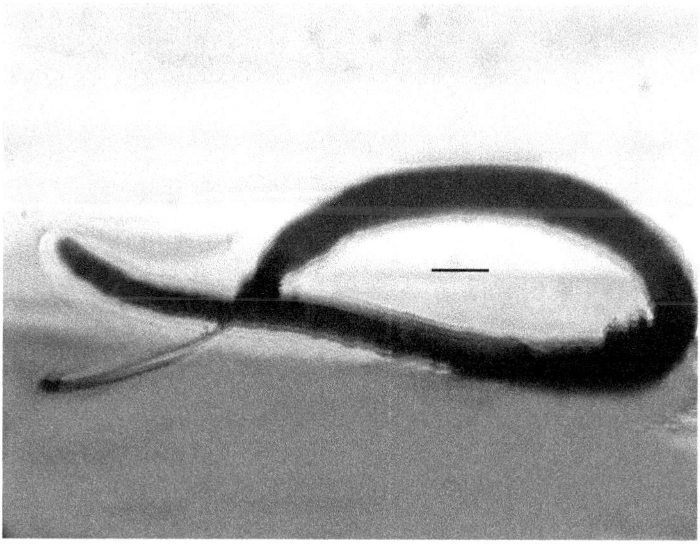

Fig. 158. *Detail of adult* Palaeodiplogaster brentiphila* *in Dominican amber (Ron Cauble collection). (Scale bar = 52 μm.)*

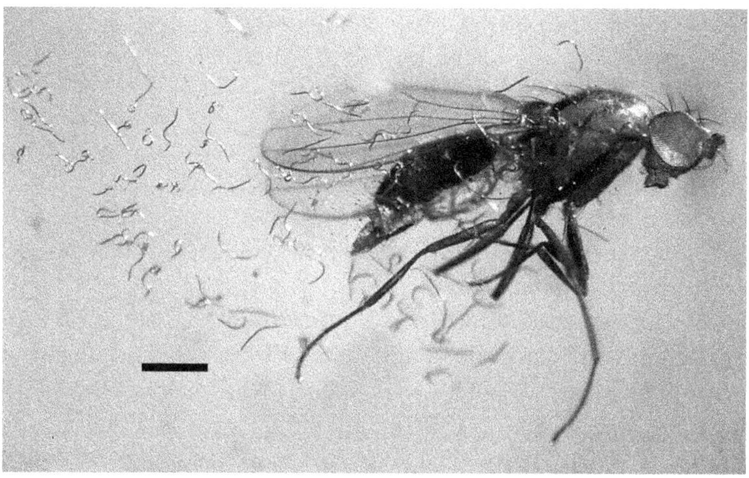

Fig. 159. *Parasitic juveniles of* Palaeoparasitylenchus dominicanus* *emerging from a drosophilid fly in Dominican amber (accession no. N-3-1) (see also Plate 18). (Scale bar = 432 μm.)*

a direct development cycle, similar to that of the other mermithids mentioned above. There are no extant records of mermithids from this family of flies.

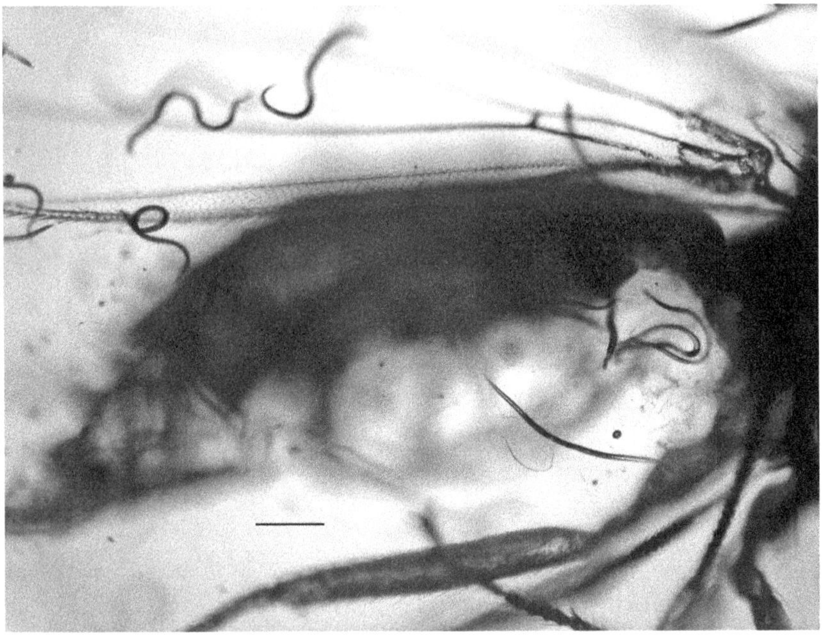

Fig. 160. *Parasitic juveniles of* Palaeoparasitylenchus dominicanus*, *some of which are emerging from the abdomen of a drosophilid fly in Dominican amber (accession no. N-3-1). (Scale bar = 100 μm.)*

Fungus gnats of the families Mycetophilidae and Sciaridae commonly occur in amber, but, in Dominican amber, mycetophilids are most abundant. While they appear to live in an ideal habitat for nematode parasites, extant mycetophilids are rarely attacked by nematodes, in contrast to the smaller sciarids.

However, iotonchid nematodes have evolved a complex life cycle involving both mycetophilids and fungi. A fossil representative of this group, *Paleoiotonchium dominicanum**, which parasitised a female Xylophilini fungus gnat in Dominican amber, may have had a similar dual host life cycle. The swollen parasitic female, as well as a number of mycetophagous juveniles, are in the body cavity of the fossil fly, while additional juveniles occur in the amber surrounding the fly (Figs 181-183; Plate 20).

The life cycle of *P. dominicanum** may have been similar to that of the extant *Iotonchium californicum* Poinar, 1991 (Poinar, 1991b). The latter has a mycetophagous cycle that occurs in mushrooms also invaded by fungus gnats. As the mushroom ages, entomogenous stages are formed

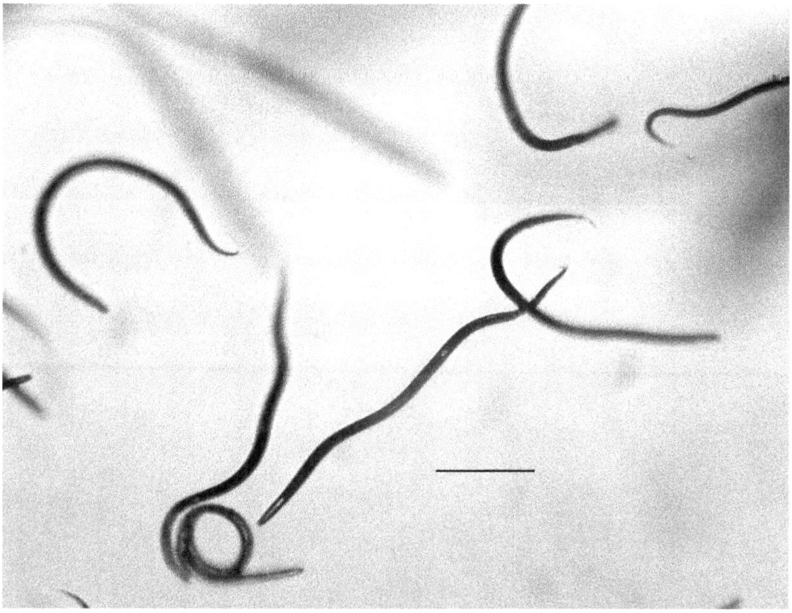

Fig. 161. *Group of parasitic juveniles of* Palaeoparasitylenchus dominicanus* *from a drosophilid fly in Dominican amber (accession no. N-3-1). (Scale bar = 79 μm.)*

within the fungal-feeding population. These infectives enter a larval fungus gnat and initiate development. Eggs from the parasitic females inside the fungus gnat develop into mycetophagous juveniles (Figs 184, 185). The parasitised female fungus gnats deposit these juveniles, along with their own eggs, on fresh mushrooms.

Another dipterous group that develops on various types of plant matter are the craneflies. Some larvae actually develop on angiosperm leaves; however, most occur in moist habitats, such as swamps, wet sand, bogs and turf. Such habitats also favour nematode survival, and the presence of *Tipulacomis dominicana** is evidence of co-habitation of free-living nematodes with cranefly larvae in the Dominican amber forest. The dauer juveniles of *T. dominicana** are adjacent to the abdomen of an adult cranefly and obviously were being transported by the insect (Figs 186-188). It is likely that *T. dominicana** developed in the same habitat as the fossil cranefly and, when that habitat began to deteriorate, formed dauer juveniles that climbed on the emerging flies. This association is probably very similar to that of *Tipulacomis balticus** men-

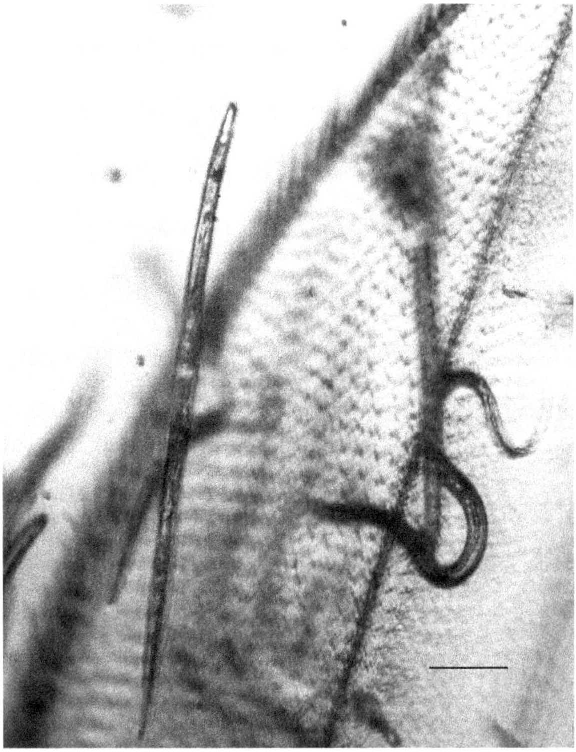

Fig. 162. *Detail of several parasitic juveniles of* Palaeoparasitylenchus dominicanus* *adjacent to the wing of a drosophilid fly in Dominican amber (accession no. N-3-1). (Scale bar = 665 μm.)*

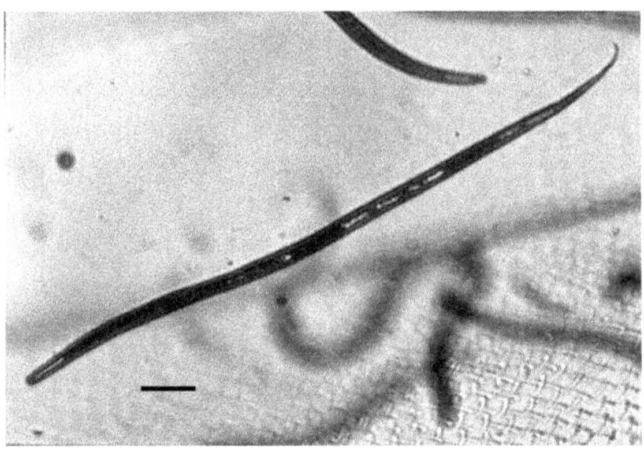

Fig. 163. *Detail of a parasitic juvenile of* Palaeoparasitylenchus dominicanus* *in Dominican amber (accession no. N-3-1). (Scale bar = 25 μm.)*

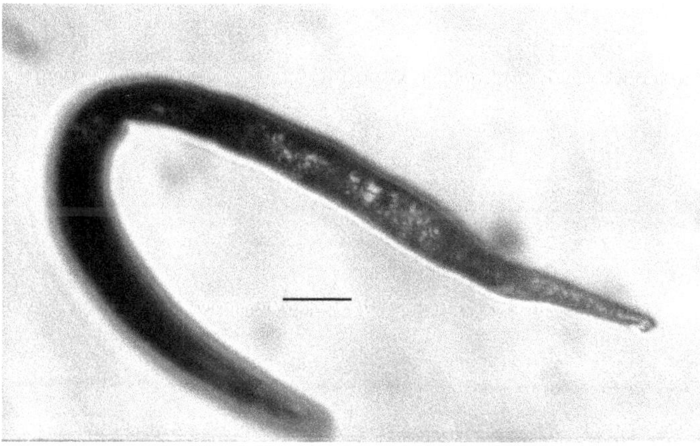

Fig. 164. *Second generation female of* Palaeoparasitylenchus dominicanus* *in Dominican amber (accession no. N-3-1). (Scale bar = 11 μm.)*

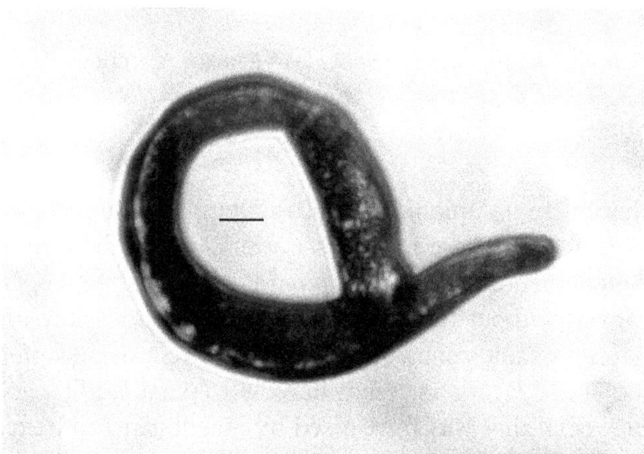

Fig. 165. *Second generation female of* Palaeoparasitylenchus dominicanus* *in Dominican amber (accession no. N-3-1). (Scale bar = 8 μm.)*

tioned earlier in Baltic amber. While there are no reports of dauer nematodes associated with adult craneflies today, Lam and Webster (1971) described two rhabditoids from larvae of *Tipula paludosa*. However, they did not comment on whether these nematodes were carried by the adult flies.

Nematode parasites of fossil herbivorous insects are rare. Thus, the discovery of five juveniles of the allantonematid *Palaeoallantonema api-*

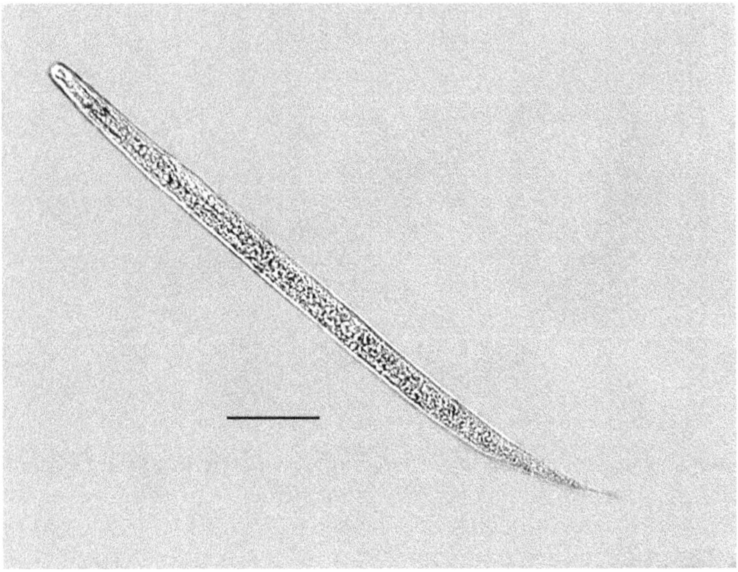

Fig. 166. *Extant parasitic juvenile of* Parasitylenchus nearcticus *Poinar, Jaenke & Dombeck, 1997, a parasite of* Drosophila *sp. in North America. (Scale bar = 31 μm.)*

*onae**, adjacent to an apionid weevil in Dominican amber, was a surprise (Figs 189-190). These juveniles are smaller than those of the extant *Apion* allantonematid, *Howardula apioni* Poinar, Laumond & Bonifassi, 1980; however, without free-living adults, it is not possible to place *P. apionae** in an extant genus. While the life cycle of *H. apioni* is unknown, both it and *P. apionae** may have had a plant-feeding generation.

Another weevil that was parasitised by allantonematid nematodes in the Dominican amber forest was a small twig and stem weevil belonging to the subfamily Zygopinae. These weevils have large eyes that often meet on the top of the head. Zygopine larvae develop in twigs, stems and wood of herbaceous and woody plants. Some of the third-stage juvenile nematodes of *Palaeoallantonema zygopinae** had emerged from the beetle, probably through the gut or reproductive system, and were trapped adjacent to their host (Figs 191, 192).

Fig wasps (Hymenoptera: Agaonidae) in Dominican amber provide indirect evidence that fig trees thrived in the forest. Some fig wasps were associated with juvenile diplogastrid nematodes described here as *Syconema dominicana**. This species is based on 59 nematodes in one

Fig. 167. *Dauer juvenile (arrow) of* Psychodorhabditis attenuatus* *adjacent to its mothfly host in Dominican amber (accession no. N-3-57A). (Scale bar = 355 μm.)*

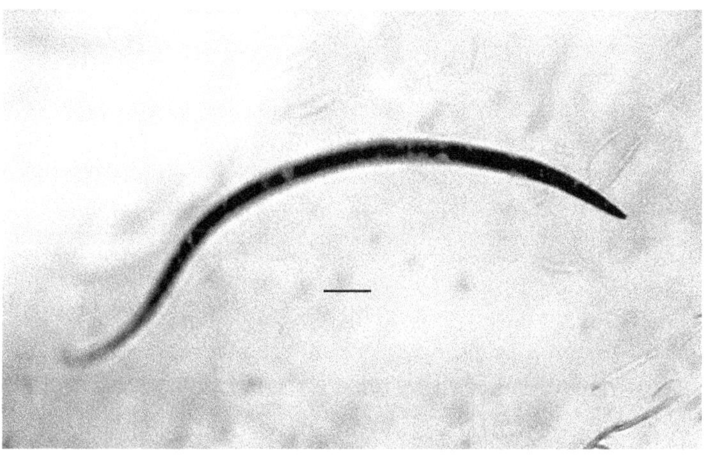

Fig. 168. *Detail of dauer juvenile of* Psychodorhabditis attenuatus* *in Dominican amber (accession no. N-3-57A). (Scale bar = 33 μm.)*

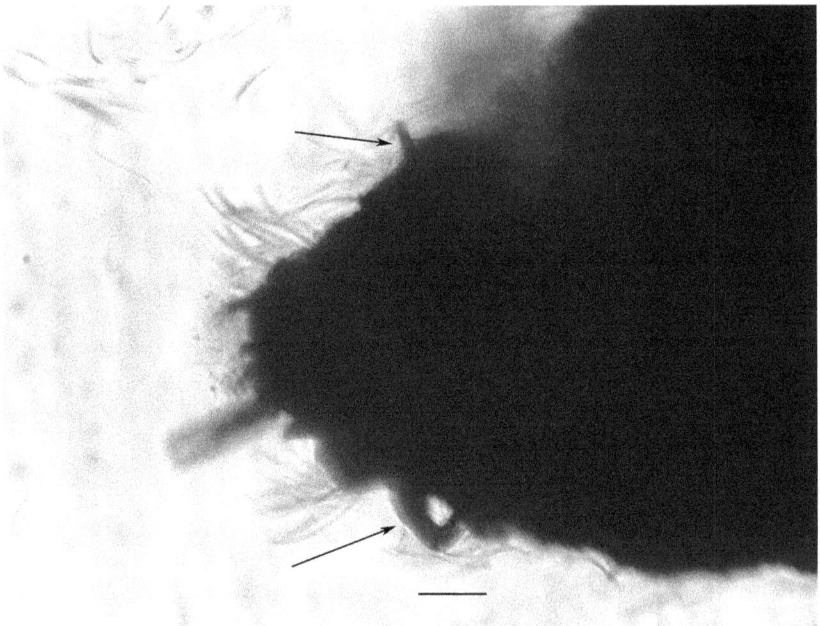

Fig. 169. *Specimen of* Psychodorhabditis attenuatus* *(arrows) coiled around the abdomen of its mothfly host in Dominican amber (accession no. N-3-57A). (Scale bar = 83 μm.)*

piece of amber, four nematodes in a second piece and five nematodes in a third (Figs 193-195; Plate 21). All of the nematodes were juveniles that had emerged, or were in the process of emerging, from the bodies of the wasps.

Two diverse genera of nematodes occur in the body cavity of extant fig wasps. One is the aphelenchoidid *Schistonchus* Cobb, 1927 and the other the diplogastrid *Parasitodiplogaster* Poinar, 1979 (Fig. 196). Measurements of the fossil *Parasitodiplogaster dominicana** overlap with the sizes of the juveniles of both genera; however, *Schistonchus* has a well developed stylet and an offset, blunt, lip region, which does not occur on the fossils. Based on their attenuated head, lack of a stylet, rounded lip region and elongated tail, the nematodes are placed in the *Parasitodiplogaster* clade. This genus parasitises fig wasps in Africa (Fig. 196) and Mesoamerica today (Poinar, 1979; Poinar & Herre, 1991).

The co-evolution of the association between fig wasps and nematodes is intriguing. Clearly it is advantageous for the nematodes since, without fig wasps, they would be unable to reach their breeding sites in the

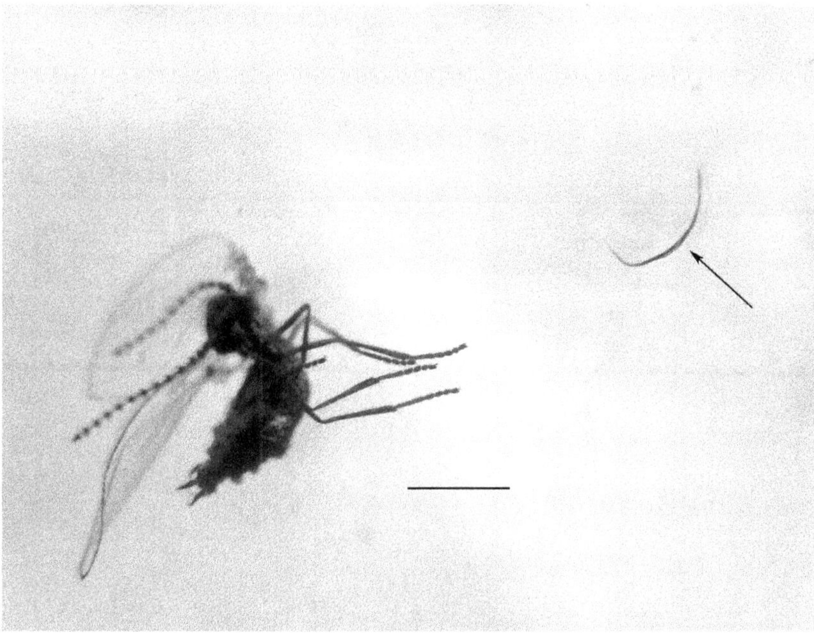

Fig. 170. *Dauer juvenile (arrow) of* Psychodorhabditis dominicanus* *with its mothfly carrier in Dominican amber (accession no. N-3-57C). (Scale bar = 565 μm.)*

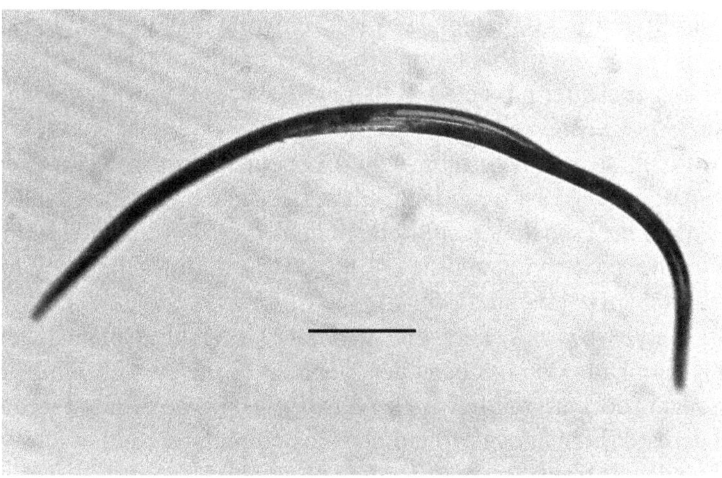

Fig. 171. *Detail of dauer juvenile of* Psychodorhabditis dominicanus* *in Dominican amber (accession no. N-3-57C). (Scale bar = 100 μm.)*

Fig. 172. *Two specimens of the mermithid* Heydenius psychodae* *with their mothfly host in Dominican amber (accession no. N-3-57B) (see also Plate 19). (Scale bar = 368 μm.)*

flowers. While the association may have evolved as one of external phoresis, the struggling female wasps entering the figs could easily injure the nematodes. By entering the body cavity of their carriers, the nematodes are better protected. Is there any advantage for the fig wasps? If both *Schistonchus* and *Parasitodiplogaster* feed on plant tissues, then it is difficult to see how they would benefit the wasps. However, if the nematodes feed on fungi and bacteria detrimental to the developing wasps, then the association could be mutualistic. The nematodes do not seem to affect the flying abilities of the wasps, and whether they lower their reproductive potential is unknown.

Fossil stylet-bearing, parasitic nematodes of higher plants are extremely rare and none has been described up to the present. This is because plant roots are rarely fossilised and it would be difficult to detect fossil nematodes in above-ground portions of plants if such tissues were ever fossilised. However, by a strange sequence of events, a section of a rootlet containing a population of stylet-bearing nematodes was preserved in Dominican amber.

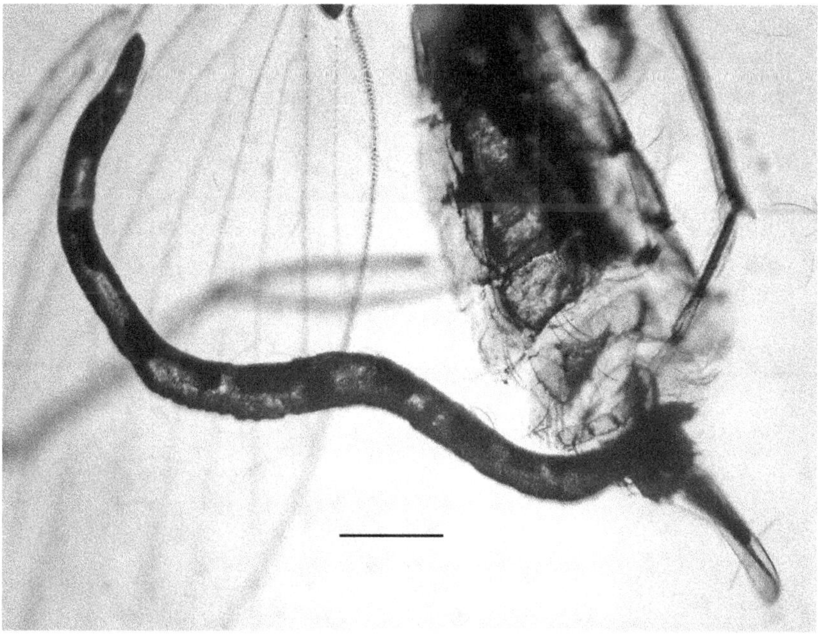

Fig. 173. *Specimen of* Heydenius psychodae* *adjacent to its mothfly host in Dominican amber (accession no. N-3-57B). (Scale bar = 175 μm.)*

The nematode is described as *Oligaphelenchoides maximus**, and is represented by numerous specimens in all developmental stages (Figs 197-201; Plate 22). The long, slender, stylet-bearing adults and males with a C-shaped terminus align the species with the Aphelenchoidea. The section of plant tissue is 14 mm long and 3 mm wide and, based on the presence of scattered vascular bundles, apparently represents a monocot rootlet. There is no evidence of fungi associated with the rootlet, and the presence of eggs, juveniles and adults indicates that *O. maximus** was developing in the plant tissue, similar to extant species of migratory root parasites of higher plants, such as members of the genus *Aphelenchoides*. Extant species of *Aphelenchoides* are known to develop in orchids (Feder, 1951), and these epiphytes, as well as bamboos, were present in the Dominican amber forest (Poinar & Poinar, 1999).

Some aphelenchoidids can survive on the external surface of insects where they feed as ectoparasites through the body wall of the host. A few that occur on bark beetles are thought to feed ectoparasitically, but it is usually only one or two juvenile stages that have this habit. However, members of the nematode family Acugutturidae Hunt, 1993

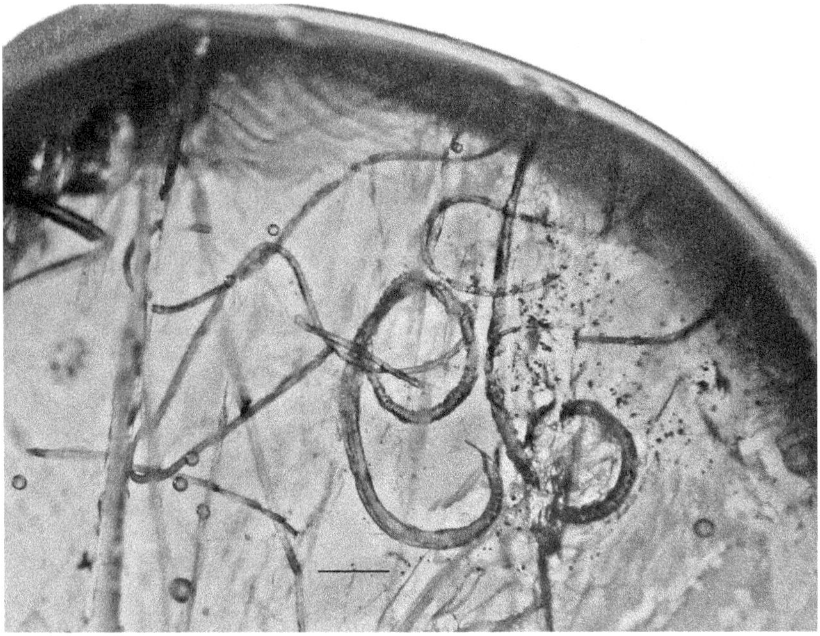

Fig. 174. *Seven specimens of the mermithid* Heydenius scatophilus* *that had emerged from a scavenger fly (in upper left corner of photo) in Dominican amber (accession no. N-3-8). (Scale bar = 448 μm.)*

complete their entire life cycle on the surface of moths and cockroaches (Hunt, 1993). Nourishment is obtained with the aid of a long stylet that passes through the body wall of the insect. Desiccation is prevented by adhering to the body of hosts living in humid tropical and subtropical climates. When conditions are favourable the nematodes can survive outside their normal range and are often found on migrating moths entering the southern United States. Several specimens of *Setonema protera** in a piece of Dominican amber are considered to belong to the Noctuidonematinae Hunt, 1993. The largest of the specimens, which is adjacent to a moth (Figs 202, 203; Plate 23A), appears to be a late stage juvenile female, while another one further away from the moth is probably an early stage juvenile female. The fossil moth also contained addition specimens of *Setonema** attached to its abdomen (Fig. 204), but details were obscured by the scales. Extant species of this subfamily occur on six families of moths, especially noctuids, and apparently transfer from moth to moth during mating (Figs 205, 206; Plate 23B).

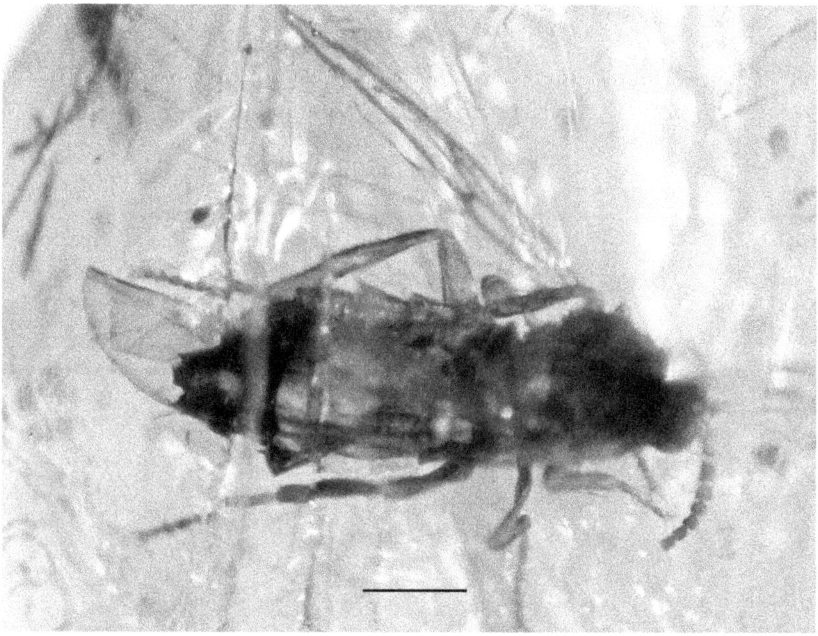

Fig. 175. *Scavenger fly host of* Heydenius scatophilus* *in Dominican amber (accession no. N-3-8). (Scale bar = 800 μm.)*

Fig. 176. *Dung beetle associated with two dauer juveniles of* Scarabinema minuta* *in Dominican amber (accession no. N-3-94). (Scale bar = 706 μm.)*

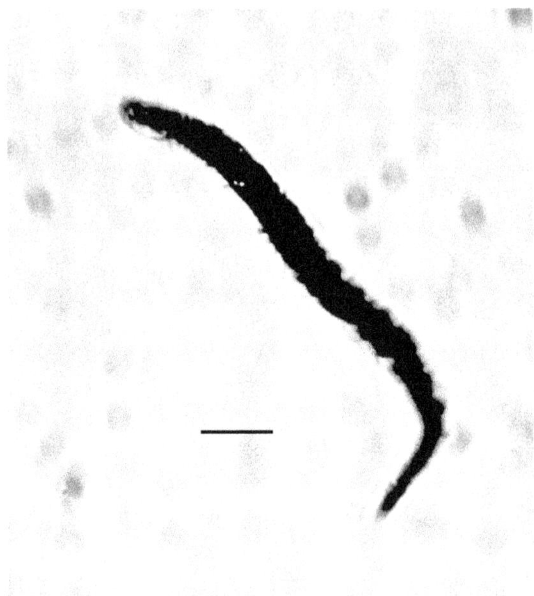

Fig. 177. *Dauer juvenile of* Scarabinema minuta* *in Dominican amber (accession no. N-3-94). (Scale bar = 15 μm.)*

The Noctuidonematinae parasitise moths of the families Noctuidae, Lasiocampidae, Notodontidae, Pyralidae, Geometridae and Sphingidae (Rogers *et al.*, 1991; Marti *et al.*, 2000). This association occurs in tropical and subtropical America as well as in Fiji and Java. Hunt has also observed moth parasitism in Aldabra (D.J. Hunt, pers. comm., June, 2010). The nematodes occur around the genitalia, in the intersegmental abdominal regions and in tympanic cavities of adult moths and are transmitted during mating. Adult fall armyworms with *Noctuidonema guyanense* show melanised spots of haemolymph from stylet penetration on the cuticle as well as cuticular aberrations and subcuticular muscular damage from the feeding activities of the nematodes. Laboratory trials showed that the nematodes reduced the longevity of infected adults as well as egg viability (Rogers *et al.*, 1993).

Another chance encounter resulted in a second fossil species of stylet-bearing plant nematode in Dominican amber. How a monocot seed became embedded in resin is unknown, but it may have occurred during the feeding activities of a herbivore (Fig. 207). At any rate, the immature seed contained a developing female of the anguinid, *Palaeoanguina dominicana** (Fig. 208). When the seed fell into the resin, the nematode

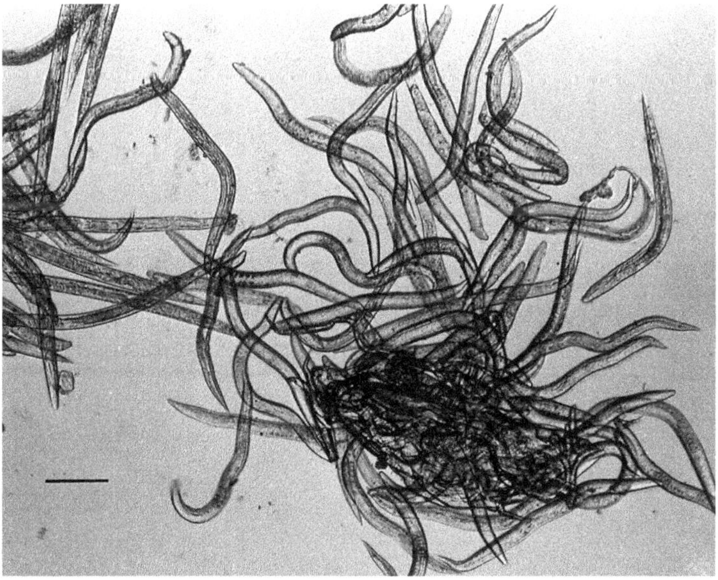

Fig. 178. *Group of dauer juveniles of the extant species* Pelodera coarctata *(Leuckart), a dung beetle associate. (Scale bar = 90 μm.)*

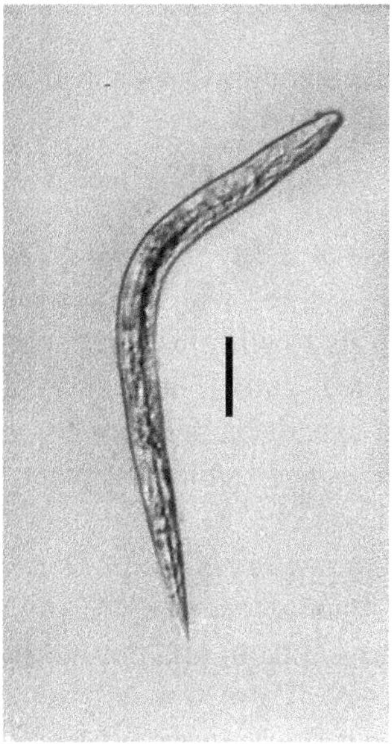

Fig. 179. *Dauer juvenile of the extant species* Pelodera coarctata *(Leuckart). (Scale bar = 15 μm.)*

Fig. 180. *Parasitic juvenile of* Heydenius dipterophilus* *adjacent to its fly host in Dominican amber (accession no. N-3-66). (Scale bar = 425 μm.)*

was obviously still active enough to escape from the plant tissue. The finely annulated cuticle, pointed head bearing a stylet, obese, curled body, and conoid tail with a peg-like tip align the nematode with the anguinids (Plate 24A).

Under natural conditions the female would have developed into an immobile reproductive sac filled with juveniles. Anguinid juveniles are well known for their ability to withstand desiccation and can survive for years in dried seed and leaf galls (Fortuner & Maggenti, 1987). While the identity of the seed cannot be determined, it may be from a grass, since bamboo seeds occur in Dominican amber (Poinar & Judziewicz, 2005).

A number of predatory beetles occur in Dominican amber and some are attacked by mermithid nematodes. An unusual discovery was a mermithid that had emerged from an adult firefly in Dominican amber (Figs 209, 210; Plate 24B). There was only a single specimen of *Heydenius lamprophilus**, but it was large, measuring slightly over 14 mm in length. The size of mermithids is usually correlated with the size of their hosts and the number of parasites present. In general, mermithids

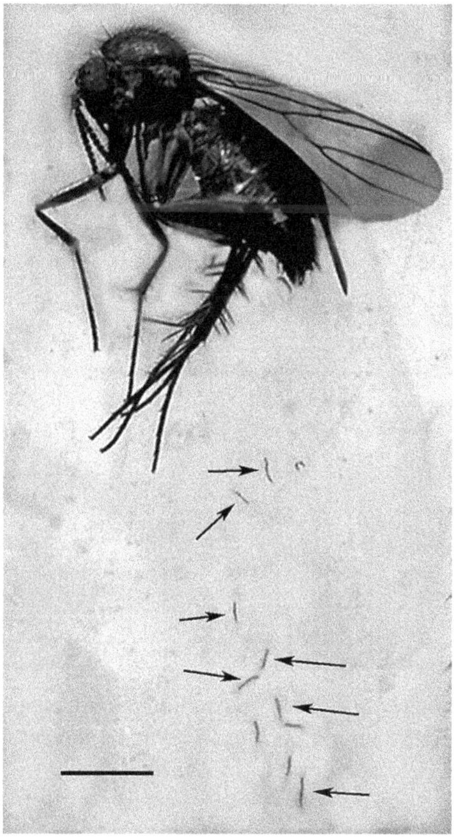

Fig. 181. *Parasitic juveniles (arrows) of* Paleoiotonchium dominicanum* *that had emerged from their mycetophilid fungus gnat host in Dominican amber (accession no. N-3-23). (Scale bar = 932 μm.)*

are at least twice the length of their hosts and have to coil in order to fit into the body cavity of their host. The exit location of the mermithid was probably a rupture located at the tip of the beetle's abdomen. While there appear to be no reports of mermithid parasites of extant Lampyridae, mermithids parasitise members of the related family Cantharidae today (Poinar, 1975).

Although fully aquatic insects are quite rare in amber, the adults of insects that have aquatic immature stages are fossilised. Chironomid midges are one of the most common groups in this category. These flies and ceratopogonids probably fell into resin deposits during mating swarms. Also in this category were dragonflies and damselflies

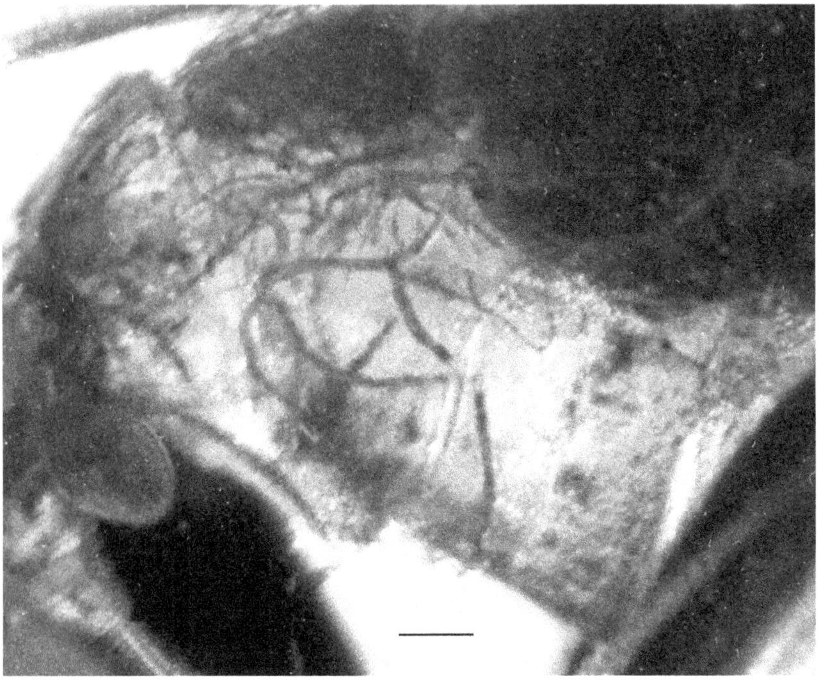

Fig. 182. *Parasitic juveniles of* Paleoiotonchium dominicanum* *inside the body cavity of a mycetophilid fungus gnat in Dominican amber (accession no. N-3-23) (see also Plate 20). (Scale bar = 75 μm.)*

(Odonata) that became entrapped when they attempted to oviposit in water collected in the bases of bromeliad leaves or in other types of phytotelmata.

The high rate of parasitism of chironomid midges by *Heydenius matutinus* in Baltic amber was previously noted. Mermithid parasitism of chironomids is not as common in Dominican amber, although a few examples of *Heydenius neotropicus** have been detected. In one piece of amber two individuals had partly emerged from their midge host (Fig. 211). These appear to be fairly mature parasitic juveniles that were probably ready to leave the insect. A separate piece of amber contains a midge parasitised by four specimens of *H. neotropicus**, three of which had completely exited from the host. The fourth specimen was still partially coiled in the body cavity of the midge (Fig. 212; Plate 25). The life cycle of *H. neotropicus** was probably direct and did not involve a paratenic host.

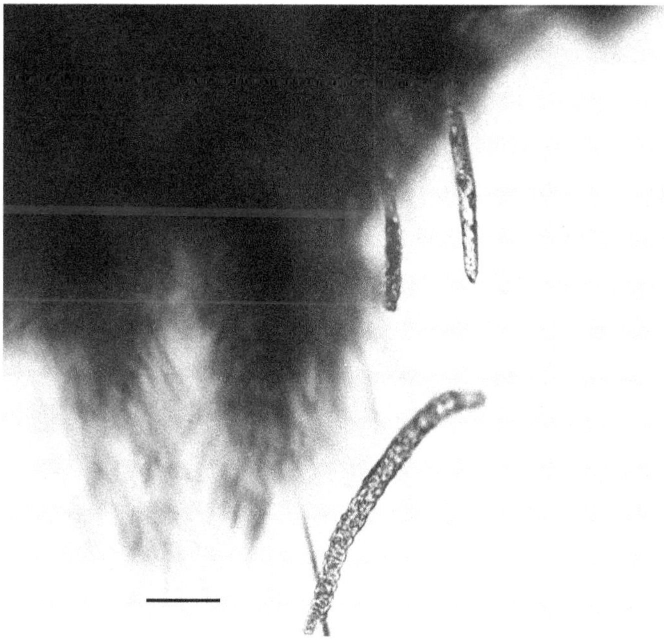

Fig. 183. *Parasitic juveniles of* Paleoiotonchium dominicanum* *emerging from the body cavity of a mycetophilid fungus gnat in Dominican amber (accession no. N-3-23). (Scale bar = 58 μm.)*

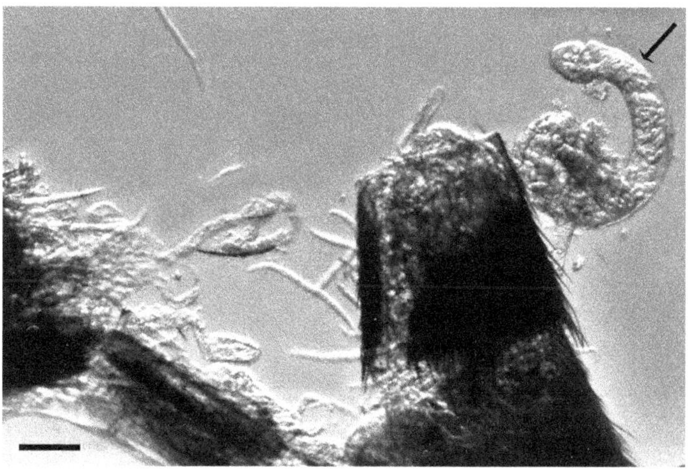

Fig. 184. *Parasitic juveniles and a mature female (arrow) of* Iotonchium californicum *Poinar, 1991 from the body cavity of the extant mycetophilid fungus gnat* Mycetophila fungorum *(DeGeer). (Scale bar = 234 μm.)*

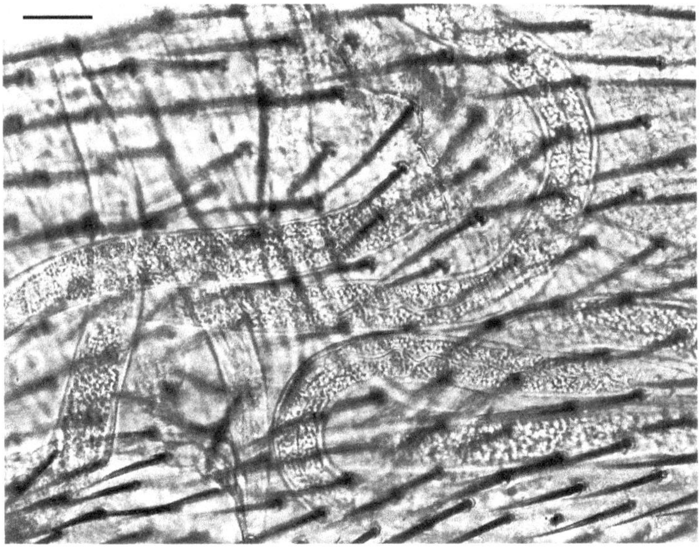

Fig. 185. *Parasitic juveniles of* Iotonchium californicum *Poinar, 1991 inside the body cavity of the extant mycetophilid fungus gnat* Mycetophila fungorum *(DeGeer). (Scale bar = 40 μm.)*

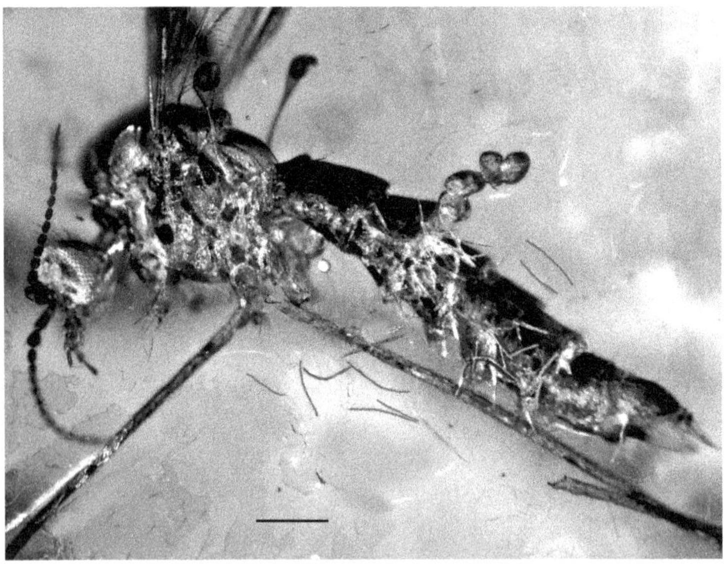

Fig. 186. *Dauer stages of* Tipulacomis dominicanus* *associated with their cranefly carrier in Dominican amber (accession no. N-3-72). (Scale bar = 480 μm.)*

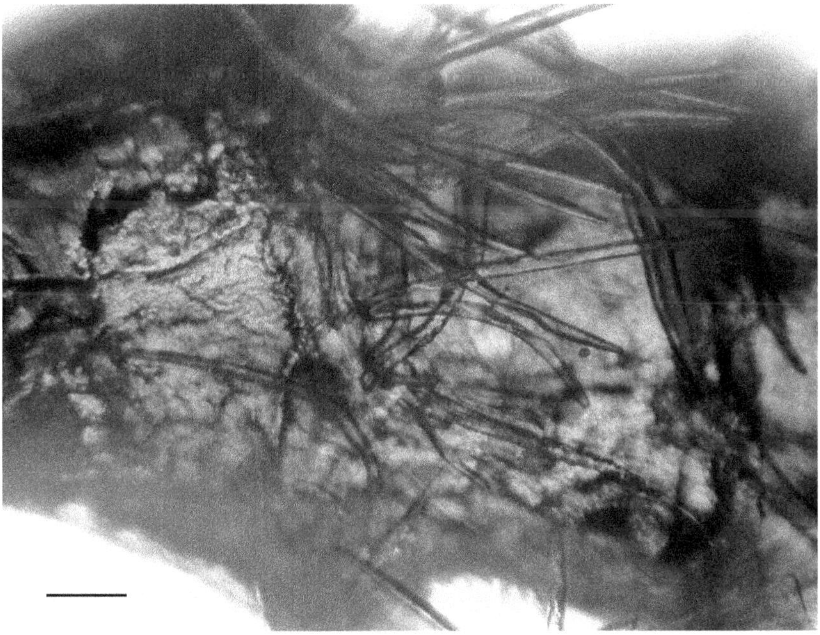

Fig. 187. *Cluster of dauer juveniles of* Tipulacomis dominicanus* *on the body of a cranefly in Dominican amber (accession no. N-3-72). (Scale bar = 88 μm.)*

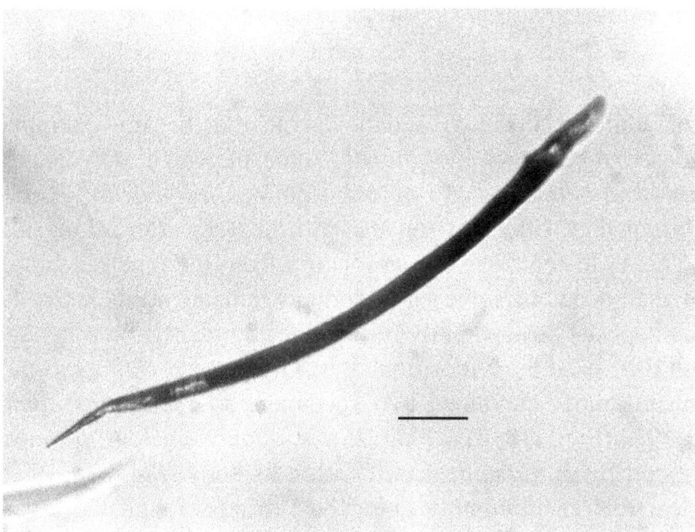

Fig. 188. *Detail of a dauer juvenile of* Tipulacomis dominicanus* *in Dominican amber (accession no. N-3-72). (Scale bar = 43 μm.)*

Fig. 189. *Parasitic juvenile (arrow) of* Palaeoallantonema apionae* *adjacent to its apionid weevil host in Dominican amber (accession no. N-3-41). (Scale bar = 396 μm.)*

Other adult insects with aquatic larvae that became entrapped in Dominican amber were mosquitoes, some of which were parasitised by mermithid nematodes. Female mosquitoes, as with the females of other biting flies, often rest on the bark of trees after taking a blood meal. This is to remove excess moisture from the ingested blood and lighten their load. In some cases, gravid females accidentally landed on resin deposits or apparently mistook resin patches for water sources that could be used as oviposition sites. One female *Culex* mosquito in Dominican amber contained two specimens of *Heydenius dominicus* Poinar, 1984 (Figs 213, 214; Plate 26). A second female *Culex* mosquito had the coils of *H. dominicus* still within its body (Fig. 215; Plate 27). It is clear that *H. dominicus* could be carried through the larval and pupal stages into the adult mosquito, similar to the biology of some extant mosquito parasites. The West African mermithid, *Empidomermis cozii* Poinar, 1977, commonly emerges from the adult stage of the

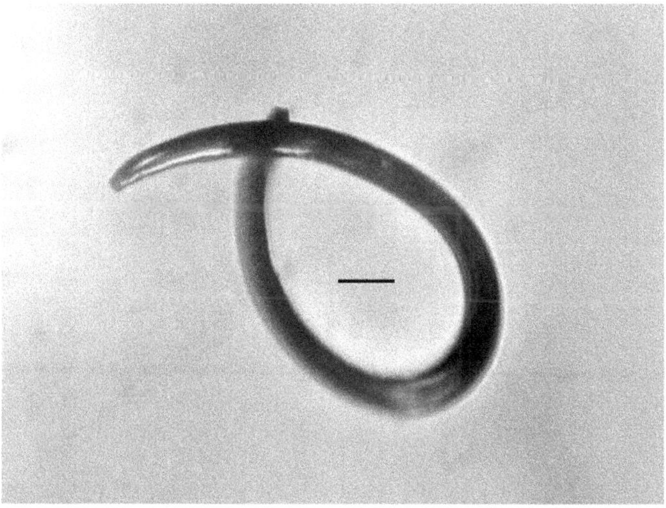

Fig. 190. *Detail of parasitic juvenile of* Palaeoallantonema apionae* *in Dominican amber (accession no. N-3-41). (Scale bar = 22 μm.)*

Fig. 191. *Parasitic juveniles of* Palaeoallantonema zygopinae* *adjacent to their zygopine weevil host in Dominican amber (accession no. N-3-106). (Scale bar = 650 μm.)*

malaria mosquito, *Anopheles funestus* Giles, 1900 (Poinar, 1977b) (Fig. 216).

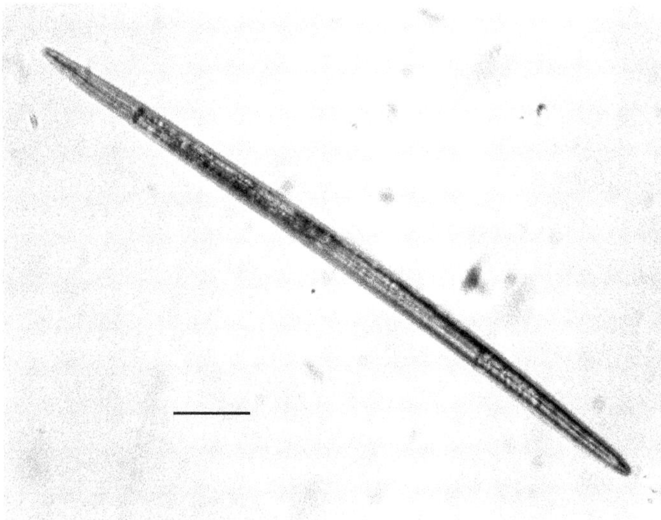

Fig. 192. *Detail of parasitic juvenile of* Palaeoallantonema zygopinae* *in Dominican amber (accession no. N-3-106). (Scale bar = 35 μm.)*

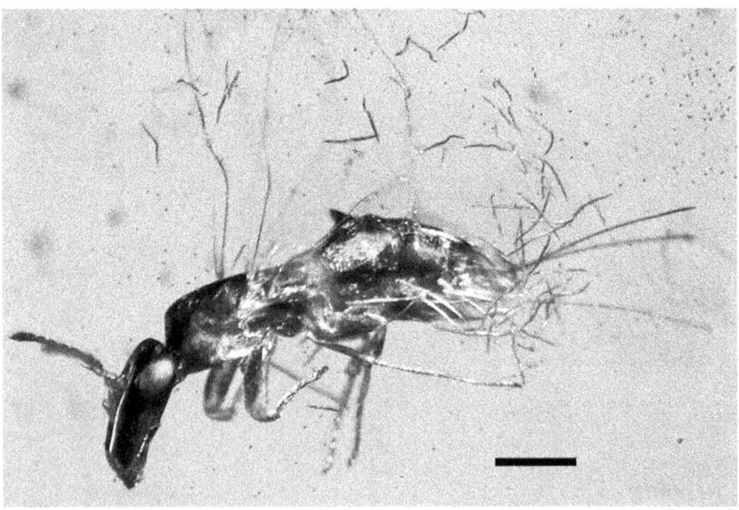

Fig. 193. *Parasitic stages of* Syconema dominicana* *emerging from a fig wasp in Dominican amber (accession no. N-3-24A) (see also Plate 21). (Scale bar = 545 μm.)*

Fossil nematode parasites of vertebrates are extremely rare and one source in amber is insects vectoring filarial parasites. In addition

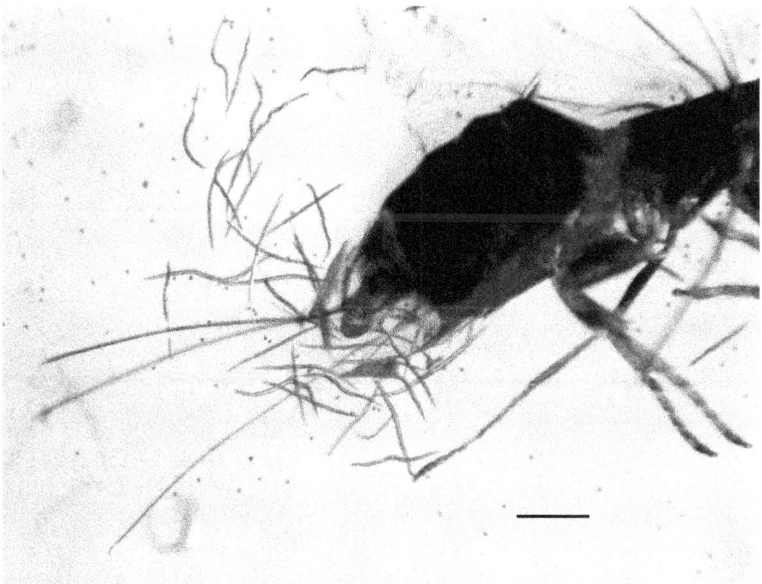

Fig. 194. *Group of* Syconema dominicana* *adjacent to a fig wasp in Dominican amber (accession no. N-3-24A). (Scale bar = 195 μm.)*

to a filarial parasite carried by a black fly in Baltic amber, two female mosquitoes in Dominican amber were associated with filarial nematodes. One filariid, *Cascofilaria dominicana**, was represented by two juveniles. The first was a pre-infective stage with features similar to members of the genera *Foleyella* Seurat, 1917 or *Waltonia* Schacher & Crans, 1973 (Fig. 217). Extant members of these genera parasitise amphibians, especially frogs, and are vectored by mosquitoes (Witenberg & Gerichter, 1944; Anderson, 2000). Pre-infectives can be active, and it is presumed that the fossil juvenile emerged prematurely from the mosquito. A cell attached to its body could represent a mosquito haemocyte. The second specimen of *C. dominicana** (Fig. 218) is considered to be an infective stage and its size is similar to that of the infective stages of the extant frog parasite, *Foleyella duboisi* (Witenberg & Gerichter, 1944). Members of *Foleyella* infect tree frogs (*Leptodactylus* spp.) in Brazil and tree frogs of the related genus *Eleutherodactylus* occur in Dominican amber (Poinar & Poinar, 1999).

A second filarial nematode adjacent to a mosquito in a separate piece of amber is described here as *Cascofilaria parvus** (Fig. 219). It is thought to represent an infective-stage juvenile although it is smaller

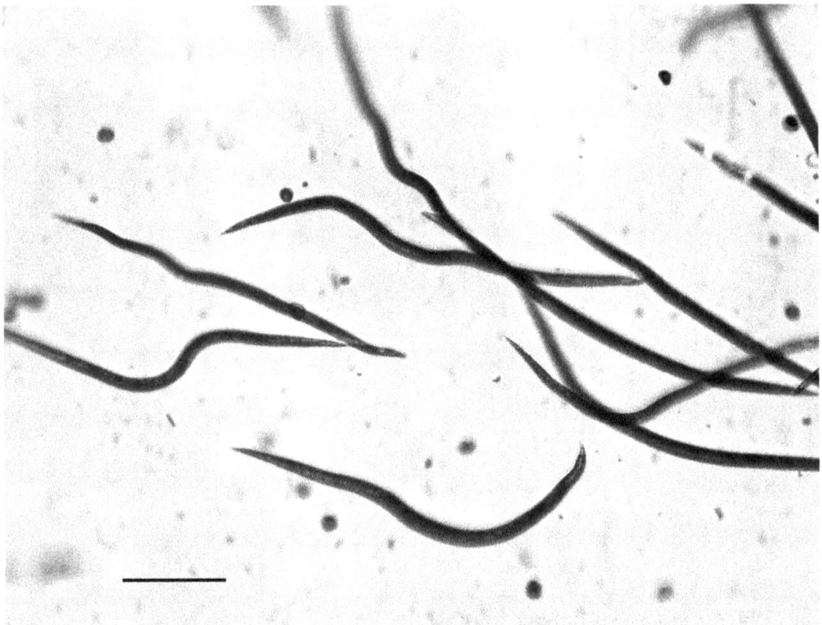

Fig. 195. *Cluster of* Syconema dominicana* *that had emerged from a fig wasp in Dominican amber (accession no. N-3-24A). (Scale bar = 80 μm.)*

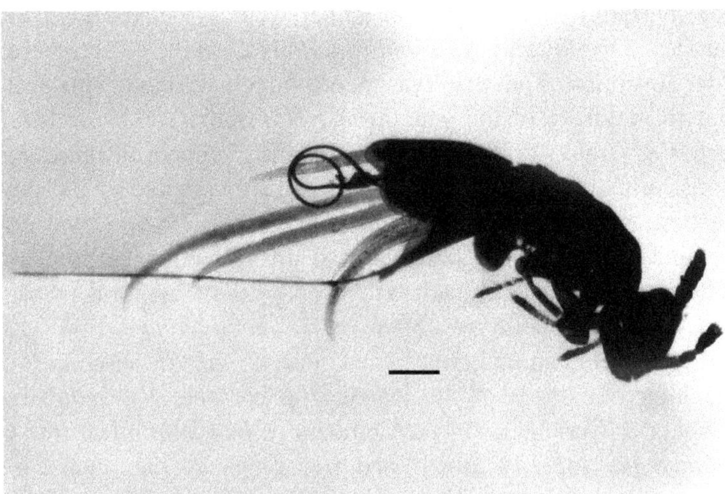

Fig. 196. *Developing stages of the extant* Parasitodiplogaster sycophilon *Poinar, 1979 in the body cavity of the fig wasp* Elisabethiella stuckenbergi *Grandi. (Scale bar = 230 μm.)*

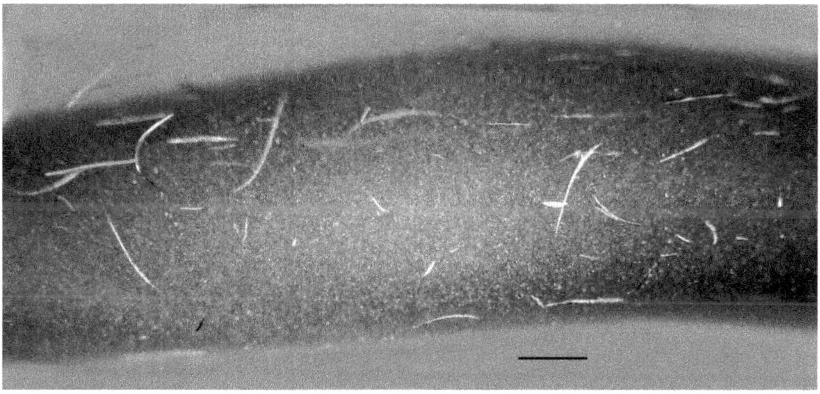

Fig. 197. *Population of the plant nematode* Oligaphelenchoides maximus* *in Dominican amber (accession no. N-3-14) (see also Plate 22). (Scale bar = 857 μm.)*

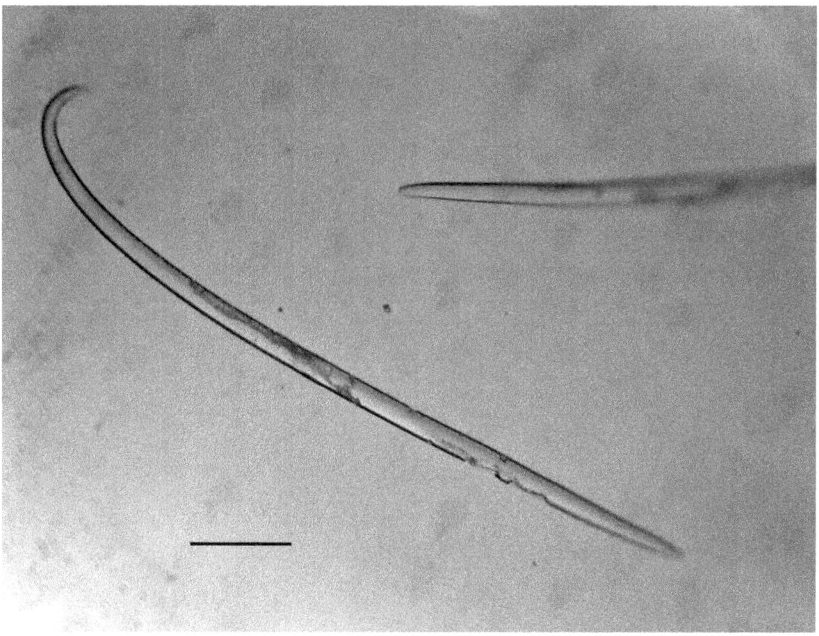

Fig. 198. *Male of* Oligaphelenchoides maximus* *in Dominican amber (accession no. N-3-14). (Scale bar = 1.2 mm.)*

than most filarial infective stages and not enough characters are present to align it with any extant genus. The tail tip bears a small knob and,

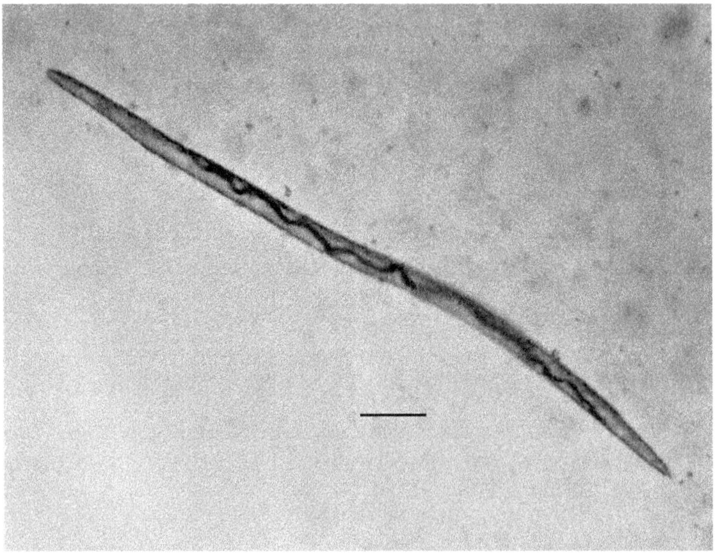

Fig. 199. *Juvenile of* Oligaphelenchoides maximus* *in Dominican amber (accession no. N-3-14). (Scale bar = 45 μm.)*

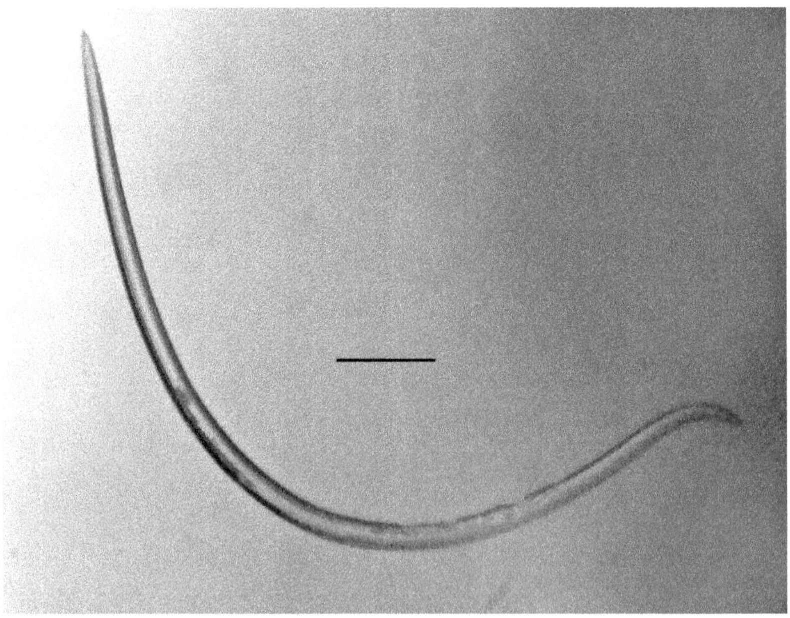

Fig. 200. *Female of* Oligaphelenchoides maximus* *in Dominican amber (accession no. N-3-14). (Scale bar = 120 μm.)*

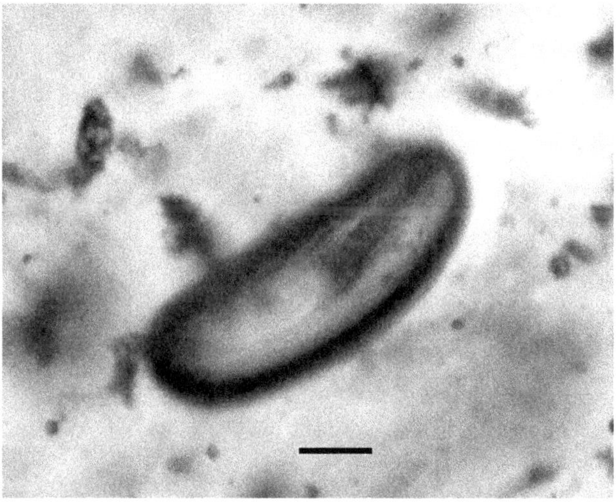

Fig. 201. *Egg of* Oligaphelenchoides maximus* *in Dominican amber (accession no. N-3-14). (Scale bar = 14 μm.)*

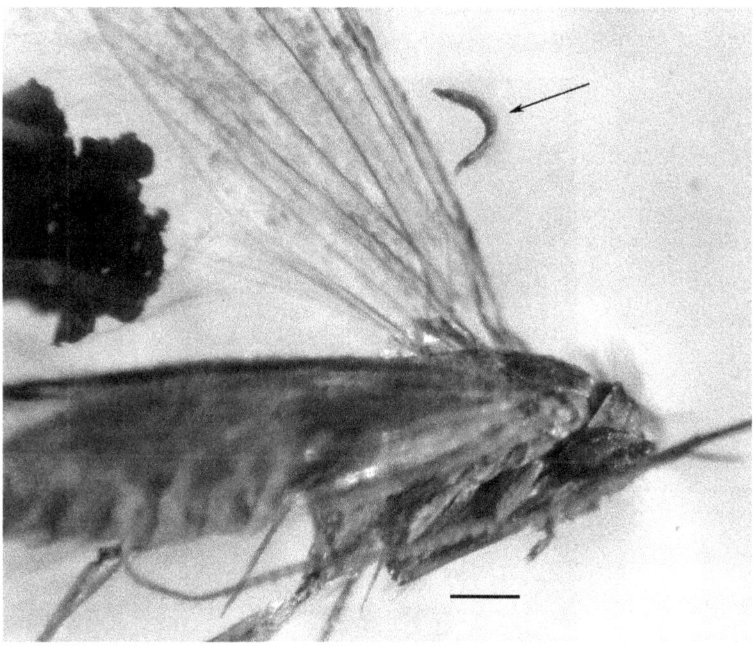

Fig. 202. *The ectoparasitic nematode* Setonema protera* *(arrow), adjacent to its moth host in Dominican amber (accession no. N-3-31) (see also Plate 23A). (Scale bar = 245 μm.)*

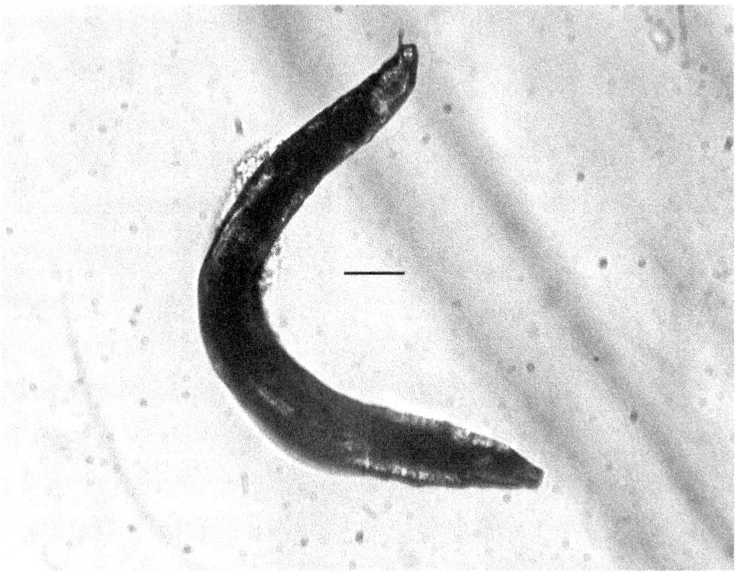

Fig. 203. *Developing female of the ectoparasitic moth nematode* Setonema protera* *in Dominican amber (accession no. N-3-31). (Scale bar = 43 μm.)*

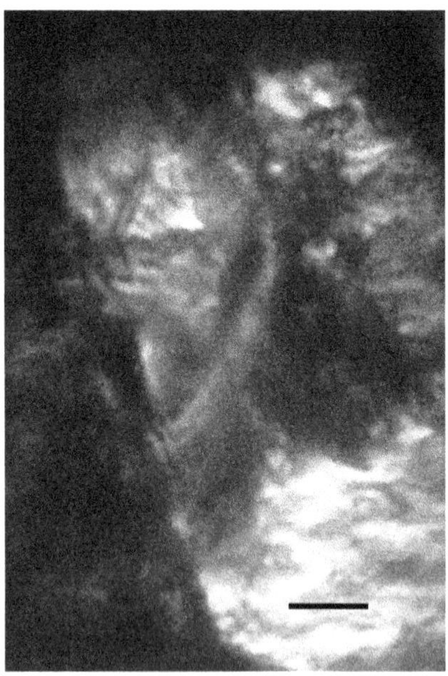

Fig. 204. *Stages of* Setonema protera* *on the body surface of a moth in Dominican amber (accession no. N-3-31). (Scale bar = 100 μm.)*

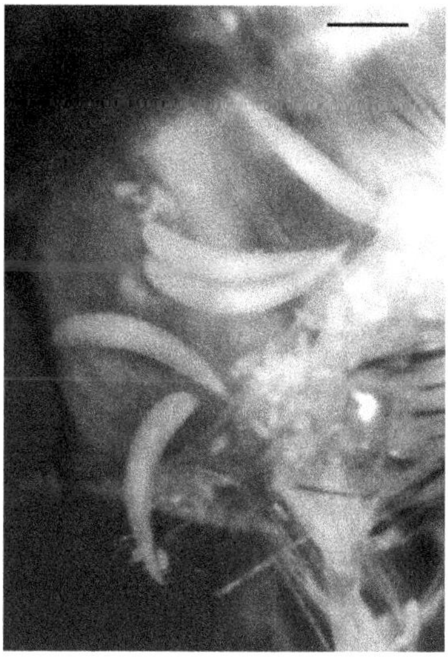

Fig. 205. *Developing females of the extant ectoparasitic moth nematode* Noctuidonema guyanense *Remillet & Silvain, 1988 on a noctuid moth (photo courtesy of M. Remillet) (see also Plate 23B). (Scale bar = 270 μm.)*

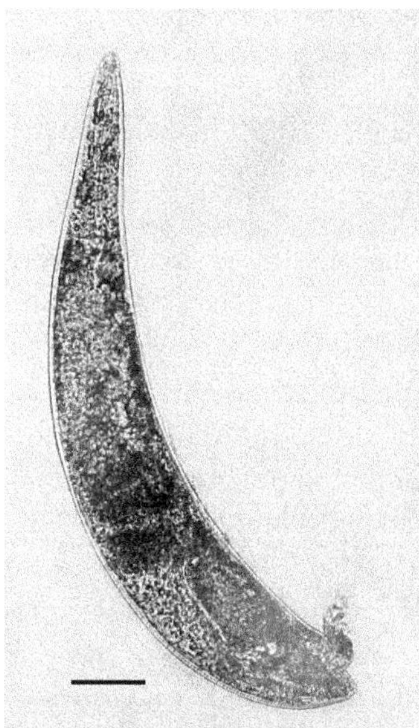

Fig. 206. *Female of the extant ectoparasitic moth nematode* Noctuidonema guyanense *Remillet & Silvain, 1988 removed from the body of a noctuid moth (photo courtesy of M. Remillet). (Scale bar = 68 μm.)*

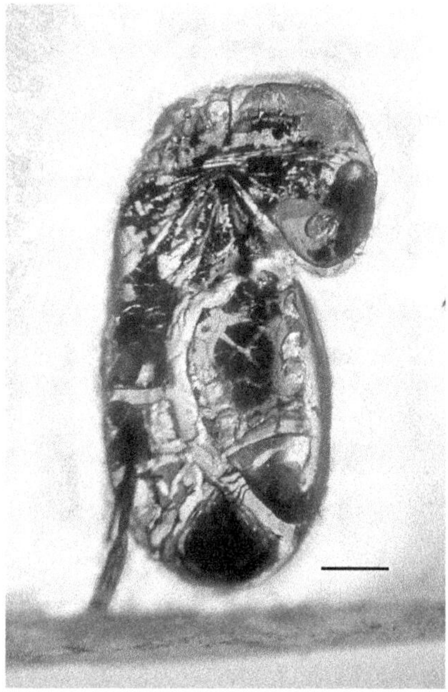

Fig. 207. *Developing monocot seed that contained* Palaeoanguina dominicana* *in Dominican amber (accession no. N-3-6A). (Scale bar = 589 μm.)*

while ornamentation of the tail tip is a key character in identifying the infective stages of filarial nematodes, a similar structure could not be found on extant filariids (Bain & Chabaud, 1986).

Another group of bloodsuckers in the Dominican amber forest comprised sand flies or phlebotomines. While the larvae are not aquatic, they occur in damp habitats where they feed on fungi and other microbes in decomposing plant matter. Since this habitat is also an attractive one for nematodes, it is not surprising that extant forms are parasitised by both tetradonematids and allantonematids. A female sand fly belonging to the genus *Lutzomyia* in Dominican amber was associated with three juvenile allantonematids described as *Palaeoallantonema phlebotomae** (Figs 220-222). Additional forms could be seen inside the body of the sand fly. Whether this species belongs to the same clade as the extant South American sand fly parasite, *Anandranema phlebotophaga* Poinar, Ferro, Morales & Tesh, 1993, could not be determined.

A number of insects survive by feeding on decaying animal and plant matter. Wood gnats or Anisopodidae are one group of such flies. The adults look like large fungus gnats with thickened antennal segments.

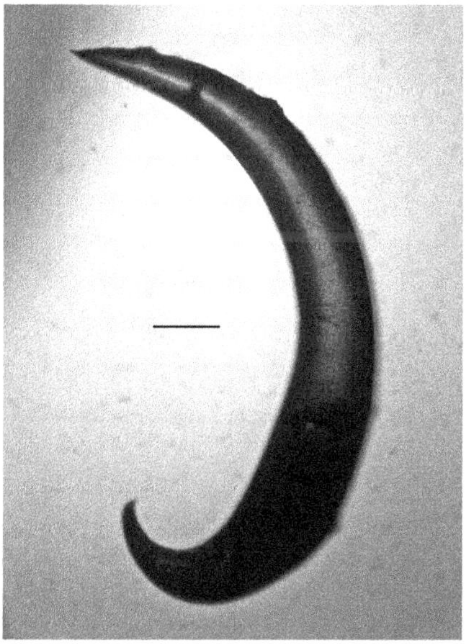

Fig. 208. *Juvenile female of* Palaeoanguina dominicana* *in Dominican amber (accession no. N-3-6A) (see also Plate 24A). (Scale bar = 128 μm.)*

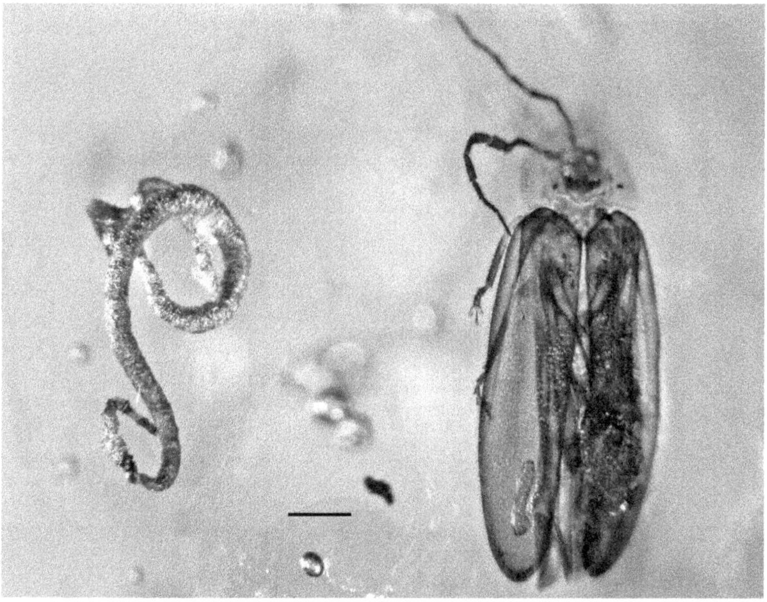

Fig. 209. Heydenius lamprophilus* *from a lampyrid beetle in Dominican amber (accession no. N-3-17) (see also Plate 24B). (Scale bar = 914 μm.)*

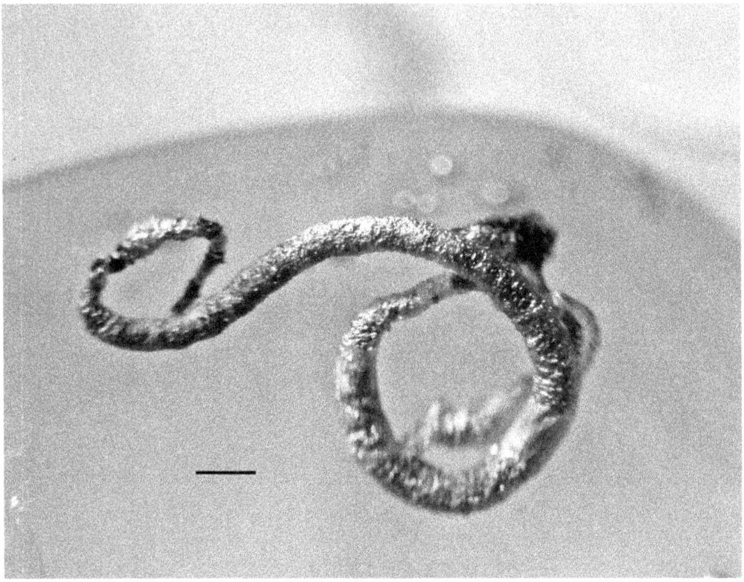

Fig. 210. *Detail of* Heydenius lamprophilus* *from a lampyrid beetle in Dominican amber (accession no. N-3-17). (Scale bar = 427 μm.)*

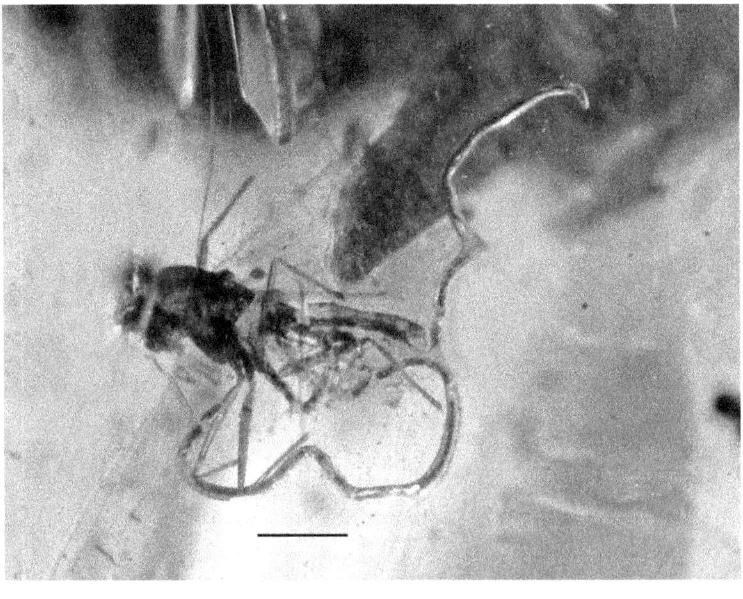

Fig. 211. *Two specimens of* Heydenius neotropicus* *from a chironomid midge host in Dominican amber (accession no. N-3-13A). (Scale bar = 700 μm.)*

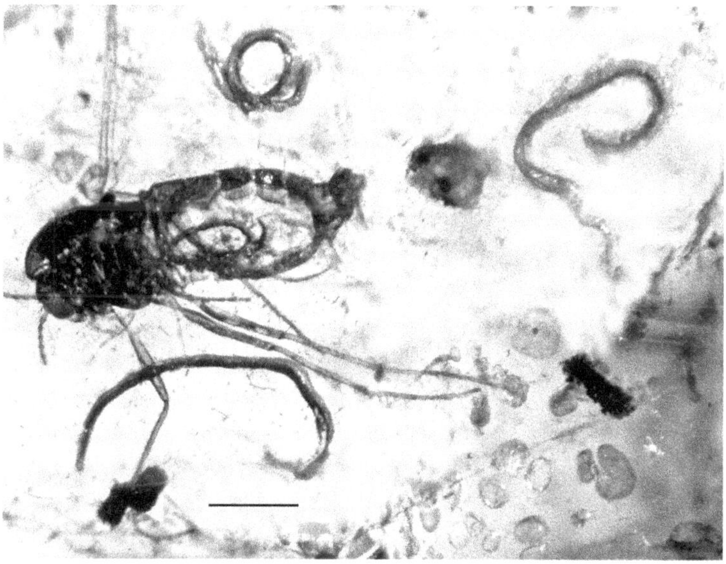

Fig. 212. *Three specimens of* Heydenius neotropicus* *adjacent to a chironomid midge in Dominican amber. A fourth specimen is still inside the fly's body (accession no. N-3-13B) (see also Plate 25). (Scale bar = 310 μm.)*

Some of these in Dominican amber were parasitised by mermithids. A mature parasitic juvenile of *Heydenius saprophilus** was near its anisopodid host (Fig. 223; Plate 28). A wide tail spike showed that it was nearly ready to leave under natural conditions (Fig. 224). It probably had a direct life cycle similar to that of other mermithid parasites of Diptera.

Insect scavengers are prey for predators that can withstand decaying environmental conditions. Rove beetles of the family Staphylinidae are attracted to decaying organic matter for the purpose of feeding on the primary decomposers. As noted in the section on Baltic amber, rove beetles are infected by allantonematid nematodes. One species in Dominican amber was also attacked by a similar nematode, which is described here as *Palaeoallantonema dominicana**. This species is represented by 44 juveniles and some other stages associated with an adult staphylinid beetle in Dominican amber (Figs 225-227; Plate 29). The size of the juveniles is significantly smaller than any nematodes recovered from extant staphylinids.

There were a number of other arthropods among the decaying vegetation in the amber forest. One group of detritivores was the millipedes, among which were several representatives of the family

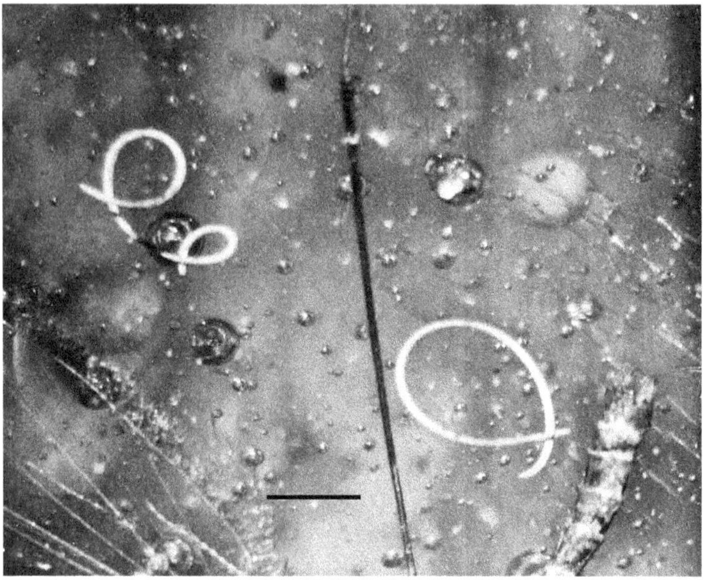

Fig. 213. *Two specimens of* Heydenius dominicus *Poinar, 1984 from a* Culex *mosquito in Dominican amber (Staatliches Museum, Stuttgart, Germany). (Scale bar = 611 μm.)*

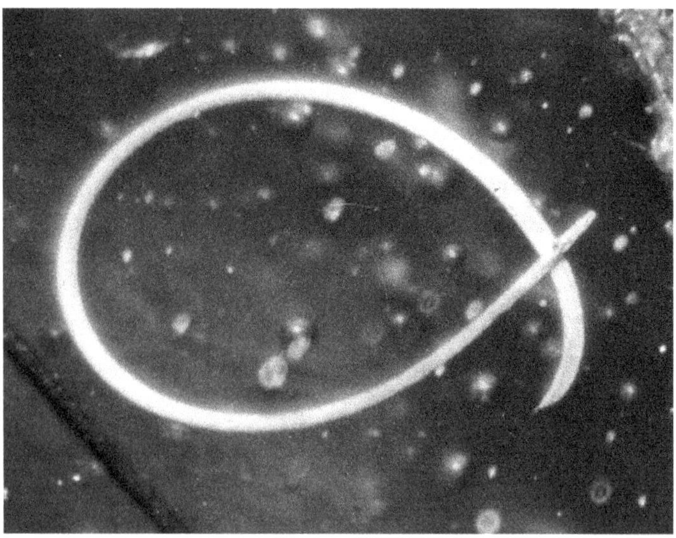

Fig. 214. *Detail of one of the specimens of* Heydenius dominicus *Poinar, 1984 in Dominican amber (Staatliches Museum, Stuttgart, Germany) (see also Plate 26). (Scale bar = 110 μm.)*

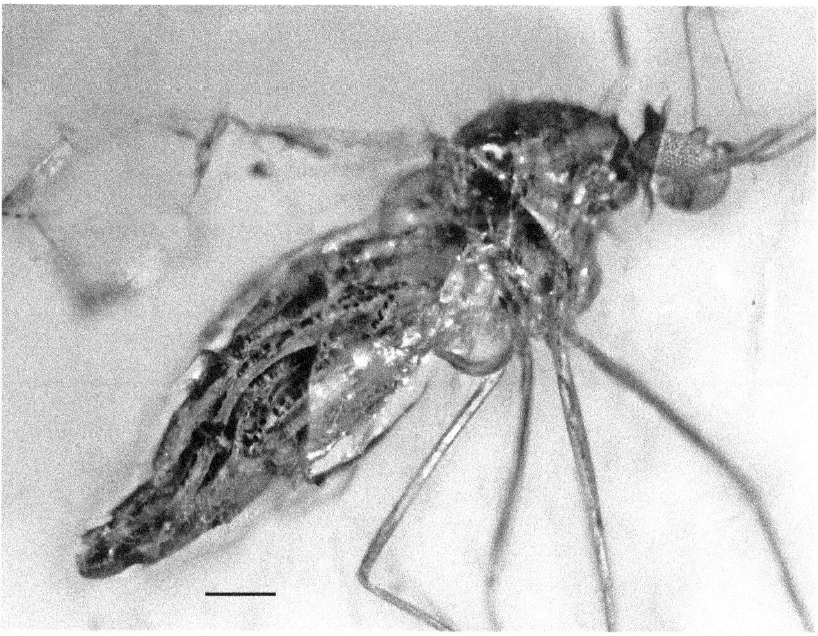

Fig. 215. *Specimen of* Heydenius dominicus *Poinar, 1984 inside the body cavity of a* Culex *mosquito in Dominican amber (accession no. N-3-42) (see also Plate 27). (Scale bar = 348 μm.)*

Polydesmidae (Santiago-Blay & Poinar, 1992). A member of the genus *Docodesmus* was associated with two juveniles of *Diplonema minuta** (Figs 228-230). The juveniles are considered to represent dauer stages that were possibly in the gut when the myriapod was captured in resin. Extant millipedes carry dauer rhabditids in their guts and body cavities (Poinar, 1986; Schulte, 1989; Sudhaus & Schulte, 1989) (Fig. 231). Under normal conditions, these dauer juveniles apparently do not mature to adults until the host dies.

Predatory centipedes also patrolled the forest floor and a member of the family Scutigeridae in Dominican amber was associated with a specimen of *Chilonema filaforma**.

The nematode, apparently a dauer juvenile, may have crawled out of the gut of the centipede when it encountered the resin (Figs 232, 233). No reports on dauer nematodes associated with extant centipedes could be found.

Also in the same habitat as centipedes and millipedes are terrestrial crustaceans belonging to the Isopoda and Amphipoda. Pill bugs are

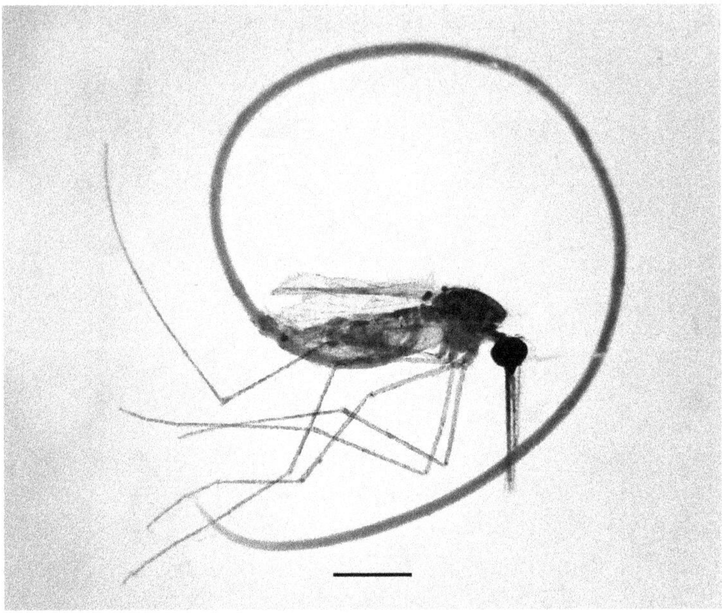

Fig. 216. *The extant mermithid* Empidomermis cozii *Poinar, 1977 emerging from its mosquito host,* Anopheles funestus *Giles, in West Africa. (Scale bar = 1.1 mm.)*

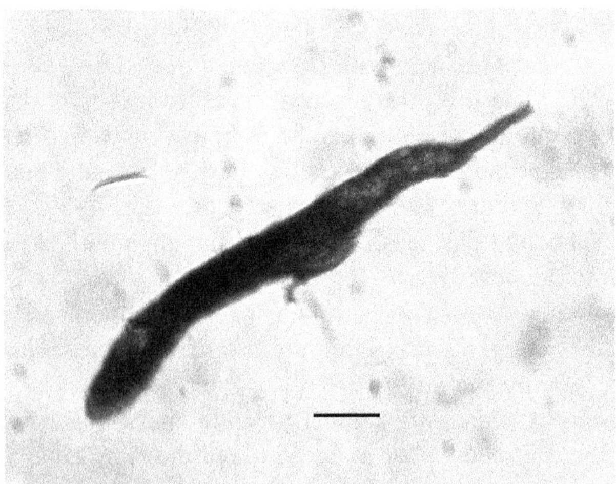

Fig. 217. *Pre-infective stage of the filariid* Cascofilaria dominicana* *adjacent to a* Culex *mosquito in Dominican amber (accession no. N-3-83). (Scale bar = 31 μm.)*

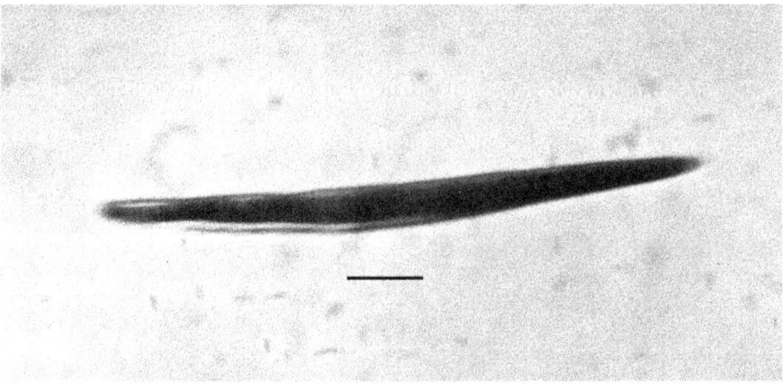

Fig. 218. *Infective stage of the filariid* Cascofilaria dominicana* *adjacent to a* Culex *mosquito in Dominican amber (accession no. N-3-83). (Scale bar = 69 μm.)*

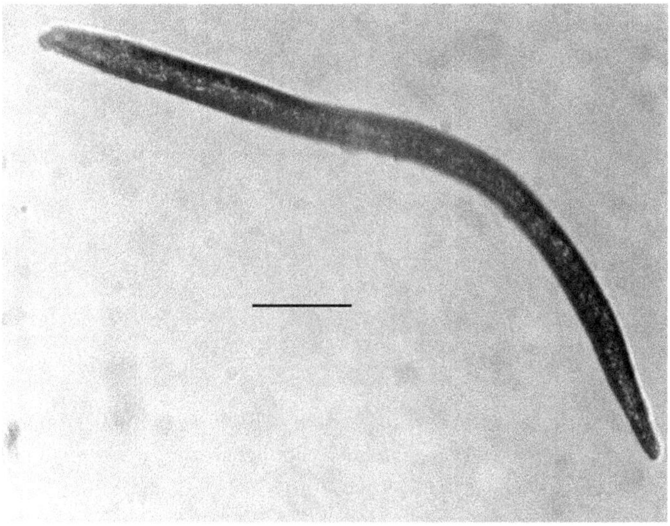

Fig. 219. *Infective stage of the filariid* Cascofilaria parva* *that was adjacent to a* Culex *mosquito in Dominican amber (accession no. Sy-1-165). (Scale bar = 26 μm.)*

the most common crustaceans in amber and one member of the family Pholisciidae was associated with nematodes described as *Crustafila confina** (Figs 234-236). The nematodes appeared to be developing in the same microhabitat as the pill bug since a range of sizes were present. One adjacent to the fossil pill bug appeared to be a dauer juvenile

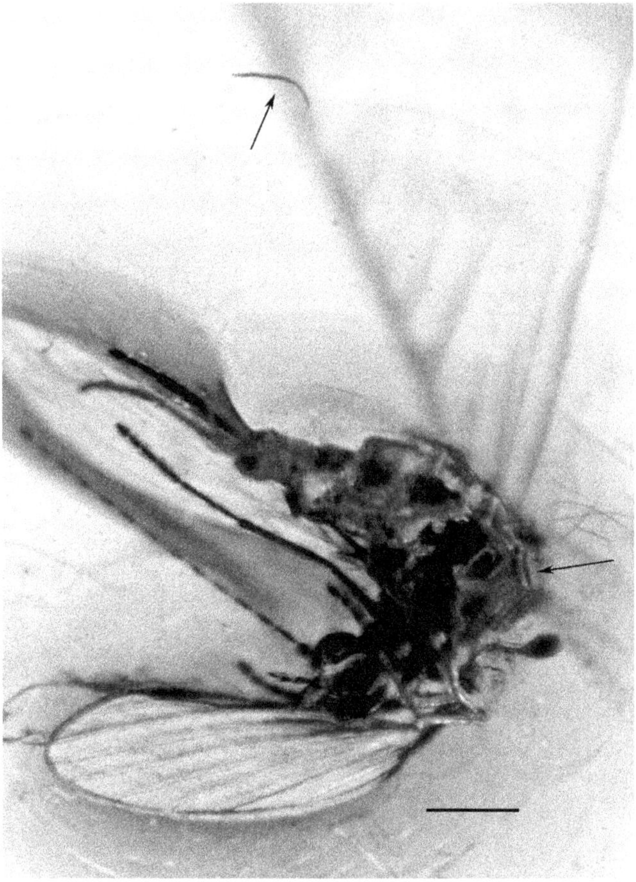

Fig. 220. *Parasitic juveniles (arrows) of the allantonematid* Palaeoallantonema phlebotomae* *adjacent to a phlebotomine sand fly in Dominican amber (accession no. N-3-93). (Scale bar = 326 μm.)*

(Fig. 236). Dauer stages of the extant nematode *Caenorhabditis vulgaris* Baird, Fitch & Emmons can be found under the dorsal carapace of pill bugs and develop on the host after it dies (Baird *et al.*, 1994). Isopods are also known to feed on nematode populations in sewage (Brown *et al.*, 1978). Nematode parasites of isopods include the mermithid *Thaumamermis cosgrovei* Poinar, 1981 and the gut-inhabiting oxyurid *Oniscicola oniscicola* Schwenk (1927).

Few nematodes have been reported from extant parasitoid wasps. That is why it was such a surprise to discover *Chalcidonema paradoxa**,

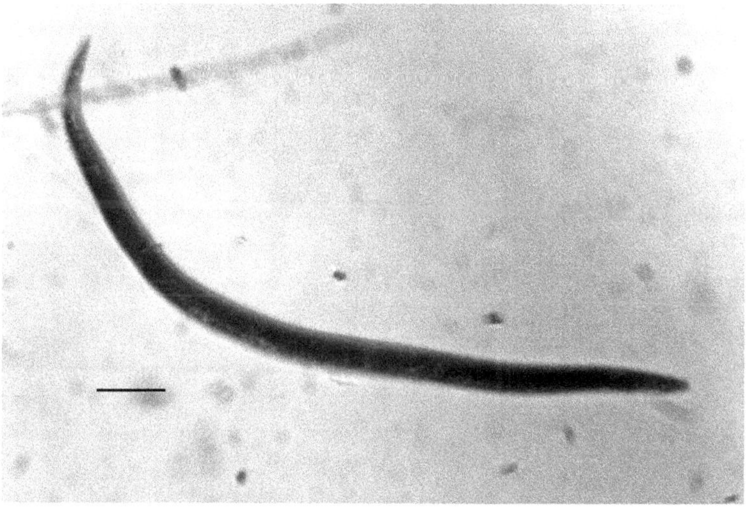

Fig. 221. *Parasitic juvenile of* Palaeoallantonema phlebotomae* *in Dominican amber (accession no. N-3-93). (Scale bar = 27 μm.)*

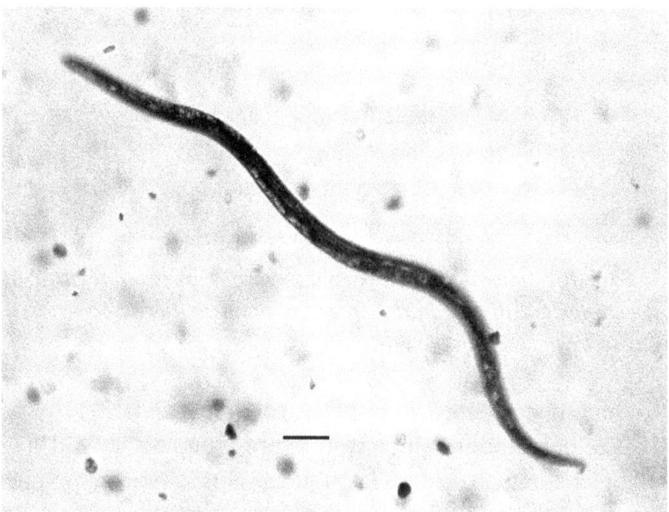

Fig. 222. *A second parasitic juvenile of* Palaeoallantonema phlebotomae* *in Dominican amber (accession no. N-3-93). (Scale bar = 22 μm.)*

associated with a minute member of the Chalcidoidea in Dominican amber. The wasp, which resembles some extant members of the family Pteromalidae, was barely 1.3 mm in length, yet it harboured nematodes,

Fig. 223. *Parasitic juvenile of* Heydenius saprophilus* *adjacent to its wood gnat host (Diptera: Anisopodidae) in Dominican amber (accession no. N-3-102) (see also Plate 28). (Scale bar = 971 μm.)*

some of which had escaped into the surrounding amber (Figs 237-239; Plate 30). The nematodes, which were represented by several stages, are tentatively placed in the Aphelenchoidea. Since two nematodes were projecting out from the body of the wasp and one was still inside, *Chalcidonema** is considered to have been a parasite.

Members of the genus *Deladenus* occur in ichneumonids and cynipids (Shepard, 1974; Poinar, 1975), and the allantonematid *Prothallonema hastatum* (Kahn, 1957) infects braconid parasites that are parasitising bark beetle hosts. In the latter case, the nematodes probably entered the developing wasp larva inside the beetle host (Khan, 1957). The fossil

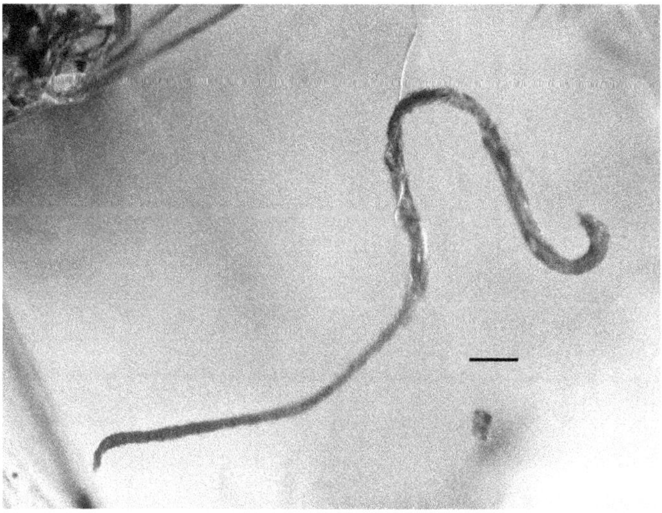

Fig. 224. *Detail of specimen of* Heydenius saprophilus* *in Dominican amber (accession no. N-3-102). (Scale bar = 283 μm.)*

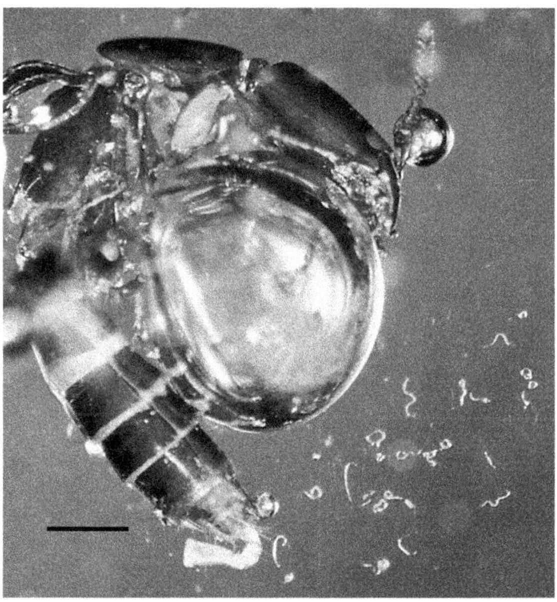

Fig. 225. *Parasitic juveniles of the allantonematid* Palaeoallantonema dominicana* *that had emerged from a rove beetle (Staphylinidae) in Dominican amber (accession no. N-3-9) (see also Plate 29). (Scale bar = 350 μm.)*

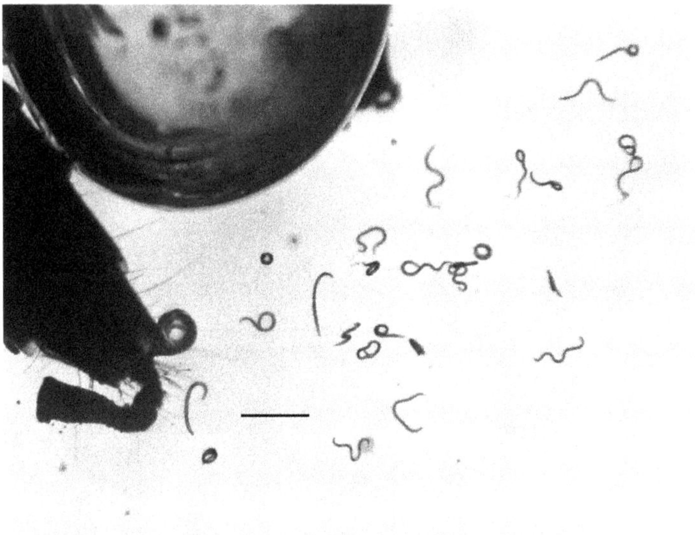

Fig. 226. *Group of parasitic juveniles of* Palaeoallantonema dominicana* *adjacent to their rove beetle host in Dominican amber (accession no. N-3-9). (Scale bar = 233 μm.)*

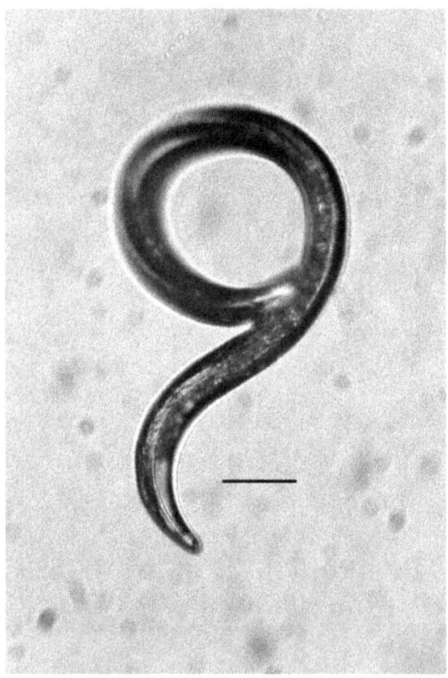

Fig. 227. *Parasitic juvenile of* Palaeoallantonema dominicana* *in Dominican amber (accession no. N-3-9). (Scale bar = 23 μm.)*

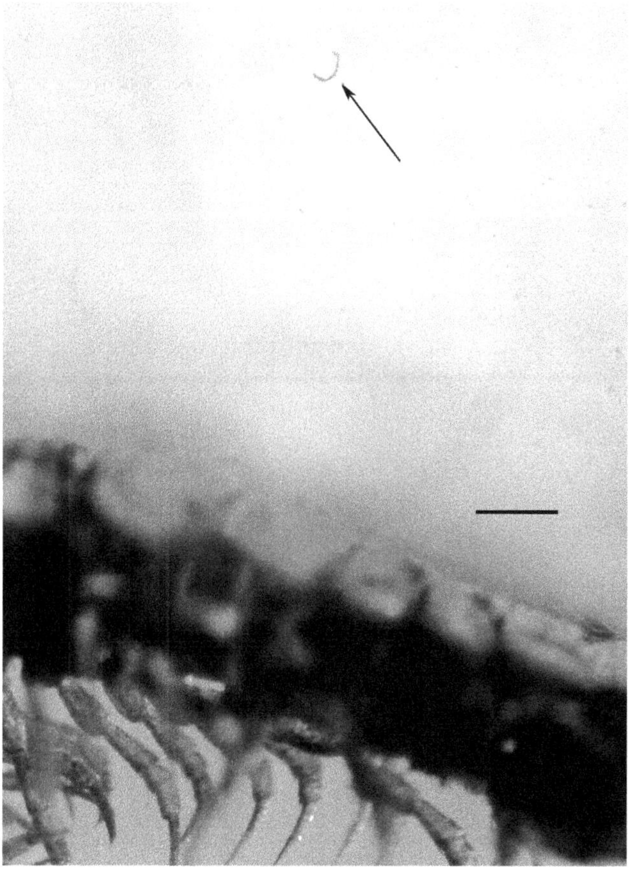

Fig. 228. Diplonema minuta* *(arrow) with its millipede host in Dominican amber (accession no. N-3-97). (Scale bar = 317 μm.)*

wasp host of *C. paradoxa** is tentatively identified as a member of the family Pteromalidae. Pteromalids are generally parasitoids of the larval and pupal stages of holometabolous insects, especially Lepidoptera and Coleoptera (Clausen, 1962). It is possible that nematode parasites of the insect chosen as host by the pteromalid entered the wasp larva during its development. The nematodes then continued their growth in the pupal and adult stage of the wasp. Another scenario is that the nematodes infected the pupal or emerging adult wasp after it left the host, but it would appear to be difficult for the nematodes to establish a successful life cycle under such conditions.

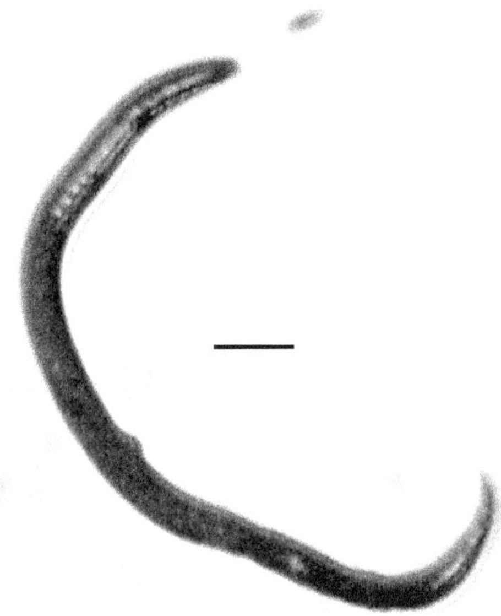

Fig. 229. *Dauer juvenile of* Diplonema minuta* *in Dominican amber (accession no. N-3-97). (Scale bar = 16 μm.)*

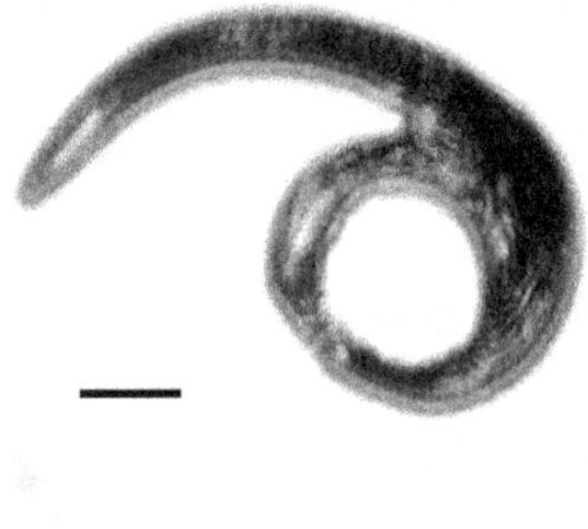

Fig. 230. *A second specimen of* Diplonema minuta* *in Dominican amber (accession no. N-3-97). (Scale bar = 14 μm.)*

It is especially interesting when natural associations involving a number of organisms are all preserved together. A piece of Dominican am-

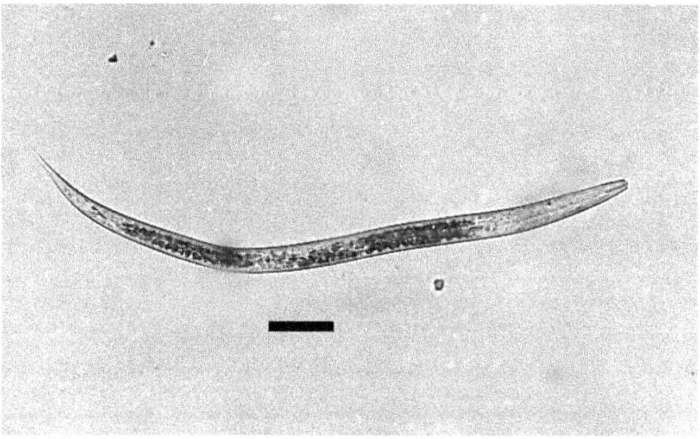

Fig. 231. *Dauer juvenile of the extant species* Rhabditis myriophila *Poinar, 1986 associated with the polydesmid millipede* Oxidis gracilis *(Koch). (Scale bar = 56 μm.)*

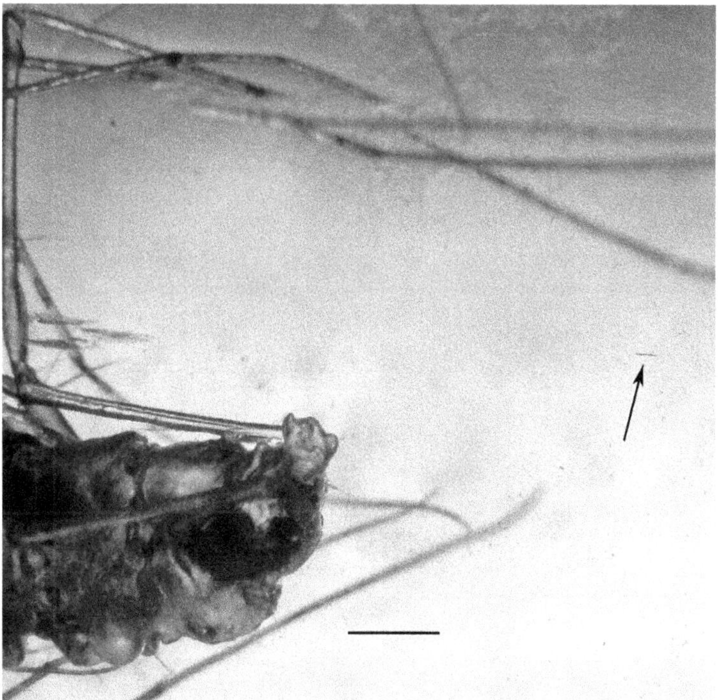

Fig. 232. *Dauer juvenile (arrow) of* Chilonema filaforma* *adjacent to a centipede in Dominican amber (accession no. N-3-96). (Scale bar = 1.6 mm.)*

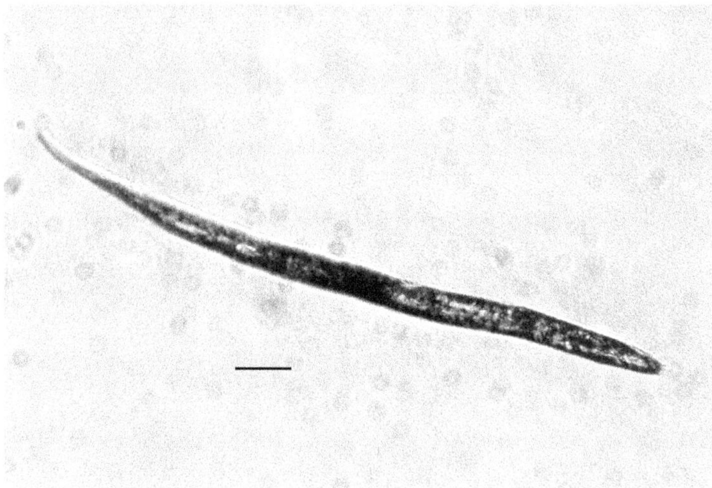

Fig. 233. *Dauer juvenile of* Chilonema filaforma* *in Dominican amber (accession no. N-3-96). (Scale bar = 20 μm.)*

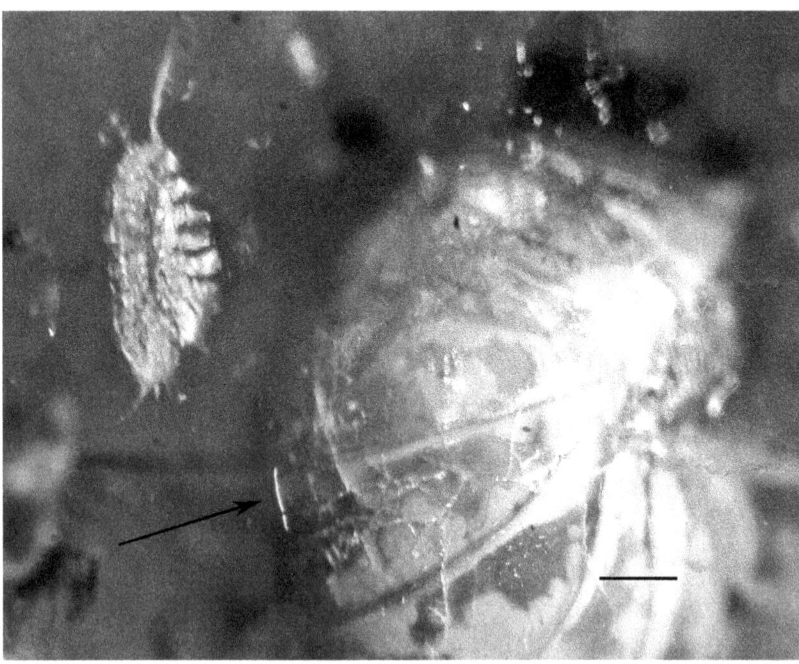

Fig. 234. Crustafila confina* *(arrow) adjacent to a pillbug in Dominican amber (accession no. N-3-98). (Scale bar = 359 μm.)*

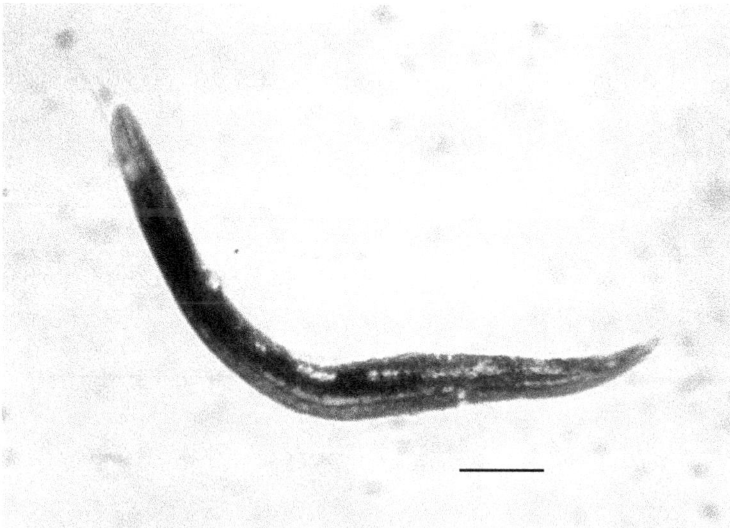

Fig. 235. *Developing stage of* Crustafila confina* *in Dominican amber (accession no. N-3-98). (Scale bar = 36 μm.)*

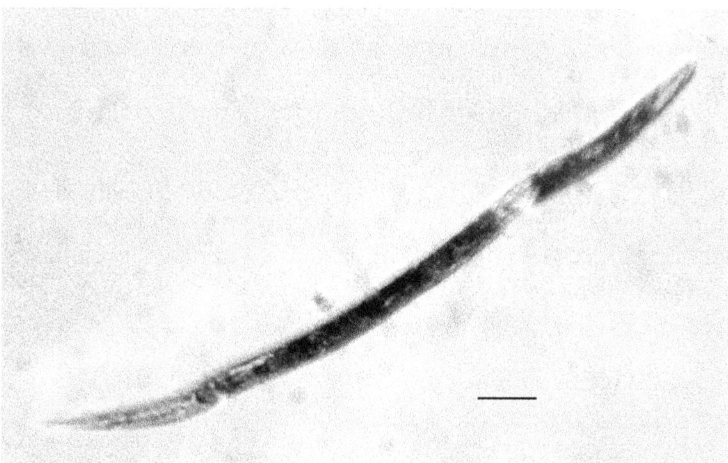

Fig. 236. *Dauer juvenile of* Crustafila confina* *in Dominican amber (accession no. N-3-98). (Scale bar = 24 μm.)*

ber contained a large population of small tardigrade-like invertebrates, along with fungal mycelium, protozoa, nematodes and a pseudoscorpion (Fig. 240). The pseudoscorpion probably fed on the tardigrade-like invertebrates as well as on the nematode, *Vetus microcosmus**. In turn, the

Fig. 237. *Stages of* Chalcidonema paradoxa* *(arrows) associated with a chalcidoid wasp in Dominican amber (accession no. N-3-99) (see also Plate 30). (Scale bar = 471 μm.)*

nematodes were apparently feeding on bacteria, which included species developing inside expired tardigrade-like invertebrates (Figs 241, 242). Scenes like this reveal a microcosm that could not have been preserved by any other means of fossilisation.

Mexican amber nematodes

Mexican amber originates from the State of Chiapas in southern Mexico and is Upper Oligocene to Lower Miocene (22-26 mya) in age. The Mexican amber forest was, in large part, similar to that of the Dominican amber forest, and the tree (*Hymenaea mexicana* Poinar & Brown, 2002) that produced the amber belonged to the same genus as the one that produced Dominican amber. While similar insect genera occur in both amber deposits, the species are different, showing that the forests were separated long enough for speciation to occur.

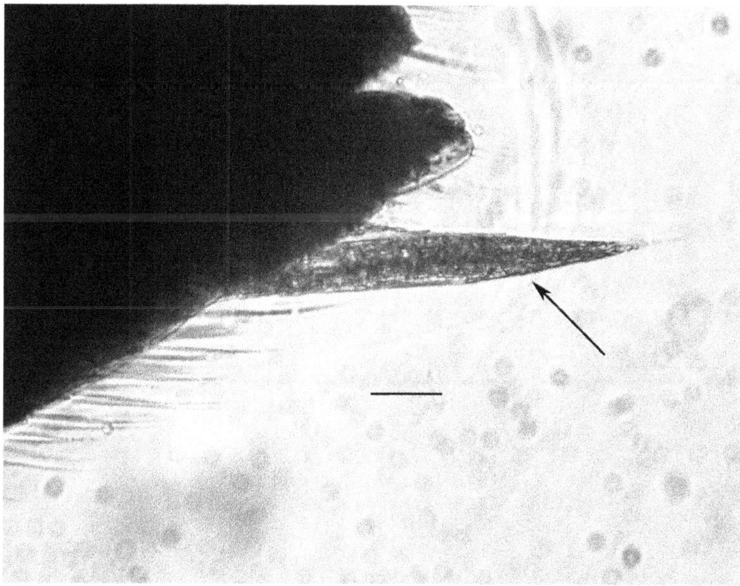

Fig. 238. *Specimen of* Chalcidonema paradoxa* *(arrow) protruding from the body of a chalcidoid wasp in Dominican amber (accession no. N-3-99). (Scale bar = 55 μm.)*

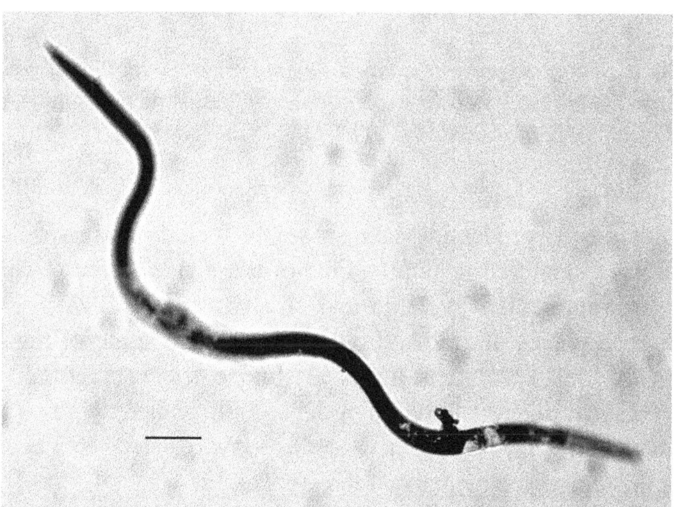

Fig. 239. *A second specimen of* Chalcidonema paradoxa* *adjacent to a chalcidoid wasp in Dominican amber (accession no. N-3-99). (Scale bar = 34 μm.)*

Fig. 240. *A pseudoscorpion dominated this microcosm with fungal mycelium, tardigrade-like invertebrates and a specimen of* Vetus microcosmus* *(arrow) in Dominican amber (accession no. ON-1-4). (Scale bar = 309 μm.)*

Free-living nematodes in Mexican amber include some species that could not be identified to family. One of these, *Vetus robustus* Poinar, 1977, was represented by large and small forms (Figs 243, 244) in two separate pieces of amber (Poinar, 1977a). On some of the larger, presumably adult, specimens, it was possible to detect six partially fused lips and a small mucron at the tip of the tail. Since there were no insects or plant material in the amber, *V. robustus* was probably an arboreal species that colonised the bark of the amber tree.

Two additional specimens in Mexican amber, described as *Oligodiplogaster antiqua* Poinar, 1977, were surrounded by a superficial dark coating (Poinar, 1977a). However, their size, indistinct lips, body curvature and long tail filament were characteristic of male diplogastrids

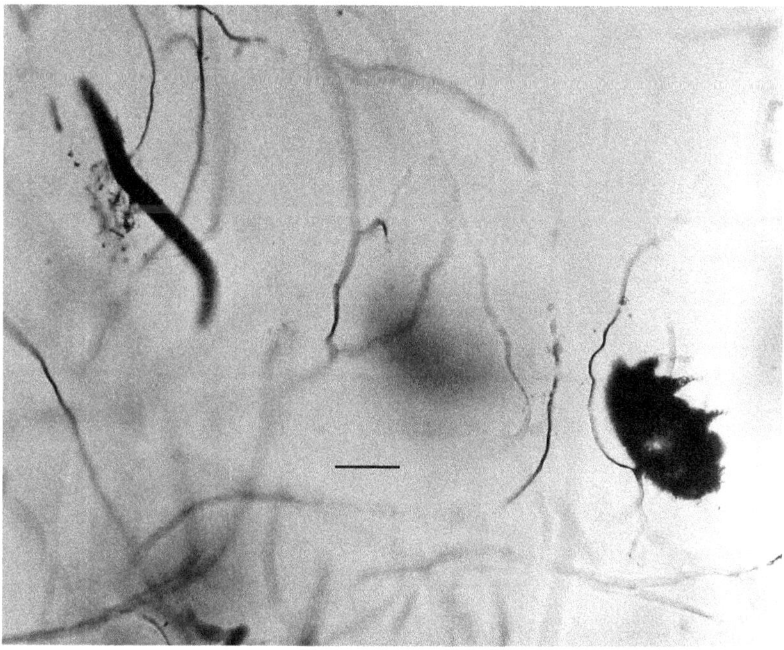

Fig. 241. *A specimen of* Vetus microcosmus* *on the left and a tardigrade-like invertebrate on the right in Dominican amber (accession no. ON-1-4). (Scale bar = 66 μm.)*

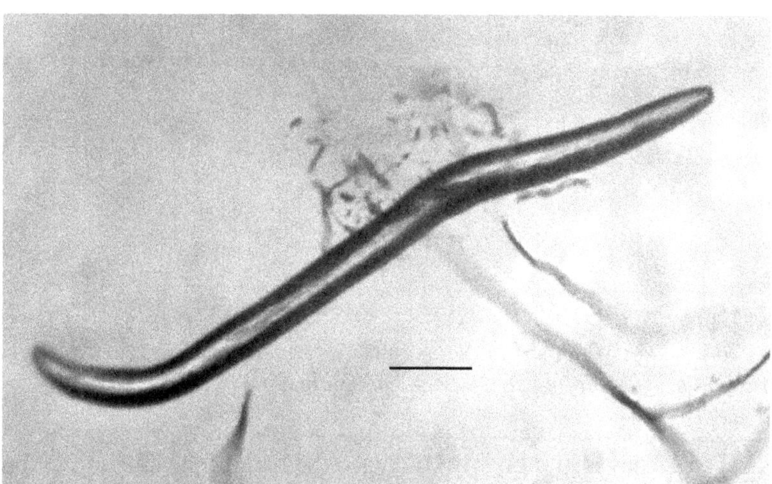

Fig. 242. *A second specimen of* Vetus microcosmus* *in Dominican amber (accession no. ON-1-4). (Scale bar = 27 μm.)*

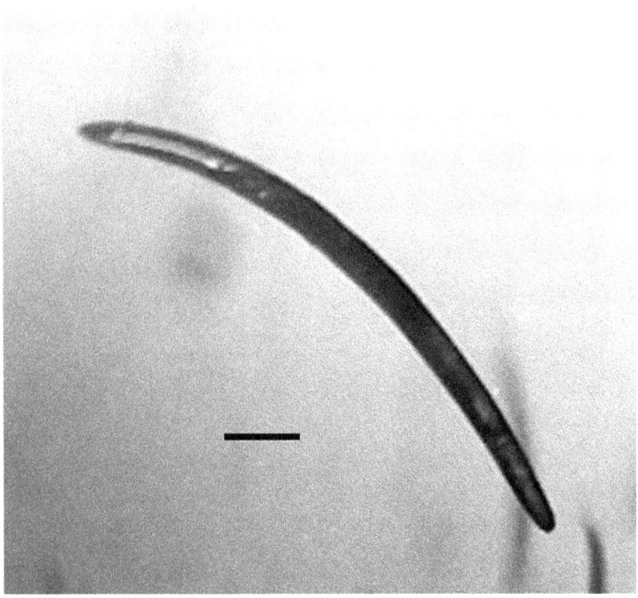

Fig. 243. *Specimen of* Vetus robustus *Poinar, 1977 in Mexican amber (Museum of Paleontology, UC Berkeley). (Scale bar = 76 μm.)*

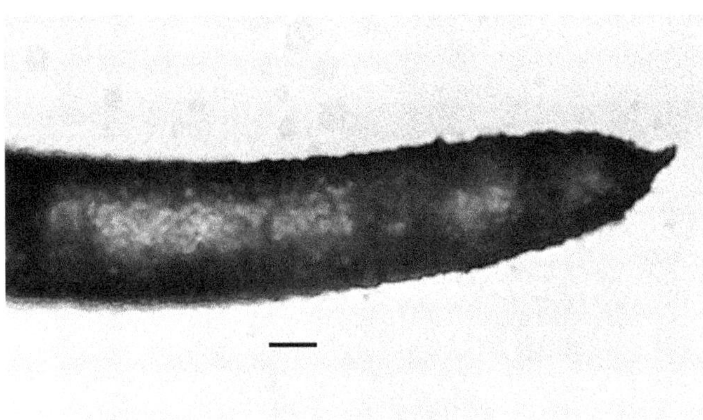

Fig. 244. *Detail of the tail of* Vetus robustus *Poinar, 1977 in Mexican amber (Museum of Paleontology, UC Berkeley). (Scale bar = 17 μm.)*

(Fig. 245). They also probably colonised the bark of the amber tree, perhaps in a decayed area, and were trapped in a resin flow.

There certainly was a variety of nematodes associated with bark beetles in the Mexican amber forest. One of these was *Scolytonema*

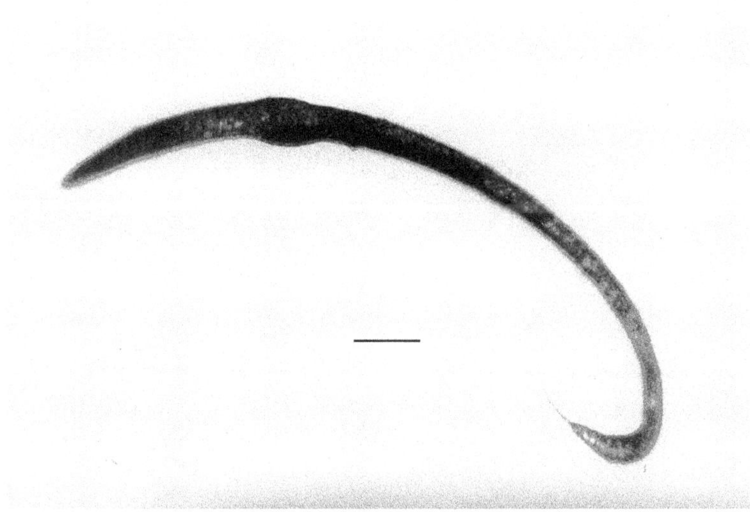

Fig. 245. *A specimen of* Oligodiplogaster antiqua *Poinar, 1977 in Mexican amber (Museum of Paleontology, UC Berkeley). (Scale bar = 38 μm.)*

*mexicana**, which was represented by 14 dauer juveniles adjacent to a platypodid beetle (Figs 246, 247). This species was determined to be a diplogastrid on the basis of its small size, long, pointed tail and general appearance.

Of particular interest in Mexican amber was the aphelenchoidid *Oligaphelenchoides atrebora* Poinar, 1977. All developmental stages (males, females, eggs and juveniles of various sizes) were present (Figs 248-251; Plate 31), as well as hyphae that served as the food source (Poinar, 1977a). The nematode colony was probably developing on fungi growing in a decayed area on the tree when a resin flow covered the microhabitat. The mycelium that the nematodes were consuming appears to be a member of the Deuteromycotina. In this microcosm were also some nematophagous fungi that had parasitised several individuals of *O. atrebora* (Jansson & Poinar, 1986) (Figs 252, 253). While there are many types of nematophagous fungi found today (Barron, 1977), this is the first case of a fossil representative with its host.

Few insect-parasitic nematodes have been recovered from Mexican amber, although the forest must have been filled with them. An interesting parasite is the tetradonematid *Palaeotetradonema phlebotomae**, represented by three immature females adjacent to a phlebotomine sand fly (Figs 254, 255). Placement in the Tetradonematidae was based on

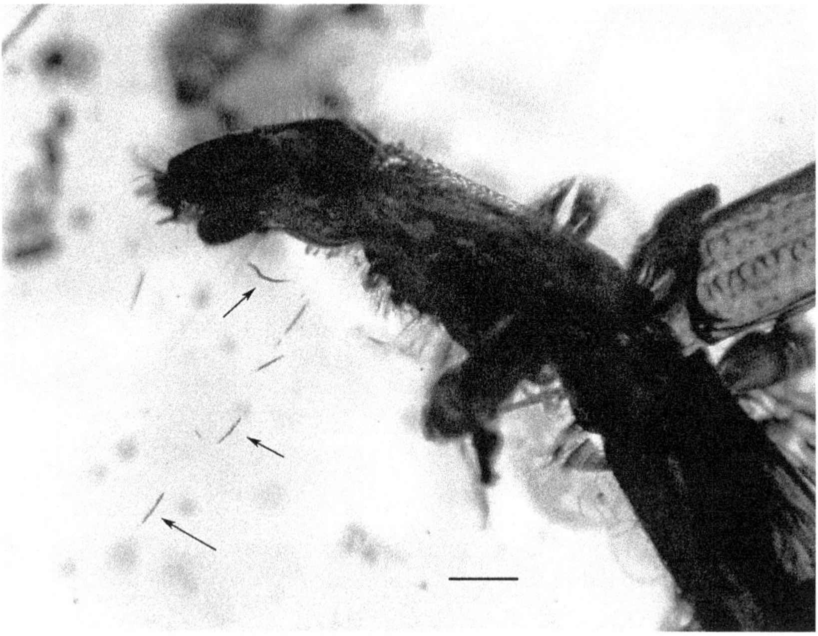

Fig. 246. *Dauer juveniles (arrows) of* Scolytonema mexicana* *associated with a platypodid beetle in Mexican amber (accession no. N-3-39). (Scale bar = 370 μm.)*

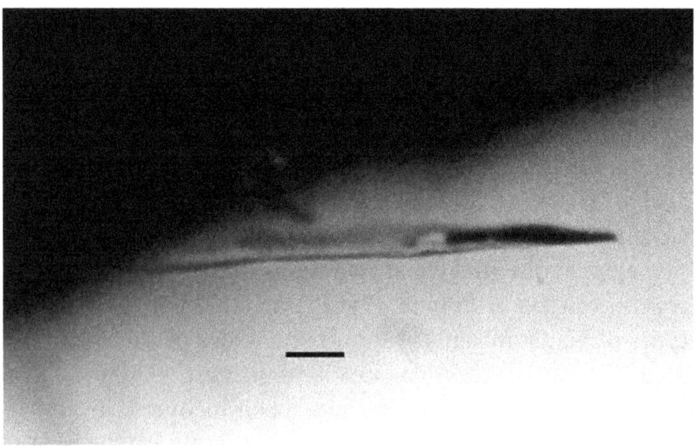

Fig. 247. *Dauer juvenile of* Scolytonema mexicana* *adjacent to the body of a platypodid beetle in Mexican amber (accession no. N-3-39). (Scale bar = 21 μm.)*

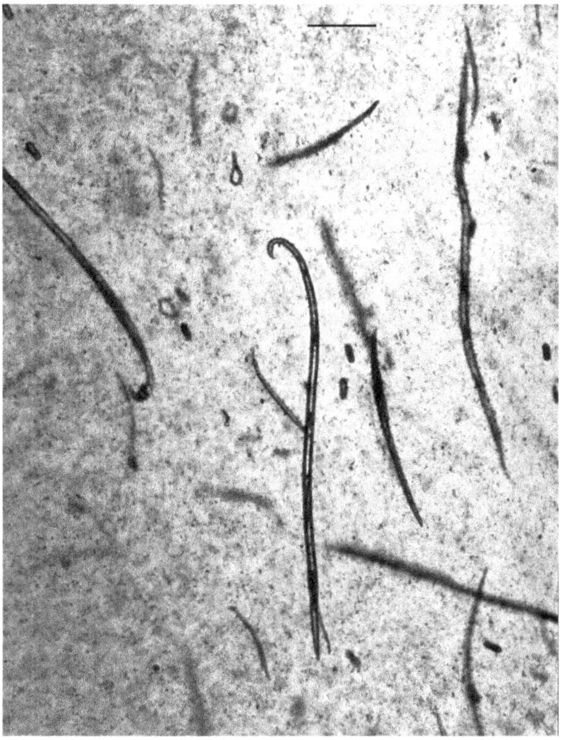

Fig. 248. *Reproducing population of* Oligaphelenchoides atrebora *Poinar, 1977 in Mexican amber (Museum of Paleontology, UC Berkeley) (see also Plate 31). (Scale bar = 176 μm.)*

one large specimen (2.7 mm in length) that contained developing ova. Similar ova in the extant sand fly tetradonematid, *Didilia ooglypta* Tang, Hominick, Killick-Kendrick, Killick-Kendrick & Page, 1993, fall within the size range of those noted in the fossil.

Pliocene nematodes

Thus far the Pliocene has yielded only a few nematode fossils. One is the mermithid *Heydenius tabanae**, the description of which is based on a single specimen emerging from an adult horsefly (Diptera: Tabanidae) in sedimentary deposits in Germany dated at approximately 2.5 million years ago (Grabenhorst, 1985). The host of *H. tabanae** was identified as the extant horsefly, *Tabanus sudeticus* Zeller.

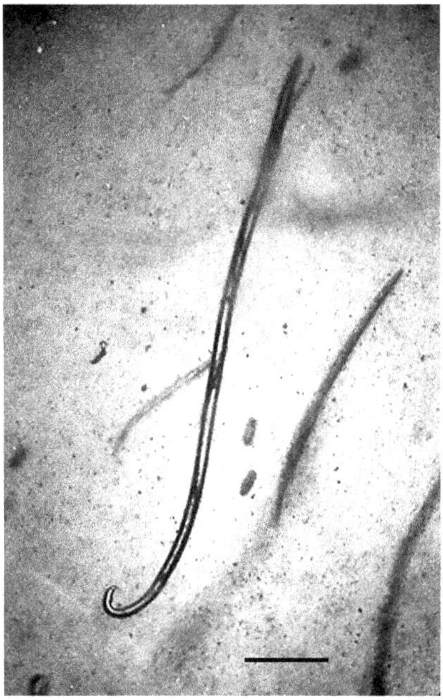

Fig. 249. *Male of* Oligaphelenchoides atrebora *Poinar, 1977 in Mexican amber (Museum of Paleontology, UC Berkeley). (Scale bar = 154 μm.)*

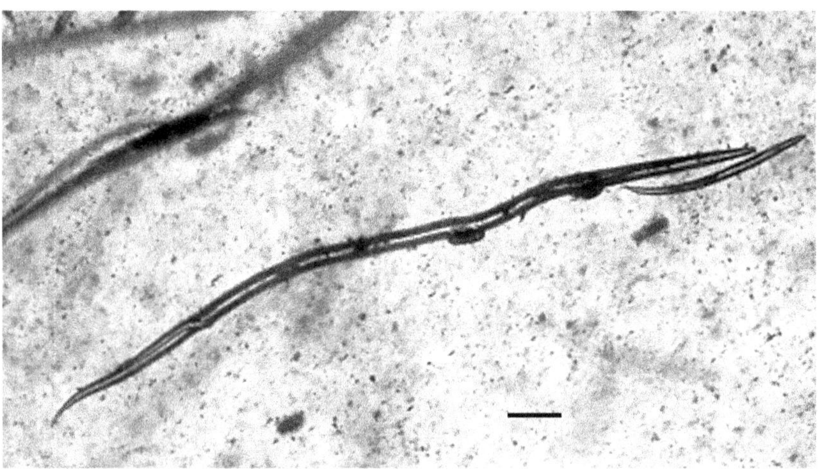

Fig. 250. *Female of* Oligaphelenchoides atrebora *Poinar, 1977 in Mexican amber (Museum of Paleontology, UC Berkeley). (Scale bar = 76 μm.)*

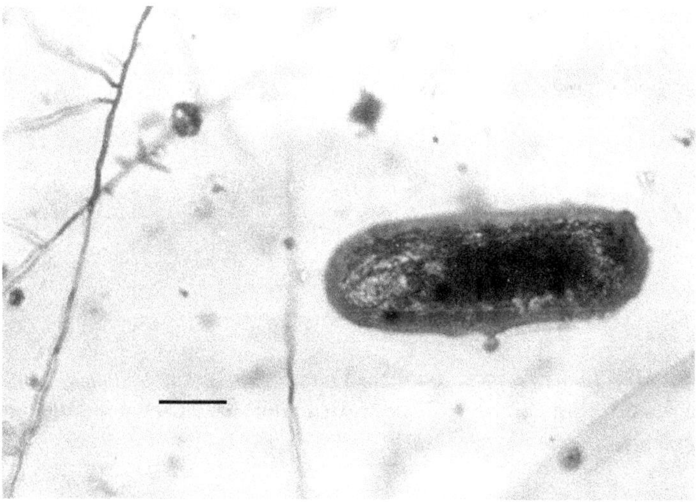

Fig. 251. *Egg of* Oligaphelenchoides atrebora *Poinar, 1977 in Mexican amber (Museum of Paleontology, UC Berkeley). (Scale bar = 13 μm.)*

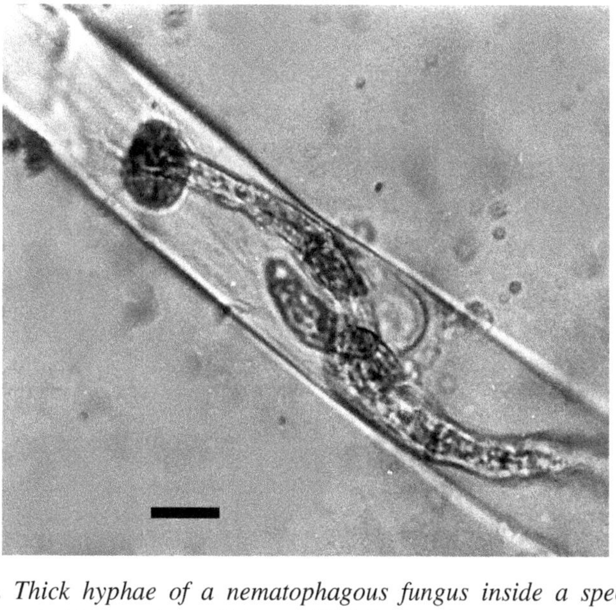

Fig. 252. *Thick hyphae of a nematophagous fungus inside a specimen of* Oligaphelenchoides atrebora *Poinar, 1977 in Mexican amber (Museum of Paleontology, UC Berkeley). (Scale bar = 9 μm.)*

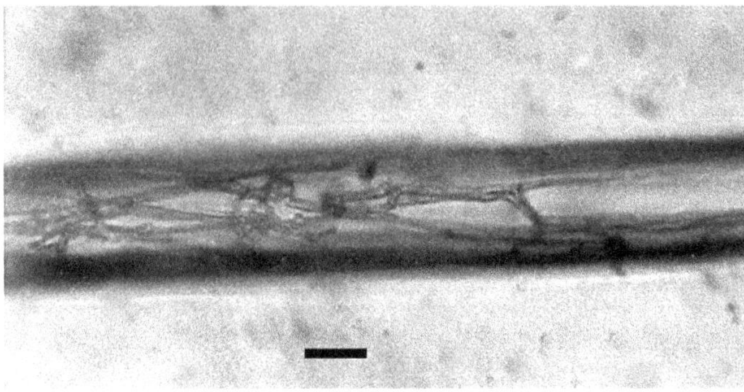

Fig. 253. *Mycelium of a nematophagous fungus inside a specimen of* Oligaphelenchoides atrebora *Poinar, 1977 in Mexican amber (Museum of Paleontology, UC Berkeley). (Scale bar = 8 μm.)*

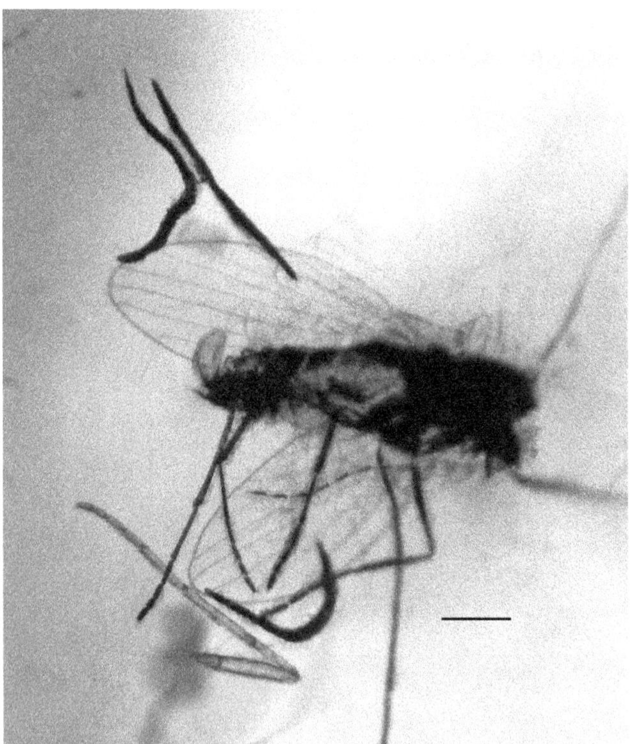

Fig. 254. *Three specimens of the tetradonematid* Palaeotetradonema phlebotomae* *adjacent to a phlebotomine sand fly in Mexican amber (accession no. N-3-49). (Scale bar = 780 μm.)*

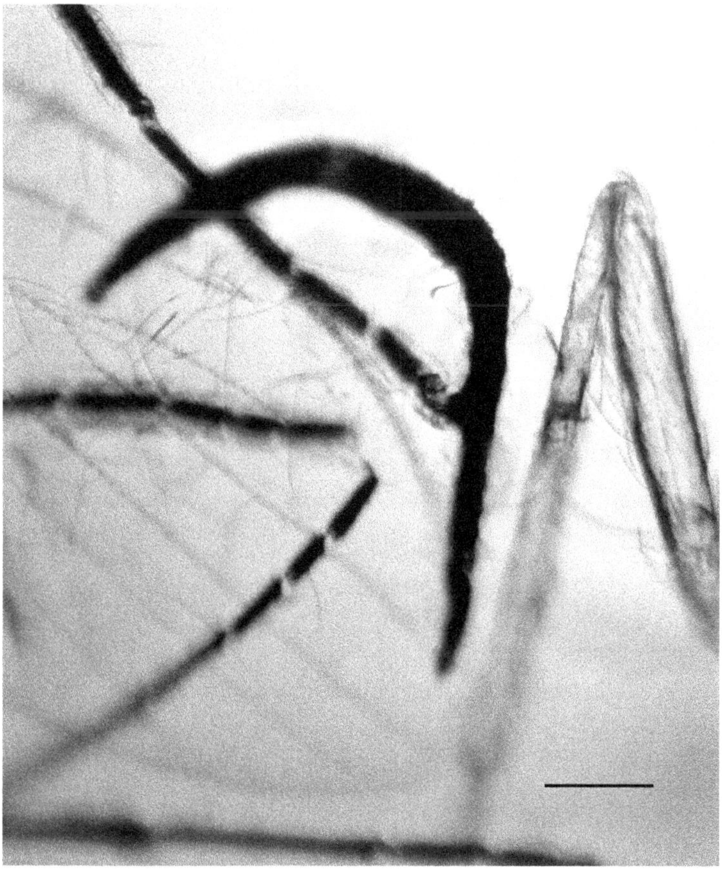

Fig. 255. *Detail of one of the three specimens of* Palaeotetradonema phlebotomae* *in Figure 252 (accession no. N-3-49). (Scale bar = 342 μm.)*

Horseflies are common hosts for mermithids since the larvae, which are the stages first infected, occur in moist habitats and other sites attractive to nematodes. Reports and descriptions of extant mermithids parasitising tabanids were summarised previously (Poinar, 1985). The interesting aspect of *H. tabanae** is that it was emerging from an adult horsefly, while all extant cases of horsefly mermithids pertain to larval parasitism. Of course, many mermithids that initially infect the larvae of holometabolous insects pass through the pupal and into the adult stage. Mermithid parasites of horseflies can undergo two types of development; direct, where the nematode infective stages enter the larva horsefly, or indirect, where the nematode infective stages first enter a paratenic host

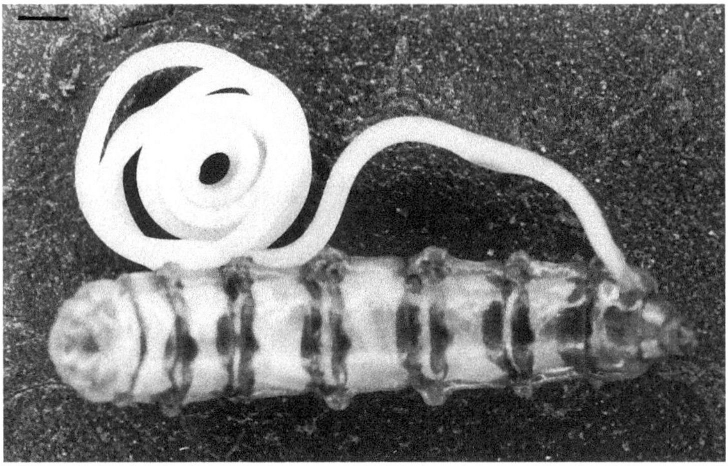

Fig. 256. *Extant mermithid parasite* Pheromermis myopis *Poinar & Lane, 1978 emerging from a larva of the horsefly* Tabanus punctifer *(Osten Sacken). (Scale bar = 1.7 mm.)*

that is subsequently eaten by a horsefly larva (Poinar, 1985). It is possible that the life cycle of *H. tabanae** was indirect, similar to that of the extant *Pheromermis myopis* Poinar & Lane, 1978, a parasite of *Tabanus punctifer* Osten Saken (Fig. 256).

Nematology Monographs & Perspectives, 2011, Vol. 9, 191-194

Nematodes from Pleistocene and Holocene deposits

The Pleistocene extends from 1.81 mya to the beginning of the Holocene, or 11 500 BP (before the present). Time periods within the Holocene include the Prehistoric (11 500 BP to 3300 BC or Before Christ), Ancient (3300 BC to 476 AD or Anno Domini), Medieval (476 AD to 1517 AD) and Modern (1517 to the present). Remains younger than the Pleistocene are often considered subfossils.

Several vertebrate-parasitic nematodes have been reported from these periods. The strongylid, *Strongylus edentatus* (Looss, 1900), was recovered from the intestine of a mummified Late Pleistocene horse (Dubinina, 1972). These 'palisade worms' are widespread in the caecum and colon of horses today. Eggs voided with the faeces hatch into rhabditiform stages that develop to third-stage infective juveniles. When these infective stages are ingested, they penetrate the gut wall and pass through a migratory stage with most ending up in the horse's liver. From there the parasites enter the colon where they mature and reproduce (Anderson, 2000).

Discovery of eggs of the horse pinworm, *Oxyuris equi* Schrank, 1788, indicates the presence of horses at ancient locations. One such site was from terp (elevated land used in ancient times during flooding) deposits in Germany dated from 100 BC to 500 AD (Jansen & Over, 1962). The same authors also reported *O. equi* from a Roman fort in Holland dated at 0-100 AD (Jansen & Over, 1966).

Pinworms assigned to the genus *Syphacia* Seurat, 1916 were reported from the fossilised remains of a 10 000 to 12 000-year-old ground squirrel (a species of *Citellus* Gueldenstaedt, 1770) in Eastern Russia (Dubinin, 1948). The fossil worms closely resemble the extant species *S. obvelata* (Rudolphi, 1802), which occurs in the intestine of rodents worldwide. While the above reports do not throw much light on the origin of specific nematode groups, they provide information on ancient hosts and past distributions.

The oldest remains of whipworms (*Trichuris* Roederer, 1761) are 30 000-year-old eggs from coprolites of the caviid rodent, *Kerodon*

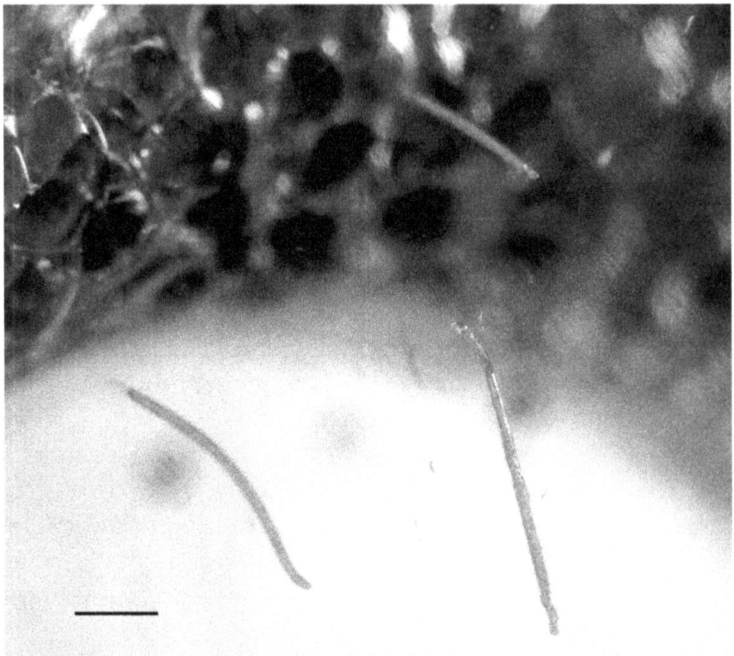

Fig. 257. *Three specimens of* Copalonema colombiana* *in Colombian copal (accession no. N-3-11). (Scale bar = 507 μm.)*

rupestris (Wied-Neuwied, 1820), in Brazil. The authors suggested that climate change was responsible for the disappearance of *Trichuris* sp. in extant populations of *K. rupestris* at the same location (Ferreira *et al.*, 1991).

Nine thousand-year-old coprolites from a Brazilian iguaniid, *Tropidurus* sp., contained eggs of an oxyurid similar to those of *Parapharyngodon sceleratus* (Travassos, 1923) that infects the extant lizard, *Tropidurus torquatus* (Wied, 1820). Based on these results, the authors concluded that the eggs in the coprolites were probably those of *P. sceleratus* (Araújo *et al.*, 1982).

Fossil nematodes from extinct animals are very rare. However, coprolites from a 1.5 million-year-old extinct *Hyaena* Brisson, 1762 yielded a number of nematode juveniles. They could not be identified, but extant hyaenids are parasitised by both *Ancylostoma* (Creplin, 1845) and *Toxascaris* Leiper, 1907 and they could have belonged to either of these genera (Ferreira *et al.*, 1993). The analysis of dried 10 000-year-old boluses of the extinct ground sloth, *Mylodon listai* Ameghino,

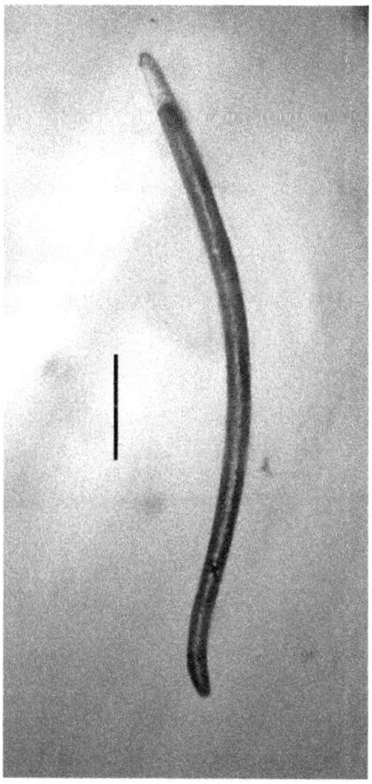

Fig. 258. *Detail of a specimen of* Copalonema colombiana* *in Colombian copal (accession no. N-3-11). (Scale bar = 304 μm.)*

1889, in Argentina yielded elliptical bodies thought to be nematode eggs (Ringuelet, 1957). Also, Shasta Ground Sloth (*Nothrotheriops shastensis* Hoffstetter, 1954) boluses (see Fig. 7) dated at 10 000 ± 180 years from Rampart Cave, Arizona, contained first-stage juveniles of an oxyurid, *Agamofilaria oxyura* Schmidt, Duszynski & Martin, 1992, and a strongyloid, *Strongyloides shastensis* Schmidt, Duszynski & Martin, 1992.

The nematodes described from amber in the present work range from 15 to 135 million years old, although nematodes also occur in semi-fossilised resin or copal. Copal deposits occur throughout the world but the most common sites are in South America, Africa and New Zealand. Tests performed thus far indicate that most of this material is under a million years of age. Samples of copal from sediments in the Magdelena Valley of Colombia, South America, gave radiocarbon dates between 10 and 500 years BP (Poinar, 1996). Most species found in this material are extant, although those from habitats recently destroyed (as in Madagascar copal) could be recently extinct. Free-living nematodes described here as *Copalonema colombiana** (Figs 257, 258) were in a piece of Colombian copal. They were probably feeding on microorganisms in fissures on the bark of the resin-producing tree, which was shown to be the extant species *Hymenaea courbaril* (Poinar, 1996).

Nematology Monographs & Perspectives, 2011, Vol. 9, 195-200

Human parasites from Pleistocene and Holocene remains

The search for human parasites has been conducted by anthropologists and archaeologists and their findings fall under the disciplines of Palaeoparasitology or Archaeoparasitology (Taylor, 1955; Kliks, 1990; Reinhard, 1992; Reinhard & Araújo, 2008). Although the earliest evidence of human nematode parasites are ascarid eggs from the Pleistocene (Table 3), these records are not old enough to shed light on the origin of nematode lineages. However, they do provide important information on the types and geographical distribution of human nematode parasites.

While some 138 species of nematodes occur in humans (Crompton, 1999), just when and how they were acquired in antiquity is unclear. Those most studied are gastrointestinal parasites whose eggs and juve-

Table 3. *Earliest physical evidence of human nematode parasites in the New and Old Worlds. Unless otherwise indicated, all dates are BP (before the present). Footnotes refer to references listed below the table.*

Nematode	Common name	Old World	New World
Ancylostoma duodenale (Dubini) (and *Necator americanus* Stiles)	Hookworm	~5600[1]	~7230[2]
Ascaris lumbricoides (L.)	White stomach worm	30 000-24 000[3]	~4277[4]
Dracunculus medinensis (L.)	Guinea worm	3991-3786[5,8]	
Enterobius vermicularis (L.)	Pinworm	2100[6]	~9837[7]
Strongyloides stercoralis (Bavay)	Threadworm	3000[8]	400-1200 AD[9]
Trichinella spiralis (Owen)	Cystworm	3200-3198[10]	
Trichuris trichiura (L.)	Whipworm	10 000-7000[11]	8000-7000[12]
Wuchereria/Brugia?	Filarial worm	3113-3085[8]	

[1]Dommelier-Espejo (2001); [2]Ferreira *et al.* (1987); [3]Bouchet *et al.* (1996); [4]Patrucco *et al.* (1983); [5]Tapp & Wildsmith (1993); [6]Wei (1973); [7]Fry & Hall (1969); Fry & Moore (1969); [8]Sandison & Tapp (1998); [9]Hall (1972); [10]Millet *et al.* (1980); [11]Evans (1996); [12]Gonçalves *et al.* (2003).

niles can be obtained from coprolites, cesspits, latrines and mummies (Sandison & Tapp, 1998; Gonçalves *et al.*, 2003). More human nematodes have been recovered from Egyptian mummies than from any other source. The preservative qualities of resins were well known to the ancient Greeks and Egyptians, who added resin to wine to prevent spoilage and on corpses for preparing them for burial. Since the common people could not afford to have the internal organs of their families removed, many cadavers, especially those from the later Dynasties, were preserved with all their organs, and such 'natural mummies' reveal a variety of parasites.

Certainly, the Egyptian embalmers encountered numerous parasitic nematodes during their preparation of cadavers for the afterlife (Plate 32A). Pages from the *Papyrus Ebers*, one of the earliest known pharmacopoeias that listed medical treatments for diseases in Ancient Egypt, include parasitic nematodes. This 3553 to 3550-year-old document offers treatment for ascarids (*Ascaris lumbricoides* L., 1758), hookworms (*Ancylostoma duodenale* Dubini, 1843) and Guinea worms (*Dracunculus medinensis* L., 1758) (Bryan, 1931; Ebbell, 1937).

Additional nematodes found in Egyptian mummies, but not mentioned in the *Papyrus Ebers*, were whipworms (*Trichuris trichiura* L., 1771), filarial nematodes (probably species of either *Wuchereria* Silvo Araujo, 1877 or *Brugia* (Brug, 1931), *Trichinella spiralis* Owen, 1835 and *Stronglyoides stercoralis* Bavay, 1876. It is curious that pinworms (*Enterobius vermicularis* L., 1758) have not been recovered from Egyptian mummies and are not mentioned in the *Papyrus Ebers.* Since the earliest record of pinworms is from the New World (Table 3), perhaps this species originated in the Americas.

Mummies show that Egyptians were parasitised by ascarids some 2200 years ago (Cockburn *et al.*, 1975) and suffered infections of *D. medinensis* 3991-3786 years ago (Tapp & Wildsmith, 1993).

Ancient Egyptians suffered from a variety of nematode parasites. A young weaver named Nakht who lived near Thebes from 3200-3198 BP had *Trichinella* cysts in his intercostal muscles (Millet *et al.*, 1980). The 3000-year-old Asru mummy had *Strongyloides stercoralis* in its intestine (Gonçalves *et al.*, 2003) and the 3113 to 3085-year-old Leeds mummy, Natsef-Amun, had filarial nematodes in his scrotum (Tapp & Wildsmith, 1993).

While mummies provide evidence of tissue parasitism, eggs and juvenile nematodes provide most evidence of human parasitism in

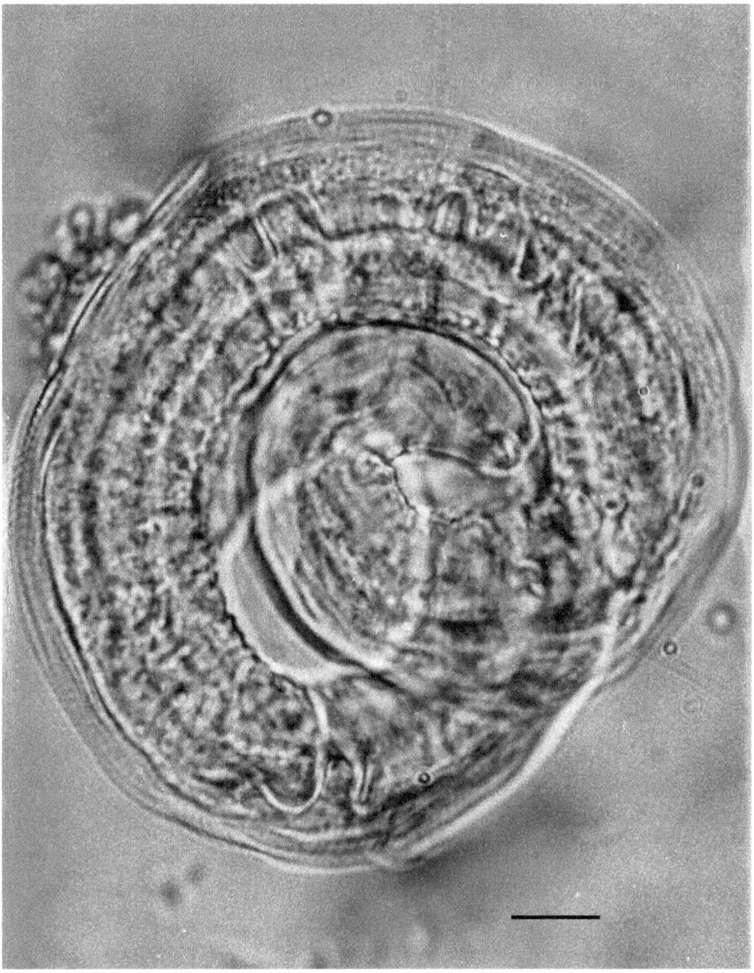

Fig. 259. *Juvenile rhabditoid nematode recovered from a 2200-year-old human coprolite at Lovelock Cave, Nevada. It appears to be a free-living juvenile of* Stronglyoides stercoralis *(Bavay), which occurs in the small intestine of humans (photo courtesy of M. Kliks) (see Dunn & Watkins, 1970). (Scale bar = 10 μm.)*

antiquity. A rhabitoid nematode recovered from a 2200-year-old human coprolite at Lovelock Cave, Nevada, was never officially described, but appears to be a free-living juvenile of *Stronglyoides stercoralis*, which parasitises the small intestine of humans (Dunn & Watkins, 1970; Kliks, 1975) (Fig. 259). If this identification is correct, it would represent the earliest record of *S. stercoralis* in the New World.

Nematode eggs are often preserved in faecal samples since they are the most resistant stage in the life cycle. Identification is based on the structure of their three primary shell layers. The innermost layer, or vitelline membrane, and the surrounding chitinous-protein layer (chorion) are both formed in the nematode's oviduct, while the outermost protein layer is applied in the uterus. The thickness and nature of these layers, together with egg size and shape, can be used to identify nematode genera and often species.

The five most commonly occurring human gastrointestinal nematode parasites are the roundworm, *Ascaris lumbricoides* (Plate 32B), whipworm, *Trichuris trichiura* (Plate 32C), pinworm, *Enterobius vermicularis* (Plate 32D), and the hookworms, *Ancylostoma duodenale* and *Necator americanus* Stiles, 1902 (Plate 32E). Pinworm eggs are flattened on one side, ascarid eggs have a mammillated outer protein coat, whipworm eggs have plug-like opercula at either pole and hookworm eggs have very thin shells.

The roundworm, *Ascaris lumbricoides*, is the most common human nematode parasite and infects nearly 1500 million people worldwide. It is not surprising that it was one of the earliest recorded parasites in the Western hemisphere. Physical evidence of its presence in human coprolites dates back some 30 000-24 000 years (Table 3).

Another human gut parasite that was probably just as common during that time period but was too inconspicuous to be noticed is the pinworm, *Enterobius vermicularis*. This helminth infects nearly a billion people at all socio-economic levels and the rate of parasitism apparently increased as humans adapted to an agricultural way of life. The oldest pinworm eggs in human coprolites were obtained from Utah and dated to 9837 BP (Fry & Hall, 1969; Fry & Moore, 1969) (Table 3). Using only molecular techniques, Iñiguez *et al.* (2006) obtained *E. vermicularis* DNA from South American pre-Columbian coprolites dated to 6110 years BP.

Hookworm ova are commonly found in human coprolites, but since the eggs of *Ancylostoma duodenale* and *Necator americanus* are similar in appearance, findings are lumped together as simply hookworms unless adults are present. Hookworms infect some 1.3 billion humans worldwide and cause a disease known as ancylostomiasis (Crompton, 1999). Both species are considered to have Old World origins, with the dispersion centre of *N. americanus* in the south-eastern portion of the Sahara and southern Asia and that of *A. duodenale* in northern Africa, southern Europe and southern Asia. The oldest records of human hookworms are 7230-year-old eggs recovered from coprolites in Brazil (Ferreira *et al.*, 1987; Montenegro *et al.*, 2006), showing that the parasites were present in South America well before the Spanish invasion.

According to the 'Clovis First' theory, hunters that came across the Bering Strait at the end of the last Ice Age, approximately 11 500-12 500 years ago, and then migrated south through the Ice-Free Corridor into North and eventually South America, represented the first peopling of the Americas. According to some, these peoples were not only responsible for introducing parasites into the New World, but they also exterminated up to 70% of the megafauna in North America (Bonnichsen & Schneider, 1999). Based on the temperatures at which hookworm larvae can develop to the infective stage in the soil (between 17°C and 35°C) and simulated palaeotemperatures from 11 000-15 000 years BP, Montenegro *et al.* (2006) concluded that humans migrating over the 'Clovis First' route across Beringia into North America could not have travelled fast enough to explain the presence of hookworms in pre-Columbian sites in the Americas. The authors suggest that hookworms had already been introduced into the New World by earlier human migrations.

The whipworm, *Trichuris trichiura*, infects slightly over one billion humans worldwide today (Crompton, 1999). The oldest record of human parasitism by this species dates back to between 7000 and 8000 years ago based on eggs from human remains in a pre-Columbian bog in Chile, suggesting that this species is endemic to the New World (Table 3), although Kliks (1990) feels that both *T. trichiura* and *E. vermicularis* were introduced by peoples migrating across the Bering strait from Siberia.

When parasitic nematodes were originally acquired by humans is unknown. Recent studies on nematode parasites of non-human primates suggest that all human intestinal nematodes could have been acquired from other primates (Dupain *et al.*, 2009; Kaur & Singh, 2009). A list of non-human primates that also host these 'human' nematode parasites is shown in Table 4. It is likely that most, if not all, of the nematode parasites that plague humans today were obtained through their distant ancestors or from other vertebrates.

Table 4. *Records of the four major human nematode parasites in other primates.*

Nematode	Primate	Geographic location*	Reference
Ascaris lumbricoides	Gorilla (*Gorilla gorilla*)	OW	Fossey (1983)
"	Howler monkey (*Alouatta* spp.)	NW	Stuart *et al.* (1998)
"	Chimpanzees (*Pan troglodytes*)	OW	Dupain *et al.* (2009)
"	Bonobos (*Pan paniscus*)	OW	Dupain *et al.* (2009)
Trichuris trichiura	Chimpanzee (*Pan troglodytes*)	OW	Dupain *et al.* (2009)
"	Bonobos (*Pan paniscus*)	OW	Dupain *et al.* (2009)
"	Baboon (*Theropithecus gelada*)	OW	Ooi *et al.* (1993)
"	Japanese Macaque (*Macaca fuscata yakui*)	OW	Hernandez *et al.* (2009)
"	Red Colobus (*Procolobus rufomitratus*)	OW	Chapman *et al.* (2009)
"	Gorilla (*Gorilla gorilla*)	OW	Fossey (1983)
Enterobius vermicularis	Chimpanzee (*Pan troglodytes*)	OW	Hasegawa (2009)
Necator americanus	Chimpanzee (*Pan troglodytes*)	OW	Huffman *et al.* (2009)
Ancylostoma duodenale	"		Dupain *et al.* (2009)
"	Bonobos (*Pan paniscus*)	OW	Huffman *et al.* (2009)
"	Howlers (*Alouetta* spp.)	NW	Thoisy *et al.* (2001)

*OW = Old World; NW = New World.

Nematology Monographs & Perspectives, 2011, Vol. 9, 201-203

Dubious nematode fossils

Some previously described nematode fossils are questionable. The specimens described by Størmer (1963) as *Scorpiophagus baculiformis* Størmer, 1963 and *S. latus* Størmer, 1963 from decomposed remains of *Gigantoscorpio willsi* Størmer, 1963, a Lower Carboniferous scorpion, fall into this category (Figs 260, 261). While working in the Nematology Department at Rothamsted Experimental Station (now Rothamsted Research) in 1964, I had the opportunity to examine specimens of *Scorpiophagus* that had been brought to that institution for examination by Dr Størmer. The *Scorpiophagus* specimens had two curious characters. One was that the posterior portion of their bodies was partly disintegrated or completely missing (Fig. 260). The other was the presence of a long pointed appendage (called 'thread' or 'marginal spine' by Størmer) emerging from one side of the head (Fig. 261). Although many specimens contained a short, narrow, darkened invagination at one end that could be considered a buccal cavity, there was no evidence of an associated alimentary tract. A careful examination of the photographs and drawings show that the lengths and diameters of *S. baculiformis* and *S. latus* overlap with those of setae and spines associated with the fossil scorpion. It is quite likely that both *S. baculiformis* and *S. latus* are actually different types of setae and/or spines that became dislodged from the scorpion during burial.

Another questionable nematode fossil is a specimen described by Voigt (1957) as a vertebrate-parasitic nematode in the muscles of a fossil buprestid beetle, *Chlorodema primordialis* Pongr., 1935 (Coleoptera: Buprestidae) from Eocene brown coal deposits near Halle, Germany. The object in question is partially coiled in a cyst-like compartment within the musculature of the beetle. Voigt thought that the beetle served as an intermediate host for a vertebrate-parasitic nematode and compared the specimen with the encysted juvenile of the lungworm parasite, *Metastrongylus* (= *Choerostrongylus*) *pudendotectus* Wostokow, 1905, which encysts in the body cavity of earthworms. However, there are no known vertebrate-parasitic nematodes that utilise members of the families Buprestidae or Cerambycidae as intermediate or paratenic hosts. The specimen shows few characters; however, if it is a nematode, it could be

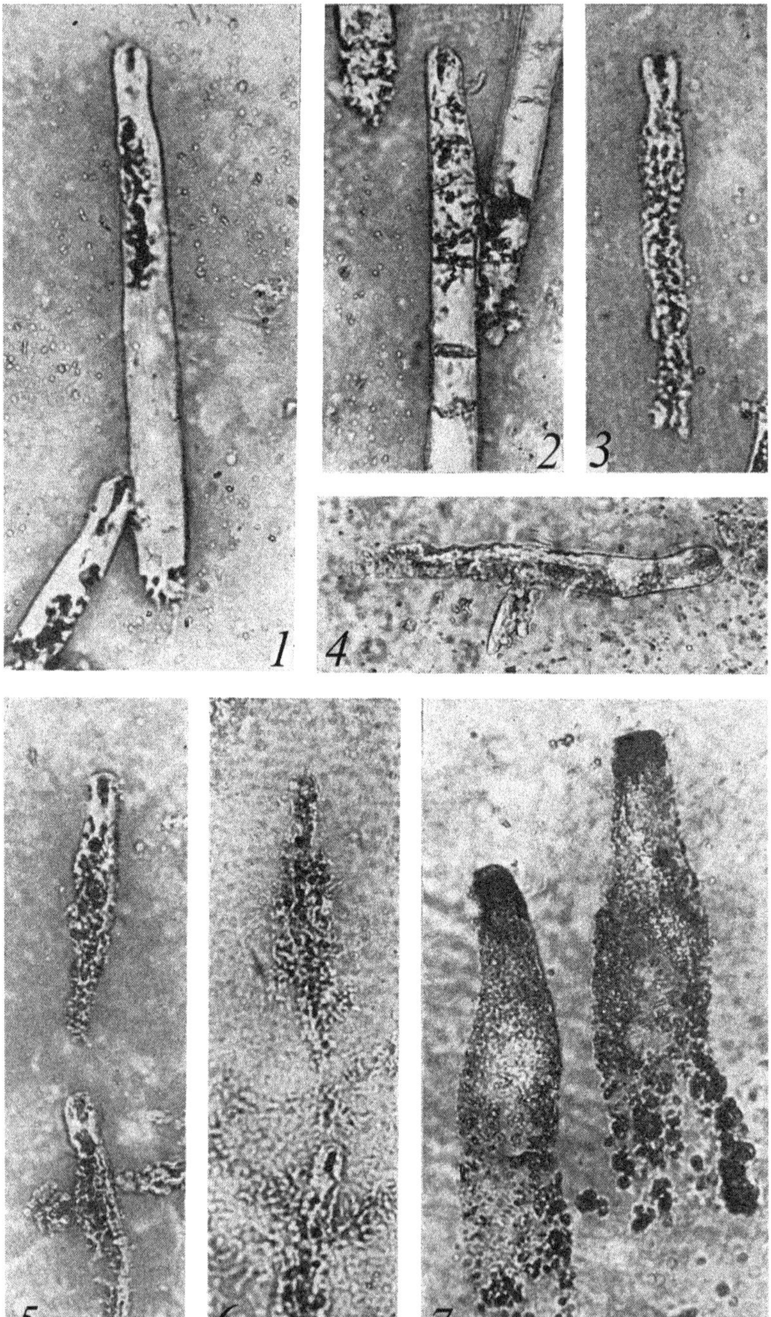
1
2
3
4
5
6
7

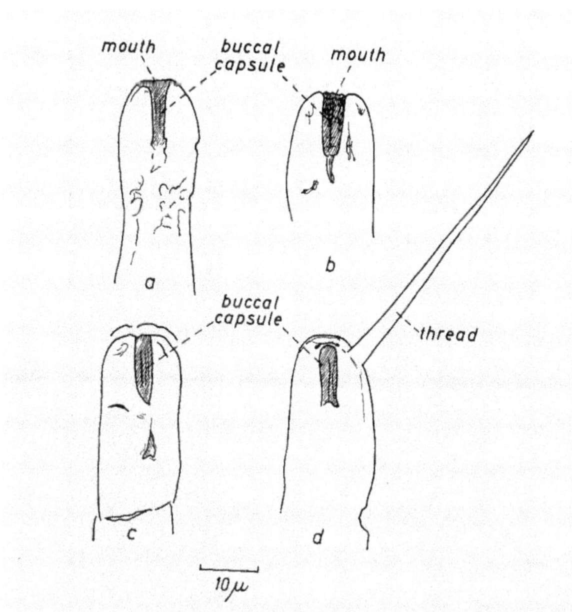

Fig. 261. Scorpiophagus baculiformis *Størmer (from Størmer, 1963, text Fig. 26, p. 68). a-c: Head with mouth and buccal capsule; d: Head with protruding 'thread' or 'marginal spine'. Specimens from tibia of the fossil scorpion,* Gigantoscorpio willsi *Størmer. (Scale bar = 10 μm.)*

a mermithid that was walled off in a host defence response. Mermithids parasitise adults of the European buprestid, *Anthaxia quadripunctata* (L., 1758) today (Dollfus & Théodoridès, 1950).

Fig. 260. *Størmer's figure of* Scorpiophagus *(from Størmer, 1963, Plate 19, p. 165) 1. Holotype of* Scorpiophagus baculiformis *Størmer. 2. Additional specimen of* S. baculiformis. *3. Specimen of* S. baculiformis *partially degraded. 4. Specimen of* S. baculiformis *with long pharynx. 5. Greatly degraded specimen of* S. baculiformis. *6. Late stage of degradation of specimen of* S. baculiformis. *7. Two specimens of* S. latus *Størmer.*

Nematology Monographs & Perspectives, 2011, Vol. 9, 205-207

Nematode artefacts

It is sometimes difficult to distinguish nematode fossils in amber from artefacts. Amongst the most common artefacts are plant hairs or trichomes, which can resemble nematodes under low power on the microscope (Fig. 262). Insect setae or hairs represent another type of artefact (Fig. 263), especially those that resemble dauer juveniles attached to the insect cuticle. Even fractures (Fig. 264) in the amber can appear as nematodes. In addition, some insect tissues, such as spermatophores, can appear as minute nematodes associated with the urogenital openings of insects (Fig. 265). Other artefacts that can resemble nematodes are extremely narrow, elongated air bubbles and isolated bits of spider webs.

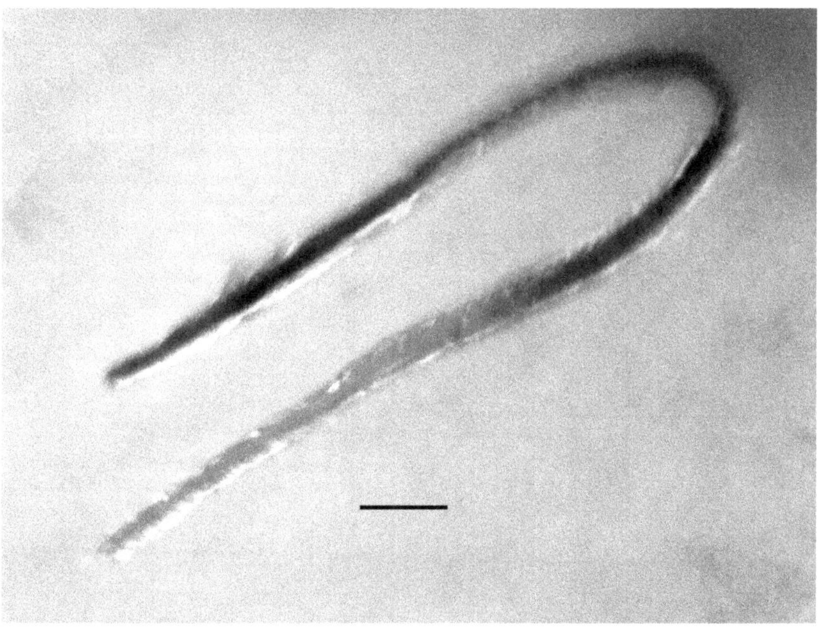

Fig. 262. *A plant hair resembling a nematode in Dominican amber. (Scale bar = 300 μm.)*

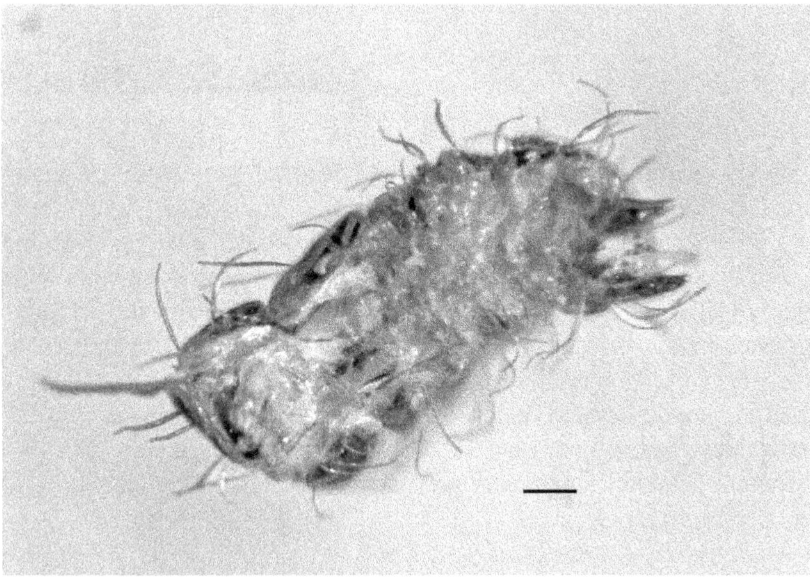

Fig. 263. *Setae resembling nematodes on a beetle larva in Dominican amber. (Scale bar = 153 μm.)*

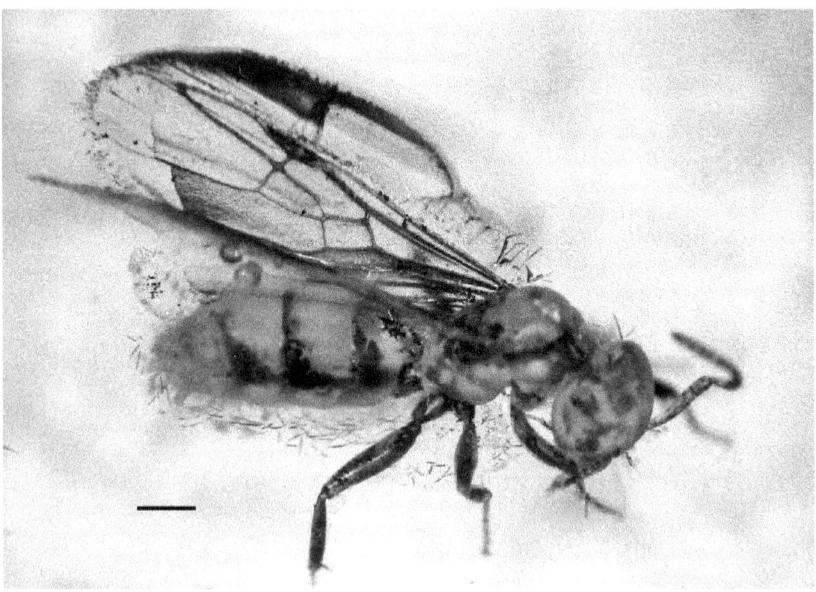

Fig. 264. *Fractures resembling nematodes surrounding a winged ant in Baltic amber. (Scale bar = 1 mm.)*

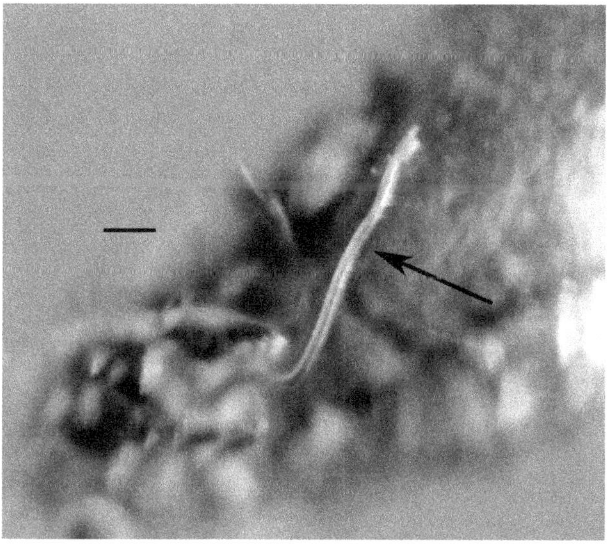

Fig. 265. *Spermatophores (arrow), like this one from a dance fly (Diptera: Empididae) in Dominican amber, can also resemble nematodes. (Scale bar = 58 μm.)*

Nematology Monographs & Perspectives, 2011, Vol. 9, 209-234

Nematode time lines

Problems with evaluating nematode fossils

Difficulties can be encountered when attempting to determine the origins of nematode lineages from fossils. The major problem is the incompleteness of the fossil record. It is clear that most of the organisms that lived in the past have never been fossilised and this is especially true for nematodes. Not only are they small and often concealed in soil or host tissues but their few hard structures are minute and difficult to recognise.

Since amber mostly traps small invertebrates that are associated with resin-producing trees, these records are biased. While amber deposits discussed here span 135 million years, many additional ones exist and they need to be examined as well. For instance, no nematodes have yet been found in Late Cretaceous or Palaeocene amber, which could help us determine whether extinction events at the end of the Cretaceous had any effect on nematodes.

The absence of fossils does not mean nematodes were not present and, when nematode remains are found, they only show that a particular lineage was present at a particular time and place. Single fossils provide no information about the global distribution of lineages, species longevity or extinctions. And the apparent absence of a nematode lineage in one location does not preclude its presence elsewhere.

There are also potential problems with dating fossils. Dating amber fossils is often indirect (based on the age of index fossils in the surrounding sediments) and much of the world's amber was re-deposited into more recent beds. Thus, dating based on trace fossils can result in younger rather than actual ages.

Longevity and extinctions of nematode lineages

Based on fossils, nematode species can survive for at least 10 million years and nematode genera for at least 40-50 million years. However, it is very difficult to determine the longevity of specific lineages since the characters needed for identification were mostly obscured during the fossilisation process. Resemblances can result from homoplasy (convergent evolution) or represent morphologically indistinguishable populations that are reproductively isolated (sibling species). Bearing these caveats in mind, Schram (1973, 1979) considered that the Carboniferous *Nemavermes mackeei* survived in offshore environments for at least 10 million years (Schram, 1973, 1979), providing an estimate of species longevity in marine nematodes. In all likelihood, the 30 000 to 24 000-year-old eggs assigned to *Ascaris lumbricoides* and the 7000 to 8000-year-old eggs identified as *Trichuris trichiura* are probably the same species that parasitise humans today and provide a minimum age of species longevity in some vertebrate lineages.

In discussing species longevity in insects, Hörnschemeyer *et al.* (2009) indicate that once a successful survival pattern has been established, evolutionary stasis may result in species persisting for millions of years. Durette-Desset (1985) felt evolutionary stasis was probably why trichostrongyloid parasites showed little evidence of speciation or diversification during the Lower Oligocene, especially in slowly evolving hosts.

Generic and family lineages are easier to trace since characters defining them are more easily detected in fossils and, if they are obscured, behavioural characters can be used, based on the principle of behavioural fixity. This principle states that the behaviour, ecology and climatic preferences of fossils will be similar to those of their present-day descendants at the generic and in some cases, the family level (Boucot, 1990; Boucot & Poinar, 2010). In the present work, feeding habits, including host selection, and habitat preference have been used to affiliate nematode fossils with extant genera and families. A number of nematodes described here in fossil group genera could actually belong to extant genera, but the absence of diagnostic generic characters make it impossible to be certain.

Based on the above, it is possible that the Early Cretaceous *Cretacimermis protus* belongs to the *Heleidomermis* lineage, which suggests a generic longevity of 100 million years. Other nematode fossils considered to belong to extant genera include *Bursaphelenchus similus**, *Cryptaphelenchus dominicus**, *Panagrolaimus neotropicus** and *Myrmeconema antiqua** in Dominican amber and *Howardula helenoschini* and *Tripius balticus** in Baltic amber. These records indicate a generic longevity of 40-50 million years for *Howardula* and *Tripius* and between 20 and 30 million years for *Bursaphelenchus*, *Cryptaphelenchus*, *Panagrolaimus* and *Myrmeconema*.

A number of nematode fossils that have been placed in collective genera could represent extant generic lineages but have been excluded due to a lack of diagnostic characters. Examples are *Heydenius araneus* Poinar, 2000, which could belong to *Aranimermis* Poinar & Benton, 1986, *Paleothelastoma** could belong to *Cephalobellus* Cobb, 1920, *Palaeocosmocerca** could be a species of *Cosmocercoides* Wilkie, 1930, *Oligaphelenchoides atrebora* could belong to *Bursaphelenchus*, *O. maximus** could be an *Aphelenchoides*, *Setonema** could be a *Noctuidonema*, *Palaeoanguina** could belong to *Anguina* Scopoli, 1777 or *Paranguina* Kirjanova, 1955, *Paleoiotonchium** could belong to *Iotonchium* Cobb, 1920, *Palaeoparasitylenchus** could belong to *Parasitylenchus* Micoletzky, 1922, *Syconema** could belong to *Parasitodiplogaster* Poinar, 1979 and *Proheterorhabditis** could belong to *Heterorhabditis*.

There have been multiple periods of mass extinctions during the course of the earth's history. These events are often accompanied by meteor impacts, volcanic eruptions, marine regression and basaltic lava flows. Changes in ocean temperatures and salinity, tidal waves and loss of necritic seas would definitely have affected marine nematodes, especially those restricted to continental shelves. Volcanic activity accompanied by excess heat, toxic fallout, changes in carbon dioxide and sulphur dioxide levels, and acid rains would have had a detrimental effect on terrestrial nematodes, especially freshwater forms, not to mention destroying plants and animals upon which nematodes were dependent.

What happened to nematodes during periods of mass extinctions, such as the one that occurred at the Cretaceous-Tertiary (K/T) boundary some 65.5 mya, is unknown. It is likely that the two species of *Ascarites* that parasitised a *Megalosaurus* dinosaur became extinct at the end of the Cretaceous, but that lineage could have continued in other reptiles that survived the K/T extinction events, eventually becoming a new species (pseudoextinction). The shape, size and shell morphology of *Ascarites* are very similar to those of modern species in the Ascarididae.

Unless they have a broad host spectrum, parasitic nematodes are eliminated when their immediate animal and plant hosts become extinct. The Devonian plant parasite, *Palaeonema*, is obviously extinct but it is impossible to determine just when that occurred. It may have become extinct along with its host plants or may have survived in other hosts for a longer period. The free-living aquatic *Nemavermes*, *Eophasma*,

Captivonema and *Palaeoenoploides* differ from any extant described nematodes and are probably extinct, although it is difficult to say just when these extinctions occurred.

It is impossible to know if the fossil species *Psocidonema dominicana**, *Chilonema filiformis**, *Chalcidonema paradoxa**, *Tipulacomis balticus**, *T. dominicus**, *Heydenius scatophilus**, *H. lamprophilus**, *H. psychodae**, *H. dipterophilus** and *H. saprophilus** that display arthropod associations that are unknown today are now extinct or have not yet been discovered. The few ecological nematode investigations in the neotropics make it difficult to compare Dominican and Mexican amber fossils with extant forms.

The greatest mass extinction may be going on right now. Human activity is destroying habitats around the world and decimating animal populations. While in the past, natural extinctions of plants and animals eliminated host-specific nematodes, humans have been doing the same for thousands of years. When a portion of the rain forest disappears, numerous species of animals and plants vanish, together with their nematode associates and parasites.

Nematode origins

Nematodes are an extremely basal clade, have no close affinity to any extant invertebrate group and the lineage from which they evolved is probably extinct. Over the years it has been proposed that nematodes originated from almost every possible group of invertebrates, ranging from echinoderms and annelids to arthropods. Twentieth century workers used comparative morphology to determine the relatedness of nematodes to other invertebrate groups (Chitwood, 1950; Paramonov, 1962; Andrássy, 1976). Chitwood (1950) compared the morphological, embryological and developmental features of nematodes with those of rotifers, gastrotrichs, echinoderms, acanthocephalans, nematomorphs, tardigrades, arthropods, annelids, turbellarians, trematodes, cestodes and nemertines. His analyses aligned nematodes with rotifers, gastrotrichs and echinoderms, and he then included the following nine phyla in the subkingdom Scolecida: Nematoda, Nematomorpha, Gastrotricha, Platyhelmintha, Nemerta, Acanthocephala, Echinoderma, Rotatoria and Entoprocta.

Based on morphology, Paramonov (1962) restricted what were then called the schizocoelids to four groups: the gastrotrichs, kinorhynchs, rotifers and nematodes. After a discussion of the characters of each of these groups, he concluded that nematodes are a sister group to marine gastrotrichs and that their common stem group was the Turbellaria.

Andrássy (1976) compared nematodes with turbellarians, trematodes, cestodes, nematomorphs, acanthocephalans, nemertines, rotifers, gastrotrichs, kinorhynchs, kamptozoans, priapuloids, siphunculoids, echiniuroids, annelids, molluscs, tentaculates, onychophorans, tardigrades, linguatuloids and arthropods. He concluded that nematomorphs (hairworms) are the closest allies of nematodes.

Unfortunately, molecular studies have not shed much more information on the question of nematode origins. A study using 146 rapidly evolving genes placed the nematodes as a sister group to the Platyhelminthes (Philippe *et al.*, 2005). However, when the analysis was based on 71 slowly evolving genes, nematodes aligned first with tardigrades, then arthropods, and after that Platyhelminthes. However, Philippe *et al.* (2005) excluded hairworms in their comparison, and when Dunn *et al.* (2008) added this group, results showed that nematodes aligned with Nematomorpha, a conclusion previously reached by Andrássy (1976), Schmidt-Rhaesa (1997) and Bleidorn *et al.* (2002).

With experimenters using different DNA sequences and models of analyses, and with the problem of long branch attraction phenomena and molecular convergence, determining the relatedness of nematodes to other groups will not be easy. Only when much more of the genome of all invertebrate representatives is available can a reasonable comparison be made. Thus the statement made by Maggenti in 1976, "...we are currently unable to delimit meaningful affinities", in reference to demonstrating phylogenetic relationships between nematodes and other invertebrates, still holds today. It is anticipated that the growing body of molecular studies will ultimately provide a greater understanding of nematode affinities.

Fossils provide a reference point for the occurrence of a particular nematode lineage at a particular time and place and these dates can be used to calibrate molecular clocks. We now have fossil lineages of the Enoplida, Mermithida, Chromadorida, Desmoscolecida, Plectida, Oxyuridomorpha, Spiruromorpha, Ascaridomorpha, Panagrolaimomorpha, Tylenchomorpha, Diplogastromorpha and Rhabditomorpha (Fig. 266).

Regarding the place of origin of nematodes, Filipjev (1934) felt that the primitive nematode was a marine enoplid and Hyman (1951) concurred. Based on the structure of the pharynx and excretory system,

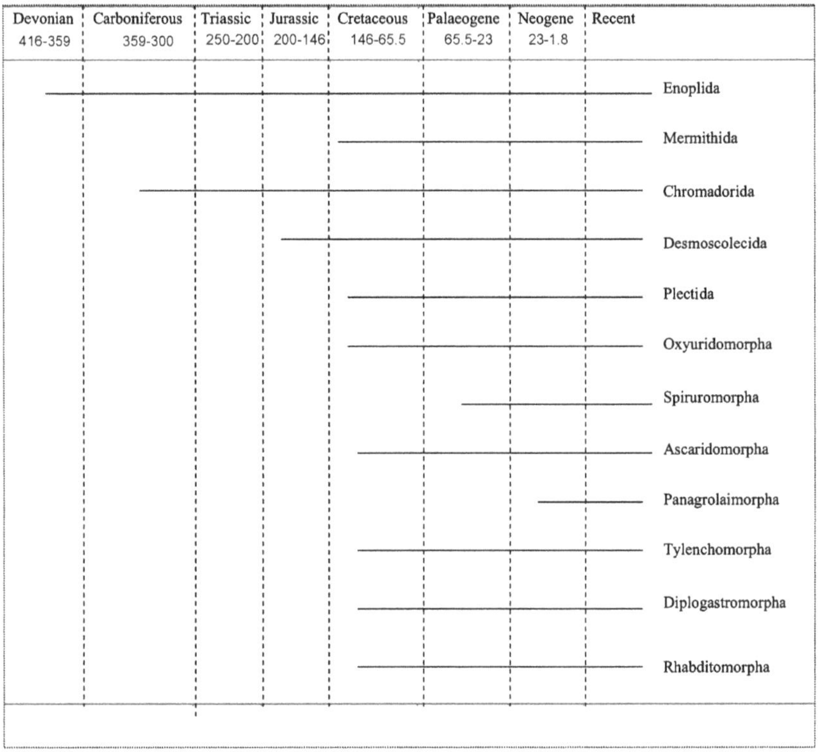

Fig. 266. *Evolutionary clock of nematode Orders and Infraorders. Palaeogene = Palaeocene, Eocene and Oligocene. Neogene = Miocene and Pliocene. Recent = Recent or the present (Holocene). Numbers refer to mya (million years ago).*

Maggenti (1963) also considered the Enoploidea as the basic nematode group and suggested that origins were in the marine intertidal zone. This seems logical since the Enoplea is obviously an ancient group and while the oldest fossil record is from the Devonian, Cambrian and Precambrian representatives can be expected. The Enoplea belongs to the archaic group 'Aphasmidia'. The aphasmids are essentially aquatic forms and include the great majority of marine nematodes. They are characterised by somatic setae, large, post-labial amphids, an excretory system lacking lateral canals and the presence of caudal and hypodermal glands, all of which can be considered plesiomorphic characters. Their counterpart is the 'Phasmidia', now limited almost entirely to terrestrial lineages.

Origin of parasitism in nematodes

Nematode parasitism of invertebrates and vertebrates probably first originated in the sea with the 'aphasmidians' and later on land with the 'phasmidians'. Parasitism is defined here as an association in which one organism takes nourishment from another organism, which is called the host. The host can be unaffected or express varying degrees of discomfort that can lead to death.

Fossil or evolutionary time lines can be established to approximate the dates when various groups of animals and plants evolved. Extrapolating from these time lines can provide dates to support various hypotheses on the origin of nematodes. Based on fossils, bacteria have existed for at least 3.5 billion years and protozoa for some 2 billion years. Assuming that nematodes arose in the sea as free-living microbotrophs in the Precambrian, they could have been parasitising marine invertebrates and vertebrates in the Cambrian and terrestrial invertebrates and vertebrates in the Ordovician. While there are no reports of marine aphasmidian plant parasites, plant parasitism on land by phasmidians could have occurred in the Silurian. Terrestrial hexapods could have been parasitised in the Devonian but insects probably not until the Carboniferous.

Since it is very likely that nematodes originated in a marine habitat, it is natural that they would first parasitise invertebrates in that environment. This scenario is supported by the discovery that representatives of the Mermithidae, Tetradonematidae, Marimermithidae, Echinomermellidae, Benthimermithidae, Monhysteridae and Leptolaimidae parasitise marine ostracods, copepods, shrimp, amphipods and isopods (Petter, 1980; Poinar *et al.*, 2002, 2009), starfish (Asteroidea) (Rubstov & Platonova, 1974; Rubstov, 1977), brittle stars (Ophiuroidea) (Ward, 1933), sea urchins (Echinoidea) (Gemmill & von Linstow, 1902; Jones & Hagen, 1987), priapulids (Rubstov, 1980), polychaetes (Petter, 1980), tubicifid oligochaetes (Hallett *et al.*, 2001), foraminifera (Hope & Tchesunov, 1999) and marine nematodes (Tchesunov, 1988; Tchesunov & Spiridonov, 1993). Some of these host groups, such as polychaetes, ostracods and foraminifera, have fossil records extending back to the Cambrian and, in the case of polychaetes, possibly the Precambrian (Lehmann & Hillmer, 1983).

It has been hypothesised that vertebrate parasitism originated on land and that those nematode parasites of marine animals originated from terrestrial lineages (Anderson, 1984, 1988). While there may be an impoverished nematode record from marine fish and mammals, and most of the existing parasites can be traced back to terrestrial lineages, this is not the case for all vertebrate parasites. Nematode parasites of elasmobranchs could represent early records of vertebrate parasitism and sharks were probably parasitised soon after they appeared in Palaeozoic seas (Carroll, 1988). Some extant bizarre elasmobranch parasites are *Phlyctainophora lamnae* Steiner, 1921, which lives between the hyomandibular arch and skull, *Ichthyostrongylus* Mawson, 1954, which occurs in the spiral valve and *Acanthocheilus* Molin, 1858, which is found in the gut. These parasites have no close ties with terrestrial lineages and their ancestors could have occurred in Late Silurian elasmobranchs. Representatives of the genus *Echinocephalus* Molin, 1858 in the Gnathostomatidae also parasitise elasmobranchs, while the remaining members of the family occur in fish, reptiles, amphibians and mammals (Yamaguti, 1961; Anderson, 2000). However, *Echinocephalus* is rather unusual and has been placed in a separate subfamily, the Echinocephalinae Yamaguti, 1961. It is difficult to determine if this lineage arose with elasmobranchs and went from sea to land or *vice versa*.

Three terrestrial groups of animal-parasitic nematodes stand out as being particularly primitive in respect to morphology, biology and host selection. These are the oxyurids, drilonematids and cosmocercoids (Osche, 1963; Inglis, 1965). The oxyurids are gut parasites of a wide range of vertebrates and invertebrates, the drilonematids are a primitive, but highly specialised, lineage confined to the coelom of earthworms and the Cosmocercoidea occurs in gastropods, amphibians and reptiles (Anderson, 2000; McClelland, 2005).

Plant parasitism evolved independently in at least four terrestrial lineages: the Enoplida, Triplonchia, Dorylaimia and Rhabditida. There is no evidence yet of any marine lineage of plant-parasitic nematodes. The few species of marine tylenchs and aphelenchoidids found today are clearly terrestrial forms that invaded the intertidal area.

Origin of invertebrate parasitism

Early nematode parasites of invertebrates were lineages of the Mermithidae, Tetradonematidae, Marimermithidae, Echinomermellidae, Benthimermithidae and Leptolaimidae that co-evolved with marine invertebrates. Of these lineages, the only ones that successfully colonised land were the Mermithidae and Tetradonematidae, and today the former is one of the most successful group of nematode parasites on land and in freshwater. Since the Mermithidae are essentially terrestrial forms, it is likely that those species attacking intertidal invertebrates, such as crustaceans and amphipods, came from the land (Poinar *et al.*, 2002).

Based in part on fossil evidence from the Early Cretaceous, chironomid midges could be the original insect host group for mermithids. There are currently at least eight genera of mermithids parasitising chironomids worldwide, which are more mermithid genera than occur in any other insect family (Poinar, 1975). Chironomid midges have a fossil record extending back to the Late Triassic (Rasnitsyn & Quick, 2002) and it is obvious from the fossil species of *Cretacimermis* and *Heydenius* that mermithid parasitism of chironomids was as extensive in the past as it is today. Also, the life cycle of mermithids parasitising chironomids is direct, which is the basic type of life cycle in this group. Even marine midges could have been parasitised since the larvae of some extant midges develop on seaweed in the intertidal zone (Schlinger, 1975).

The Mermithidae and Tetradonematidae are probably not closely related to the Marimermithidae and Echinomermithidae, and the latter family evolved from free-living enoplids (Poinar *et al.*, 2011). Superficial resemblances of the two groups are certainly due to convergence, as with the Rhaptothyreidae Hope & Murphy (1969). Members of the latter family also have elongate, narrow bodies, but are free-living with a vestigial digestive system that has been replaced with symbiotic microorganisms.

From chironomid midges, mermithids could have radiated into other Diptera with aquatic or semi-aquatic larvae. Two likely candidates are biting midges (Ceratopogonidae) and horseflies (Tabanidae). The fossil *Cretacimermis protus* shows that biting midges were hosts to mermithids in the Early Cretaceous when ceratopogonids first appeared in the fossil record. While the fossil *Heydenius tabanae** is only from the Pliocene, horseflies have a fossil record extending back to the Jurassic. Terrestrial flies parasitised by mermithids in the Tertiary included scavenger flies (Scatopsidae), gall gnats (Cecidomyiidae), mothflies (Psychodidae), woodgnats (Anisopodidae), scuttle flies (Phoridae), fruit flies (Drosophilidae) and sciarid fungus gnats (Sciaridae). These insect families all have fossil records dating back to the Jurassic (Rasnitsyn & Quicke, 2002).

Other terrestrial insect groups could also have been early hosts for mermithids. One of these is the Hemiptera and, while *Heydenius brownii* is only Eocene in age, true bugs have a fossil record dating back to the Early Permian (Rasnitsyn & Quicke, 2002). Planthoppers and other Sternorrhyncha continue to be parasitised by mermithids throughout the world today (Poinar & Linares, 1985; Choo & Kaya, 1993; Helden, 2008).

It is likely that some ancient terrestrial invertebrates such as scorpions and millipedes are only accidental hosts of mermithids (Poinar & Stockwell, 1988) (Fig. 99) (Hagmeier, 1912), which brings up the issue of host specificity. Two extant mermithid genera, *Heleidomermis* and *Hexamermis* Steiner, 1924, demonstrate the opposite extremes in host selection. Members of the genus *Heleidomermis* only parasitise biting midges of the family Ceratopogonidae. However, *Hexamermis* has a wide host range including many insect families and *H. albicans* (von Siebold, 1854) has been reported from insects, myriapods and snails (Hagmeier, 1912).

Such broad host ranges make determining evolutionary clocks based on fossil host records difficult and a judgment call has to be made to decide if a particular host group is ancient or represents accidental parasitism. In such cases where a range of host groups are widely parasitised by a single nematode lineage, the earliest fossil record of the accumulated hosts is used to estimate the origin of parasitic nematode clades (Table 5).

The family Tetradonematidae may have a marine origin since *Nematimermis enoplivora* Tchesunov & Spiridonov (1993), parasitic in marine nematodes, possesses a 'tetrad' and matures to the adult stage in its hosts. Both of these features are characters of tetradonematids. Terrestrial lineages of the Tetradonematidae extend back to the Early Tertiary *Palaeotetradonema sciarae** infecting sciarid fungus gnats. If sciarids were ancient hosts of tetradonematids then this parasite lineage could have evolved by the Jurassic (Table 5).

As mentioned in the previous section, oxyurids, drilonematids and cosmocercoids are considered to represent three groups of particularly primitive terrestrial lineages. It is likely that oxyurids parasitised invertebrates before vertebrates, and while *Paleothelastoma tipulae** establishes the oxyurids as parasites of Diptera in the Early Cretaceous, these nematodes could have occurred in Silurian millipedes (Table 5). Other early hosts for oxyurids could have been isopods (Schwenk, 1927) and annelids (Poinar, 1978b; Yeates *et al.*, 1998). A single report of a *Thelastoma* sp. from the hindgut of a scorpion (Barus & Koubková, 2002) is interesting. Oxyurids do not normally occur in

Table 5. *Estimated dates of the origin of invertebrate-parasitic nematode families inferred from the earliest fossil record of their host group. Invertebrate host records were obtained from Yamaguti (1961), Shepard (1974), Poinar (1975, 1979) and Anderson (2000). Fossil host data were taken from Lehmann and Hillmer (1983), Gray (1988), and Rasnitsyn and Quicke (2002). Nematode classification follows De Ley and Blaxter (2004).*

Nematode taxa	Earliest host fossil record
Class Enoplea Inglis, 1983	
Order Mermithida Hyman, 1951	
Mermithidae Braun, 1883	Permian (terrestrial)
	Triassic (marine)
Tetradonematidae Cobb, 1919	Jurassic (terrestrial)
	Cambrian (marine)
Order Benthimermithida Tchesunov, 1995	
Benthimermithidae Petter, 1980	Cambrian
Order Marimermithida Rubstov, 1980	
Marimermithidae Rubstov & Platonova, 1974	Ordovician
Order Echinomermellida **n. ord.**	
Echinomermellidae **n. fam.**	Ordovician
Order Dioctophymatida Baylis & Daubney, 1926	
Soboliphymatidae Petrov, 1930	Carboniferous
Dioctophymatidae Castellani & Chambers, 1910	Carboniferous
Superfamily Trichinelloidea Ward, 1907	
Trichuridae Railliet, 1915	Carboniferous
Cystoopsidae Skrjabin, 1923	Eocene
Order Muspiceida Bain & Chabaud, 1959	
Robertdollfusiidae Chabaud & Campana, 1950	Jurassic
Order Plectida Malakhov, 1982	
Leptolaimidae Örley, 1880	Cambrian
Order Monhysterida Filipjev, 1929	
Monhysteridae de Man, 1876	Eocene
Order Rhabditida Chitwood, 1933	
Suborder Spirurina Railliet & Henry, 1915	
Superfamily Dracunculoidea Leiper, 1912	
Dracunculidae Leiper, 1912	Miocene
Philometridae Baylis & Daubney, 1926	Miocene
Anguillicolidae Yamaguti, 1935	Miocene
Micropleuridae Travassos, 1960	Miocene
Infraorder Gnathostomatomorpha De Ley & Blaxter, 2002	

Table 5. *(Continued).*

Nematode taxa	Earliest host fossil record
Superfamily Gnathostomatoidea Railliet, 1895	
Gnathostomatidae Railliet, 1895	Miocene
Infraorder Oxyuridomorpha De Ley & Blaxter, 2002	
Superfamily Thelastomatoidea Travassos, 1929	
Thelastomatidae Travassos, 1929	Carboniferous
Travassosinematidae Travassos, 1929	Silurian
Hystrignathidae Travassos, 1929	Silurian
Protrelloididae Chitwood, 1932	Carboniferous
Infraorder Rhigonematomorpha De Ley & Blaxter, 2002	
Superfamily Rhigonematoidea Artigas, 1930	
Rhigonematidae Artigas, 1930	Silurian
Ichthyocephalidae Travassos, 1930	Silurian
Superfamily Ransomnematoidea Travassos, 1930	
Ransomnematidae Travassos, 1930	Silurian
Carnoyidae Travassos & Kloss, 1960	Silurian
Hethidae Travassos & Kloss, 1960	Silurian
Infraorder Spiruromorpha De Ley & Blaxter, 2002	
Superfamily Camallanoidea Railliet & Henry, 1915	
Camallanidae Railliet & Henry, 1915	Miocene
Superfamily Physalopteroidea Leiper, 1908	
Physalopteridae Leiper, 1908	Carboniferous
Superfamily Rictularioidea Railliet, 1916	
Rictulariidae Railliet, 1916	Silurian
Superfamily Thelazioidea Skrjabin, 1915	
Thelaziidae Skrjbin, 1915	Carboniferous
Rhabdochonidae Skrjabin, 1946	Carboniferous
Pneumospiruridae Wu & Hu, 1938	Palaeocene
Superfamily Spiruroidea Railliet & Henry, 1915	
Hartertiidae Quentin, 1970	Cretaceous
Gongylonematidae Sobolev, 1949	Jurassic
Spiruridae Railliet & Henry, 1915	Carboniferous
Spiroceridae Chitwood & Wehr, 1932	Permian
Superfamily Habronematoidea Ivaschkin, 1961	
Habronematidae Ivaschkin, 1961	Permian
Hedruridae Railliet, 1916	Triassic
Tetrameridae Travassos, 1914	Permian
Cystidicolidae Skrjabin, 1946	Carboniferous

Table 5. *(Continued).*

Nematode taxa	Earliest host fossil record
Superfamily Acuarioidea Railliet, Henry & Sisoff, 1912	
Acuariidae Railliet, Henry & Sisoff, 1912	Carboniferous
Superfamily Filarioidea Chabaud & Anderson, 1959	
Filariidae Chabaud & Anderson, 1959	Miocene
Onchocercidae Leiper, 1911	Jurassic
Superfamily Aproctoidea Skrjabin & Schikhobalova, 1945	
Aproctidae Skrjabin & Schikhobalova, 1945	Permian
Superfamily Diplotriaenoidea Anderson, 1958	
Diplotriaenidae Anderson, 1958	Permian
Oswaldofilariidae Chabaud & Choquet, 1953	Cretaceous
Infraorder Ascaridomorpha De Ley & Blaxter, 2002	
Superfamily Cosmocercoidea Travassos, 1925	
Cosmocercidae Travassos, 1925	Carboniferous
Kathlaniidae Travassos, 1925	Jurassic
Superfamily Ascaridoidea Blanchard, 1849	
Anisakidae Skrjabin & Karokhin, 1945	Ordovician
Ascarididae Blanchard, 1849	Permian
Superfamily Subuluroidea Yorke & Maplestone, 1926	
Maupasinidae Inglis, 1959	Permian
Subuluridae Yorke & Maplestone, 1926	Permian
Suborder Tylenchina Thorne, 1949	
Infraorder Panagrolaimomorpha De Ley & Blaxter, 2002	
Superfamily Strongyloidoidea Chitwood & McIntosh, 1934	
Steinernematidae Filipjev, 1934	Permian
Alloionematidae Chitwood & McIntosh, 1934	Carboniferous
Infraorder Tylenchomorpha De Ley & Blaxter, 2002	
Superfamily Aphelenchoidea Fuchs, 1937	
Aphelenchoididae Skarbilovich, 1947	Cretaceous
Entaphelenchidae Nickle, 1970	Triassic
Acugutturidae Hunt, 1980	Carboniferous
Superfamily Sphaerularoidea Lubbock, 1861	
Sphaerulariidae Lubbock, 1861	Jurassic
Neotylenchidae Thorne, 1941	Jurassic
Iotonchidae Goodey, 1935	Cretaceous
Allantonematidae Pereira, 1932	Permian
Superfamily Myenchoidea Pereira, 1931	
Myenchidae Pereira, 1931	Jurassic
Infraorder Drilonematomorpha De Ley & Blaxter, 2002	

Table 5. *(Continued).*

Nematode taxa	Earliest host fossil record
Superfamily Drilonematoidea Pierantoni, 1916	
Drilonematidae Pierantoni, 1916	Carboniferous
Ungellidae Chitwood, 1950	Carboniferous
Homungellidae Timm, 1966	Carboniferous
Pharyngonematidae Chitwood, 1950	Carboniferous
Creagrocercidae Baylis, 1943	Carboniferous
Suborder Rhabditida Chitwood, 1933	
Infrorder Diplogastromorpha De Ley & Blaxter, 2002	
Superfamily Diplogastroidea Micoletzky, 1922	
Diplogastridae Micoletzky, 1922	Jurassic
Parasitodiplogastridae **n. grad.**	Eocene
Infraorder Rhabditomorpha De Ley & Blaxter, 2002	
Superfamily Rhabditoidea Örley, 1880	
Carbonematidae Stammer & Wachek, 1952	Jurassic
Agfidae Dougherty, 1955	Carboniferous
Syrphonematidae Laumond & Lyon, 1971	Cretaceous
Rhabditidae Örley, 1880	Triassic
Superfamily Strongyloidea	
Syngamidae Leiper, 1912	Carboniferous
Superfamily Metastrongyloidea Leiper, 1908	
Metastrongylidae Leiper, 1908	Carboniferous
Protostrongylidae Kamenskii, 1905	Carboniferous
Crenosomatidae Skrjabin, 1933	Carboniferous
Angiostrongylidae Böhm & Gebauer, 1934	Carboniferous
Filaroididae Skrjabin, 1933	Carboniferous
Skrjabingylidae Skrjabin, 1933	Carboniferous
Pseudaliidae Railliet, 1916	Carboniferous
Superfamily Heterorhabditoidea **n. grad.**	
Heterorhabditidae Poinar, 1976	Permian

invertebrate predators, especially arachnids, so *Thelastoma euscorpii* Barus & Koubková, 2002, may be a parasite of a prey species eaten by the scorpion.

The cosmocercoids may have evolved in invertebrates and *Palaeocosmocerca burmanica** shows that a snail-infecting lineage was already in existence by the Early Cretaceous. Infected snails eaten by amphibians could have initiated the vertebrate-parasitic taxa; however, it has been suggested that amphibians may have been the original hosts (Vanderburgh & Anderson, 1987).

The drilonematids represent a primitive, but highly specialised, lineage now confined to the coelom of earthworms. These nematodes could have evolved from oxyurids. This scenario is based on the life cycle of *Mesidionema praecomasculatis* Poinar, 1978, an earthworm parasite that has morphological and biological features of both oxyurids and drilonematids (Poinar, 1978a). The females produce eggs in the earthworm's coelom but enter the host's intestine to deposit them. The voided eggs develop into juvenile females, which then penetrate into the coelom of new earthworm hosts where they mature to adults. Males are only produced in the coelom. This migrational pattern from the coelom to the gut links gut-inhabiting oxyurids with coelom-inhabiting drilonematids.

Parasitism by rhabditids is usually initiated by a resistant third-stage infective juvenile. The first fossil record of this stage is with the Early Cretaceous *Proheterorhabditis burmanicus**. Its host has some characteristics of a rove beetle and it is possible that the Staphylinomorpha, with a fossil record extending back to the Triassic, was one of the original host groups of heterorhabitids. Natural insect hosts of extant heterorhabditids are Lepidoptera and especially Coleoptera (*ca* four times more reports in beetles than in moths; Poinar, 1990b). Heterorhabditids are thought to have evolved from free-living, intertidal rhabditids (Poinar, 1993) and their early hosts were probably beach-dwelling crustaceans (isopods, amphipods, *etc.*) and intertidal insects. Rove beetles are fairly common on beaches today (Doyen, 1975) and probably were so in ancient times. If heterorhabditids co-evolved with beetles, they could have originated in the Permian with the Cupedomorpha (Rasnitsyn & Quick, 2002).

Insect-parasitic tylenchs and aphelenchoidids evolved from mycetophagous lineages that developed the ability to attack insects in their immediate environment. The dauer juveniles of *Cryptaphelenchus dominicus** and their phoretic association with a platypodid beetle in Dominican amber could represent a case of dual fungal and insect parasitism. Some stages of extant *Cryptaphelenchus* Fuchs, 1937 species are considered to be body cavity parasites of wood-boring beetles (Hunt, 1993).

Two fossils reported here probably represent species with both mycetophagous and entomogenous life cycles. The stages of *Vetus elongatus** closely resemble the entomogenous females of *Deladenus* *siricidicola* Bedding, 1968, a species with both fungal-feeding and insect-parasitic life cycles (Bedding, 1972b). Unfortunately, possible insect hosts of *V. elongatus** are unknown.

It is likely that *Paleoiotonchium dominicanum** had a similar dual life cycle alternating between mushrooms and fungus gnats. The stages emerging from the fossil fly would develop into mycetophagous adults. While there is no evidence of fungal fruiting bodies in the amber piece with *Paleoiotonchium**, mushrooms and other fungi occur in Dominican amber (Poinar & Poinar, 1999; Boucot & Poinar, 2010). The fossil representatives of *Howardula*, *Palaeoallantonema**, *Palaeoparasitylenchus** and *Tripius* show that the typical allantonematid-sphaerulariid life cycle involving only insect hosts existed by the beginning of the Tertiary.

Ectoparasitism by nematodes is a rare event and thus far appears to be limited to parasites of insects. The discovery of *Setonema protera** adjacent to its moth host in Dominican amber shows that the Noctuidonematinae lineage was already established by the mid-Tertiary. Two extant generic lineages in this subfamily, namely *Noctuidonema* Remillet & Silvain, 1988 and *Vampyronema* Hunt, 1993, attack moths in tropical and subtropical regions. Evidence of ectoparasitism by *Setonema** are spots of melanised haemolymph on the cuticle of the fossil moth. Such signs on extant moths parasitised by *Noctuidonema guyanense* are considered stylet penetration points (Rogers *et al.*, 1993).

Origin of vertebrate parasitism

Vertebrates originated in the seas and their earliest remains occur in shallow, near shore Cambrian and Ordovician deposits. There is no evidence of their presence in freshwater deposits prior to the Early Devonian (Carroll, 1988). In the Silurian seas, nematodes probably parasitised the primitive-jawed acanthodians as well as elasmobranchs and the ancestors of coelacanths.

The oxyurids were possibly the earliest nematode parasites of terrestrial vertebrates, but, unfortunately, no ancient fossils have been recovered. Pinworms in human and animal remains are all from Holocene and Pleistocene deposits (Jansen & Over, 1962, 1966). The oldest oxyurids are 10 000 to 12 000-year-old specimens from ground squirrels (Dubinin, 1948), 10 000 (± 180)-year-old juveniles of *Agamofilaria oxyura* from Shasta Ground Sloth boluses (Schmidt *et al.*, 1992) and 10 000-year-old *Enterobius vermicularis* eggs from human coprolites (Fry & Hall, 1969).

The oldest fossil nematode parasites of any vertebrate are Early Cretaceous ascarid eggs (*Ascarites* sp.) recovered from a dinosaur coprolite. The chances of discovering fossil vertebrate nematodes other than in coprolites or corpses are slight. The three records of filarial nematodes reported here in 20 to 50 million-year-old amber were carried to the sites by insect vectors. They establish the presence of filariids in the Tertiary and show what types of insect vectors were transmitting these parasites. The next oldest records of vertebrate parasitism are 30 000-year-old eggs of *Ascaris lumbricoides* from human coprolites (Patrucco *et al.*, 1983) and 30 000-year-old whipworm eggs from coprolites of a caviid rodent (Ferreira *et al.*, 1991).

The earliest fossil records of any vertebrate parasite are isolated pentastomid larvae from the Cambrian (Waloszek & Müller, 1994) and Ordovician (Andres, 1989). The earliest fossil record of a case of actual vertebrate parasitism is of helminths (cestodes and acanthocephalans) in Devonian fish (Upeniece, 2001). If these parasites already existed by the Devonian, there is no reason why nematode parasites would not also have been present. However, with the absence of fossils, we must turn to the fossil record of the hosts of extant parasites for estimating the origin of most vertebrate-parasitic nematode lineages (Table 6).

ORIGIN OF PLANT PARASITISM

Plant parasitism by nematodes may have existed in the Silurian or even the Ordovician since the earliest known fossils are from the Devonian (Poinar *et al.*, 2008). While there are no known enoploid plant parasites today, the type of plant parasitism exhibited by *Palaeonema* was similar to that of extant dyssaprobes in the families Panagrolaimidae and Cephalobidae. Dyssaprobes do not possess stylets but use their cuticularised stoma and buccal armature to enter and break open plant cells in roots, stems and leaves of healthy plants (Christie, 1933; Filipjev & Schuurmans Stekhoven, 1959; Paramonov, 1962).

At least four lineages of extant marine nematodes bear stylets, the Camacolaimidae (Weiser, 1956), Siphonolaimidae (Weiser, 1956), Leptolaimidae (Hope & Tchesunov, 1999) and Thoracostomopsidae (Ditlevsen, 1919); however, there is no evidence that these stylets are used to obtain nourishment from plants. Some marine dorylaims such as *Dolicholaimus* de Man, 1880 and *Syringolaimus* de Man, 1888 have a thickened stoma or modified teeth that could be used to obtain nutrients from seaweeds but their feeding habits are unknown (Filipjev, 1918; Weiser, 1953).

All stylet-bearing plant parasites found in the sea thus far are terrestrial lineages that invaded littoral and sublittoral habitats. These include the tylench genera *Halenchus* Cobb, 1933 (Fortuner & Maggenti, 1987) and *Hirschmanniella* Luc & Goodey, 1964 (Luc, 1987), and the aphelenchoidid species *Aphelenchoides marinus* Timm & Franklin, 1969 and

Table 6. *Estimated dates of the origin of vertebrate parasitic nematode families inferred from the earliest fossil record of their host group. Vertebrate hosts were taken from Yamaguti (1961), Anderson (2000) and McClelland (2005). Vertebrate host origins are from Carroll (1988). Nematode classification follows Anderson (2000) and De Ley and Blaxter (2004).*

Nematode taxa	Earliest host fossil record
Order Enoplida Filipjev, 1929	
Superfamily Dioctophymatoidea Castellani & Chalmers, 1910	
Soboliphymatidae Petrov, 1930	Carboniferous
Dioctophymatidae Castellani & Chalmers, 1910	Carboniferous
Superfamily Trichinelloidea Ward, 1907	
Trichuridae Railliet, 1915	Carboniferous
Trichinellidae Ward, 1907	Palaeocene
Cystoopsidae Skrjabin, 1923	Cretaceous
Superfamily Muspiceoidea Bain & Chabaud, 1959	
Muspiceidae Bain & Chabaud, 1959	Palaeocene
Robertdollfusidae Chabaud & Campana, 1950	Cretaceous
Order Rhabditida Örley, 1880	
Superfamily Rhabditoidea Chitwood, 1933	
Rhabdiasidae Railliet, 1916	Carboniferous
Strongyloididae Chitwood & McIntosh, 1959	Palaeocene
Longibuccidae Poinar, Jackson, Bell & Wahid, 2003	Eocene
Order Strongylida Diesing, 1851	
Superfamily Diaphanocephaloidea Travassos, 1920	
Diaphanocephalidae Travassos, 1920	Permian
Superfamily Ancyclostomatoidea Nicoll, 1927	
Ancyclostomatidae Nicoll, 1927	Palaeocene
Superfamily Strongyloidea Baird, 1853	
Strongylidae Baird, 1853	Permian
Chabertiidae Railliet & Henry, 1909	Palaeocene
Syngamidae Leiper, 1912	Jurassic

Table 6. *(Continued).*

Nematode taxa	Earliest host fossil record
Superfamily Trichostrongyloidea Leiper, 1912	
Strongylacanthidae York & Maplestone, 1926	Eocene
Amidostomatidae Travassos, 1919	Cretaceous
Trichostrongylidae Leiper, 1912	Jurassic
Dictyocaulidae Skrjabin, 1933	Eocene
Molineidae Cameron, 1923	Carboniferous
Herpetostrongylidae Baylis, 1931	Palaeocene
Ornithostrongylidae Travassos, 1931	Cretaceous
Heligmosomidae Travassos, 1914	Palaeocene
Helegmonellidae Mönnig, 1927	Palaeocene
Superfamily Metastrongyloidea Leiper, 1908	
Metastrongylidae Leiper, 1908	Oligocene
Protostrongylidae Kamenskii, 1905	Oligocene
Crenosomatidae Skrjabin, 1933	Cretaceous
Angiostrongylidae Böhm & Gebauer, 1934	Cretaceous
Filaroididae Skrjabin, 1933	Palaeocene
Skrjabingylidae Skrjabin, 1933	Oligocene
Pseudaliidae Railliet, 1916	Oligocene
Order Oxyurida Weinland, 1858	
Superfamily Oxyuroidea Cobbold, 1864	
Pharyngodonidae Travassos, 1920	Carboniferous
Oxyuridae Cobbold, 1864	Palaeocene
Heteroxynematidae Skrjabin & Shikhobalova, 1948	Palaeocene
Order Ascaridida Yamaguti, 1961	
Superfamily Cosmocercoidea Travassos, 1925	
Cosmocercidae Travassos, 1925	Permian
Kathlaniidae Travassos, 1918	Permian
Atractidae Travassos, 1919	Carboniferous
Superfamily Seuratoidea Railliet, 1906	
Seuratidae Railliet, 1906	Permian
Quimperiidae Baylis, 1930	Cretaceous
Cucullanidae Cobbold, 1854	Triassic
Superfamily Heterakoidea Railliet & Henry, 1912	
Heterakidae Railliet & Henry, 1912	Carboniferous
Ascaridiidae Travassos, 1919	Eocene

Table 6. *(Continued).*

Nematode taxa	Earliest host fossil record
Superfamily Ascaridoidea Baird, 1853	
Anisakidae Skrjabin & Karokhin, 1945	Devonian
Ascarididae Blanchard, 1849	Permian
Acanthochelidae Molin, 1858	Devonian
Superfamily Subuluroidea Yorke & Maplestone, 1926	
Maupasinidae Inglis, 1959	Oligocene
Subuluridae Yorke & Maplestone, 1926	Jurassic
Order Spirurida Diesing, 1861	
Superfamily Camallanoidea Railliet & Henry, 1915	
Camallanidae Railliet & Henry, 1915	Carboniferous
Superfamily Dracunculoidea Leiper, 1912	
Dracunculidae Leiper, 1912	Permian
Philometridae Baylis & Daubney, 1926	Triassic
Anguillicolidae Yamaguti, 1935	Triassic
Micropleuridae Travassos, 1960	Triassic
Daniconematidae Anderson, 2000	Cretaceous
Superfamily Gnathostomatoidea Railliet, 1895	
Gnathostomatidae Railliet, 1895	Devonian
Superfamily Physalopteroidea Leiper, 1908	
Physalopteridae Leiper, 1908	Carboniferous
Superfamily Rictularioidea Railliet, 1916	
Rictulariidae Railliet, 1916	Palaeocene
Superfamily Thelazioidea Skrjabin, 1915	
Thelaziidae Skrjabin, 1915	Jurassic
Rhabdochonidae Skrjabin, 1946	Triassic
Pneumospiruridae Wu & Hu, 1938	Palaeocene
Superfamily Spiruroidea Railliet & Henry, 1915	
Hartertiidae Quentin, 1970	Jurassic
Gongylonematidae Sobolev, 1949	Jurassic
Spiruridae Railliet & Henry, 1915	Cretaceous
Spiroceridae Chitwood & Wehr, 1932	Jurassic
Superfamily Habronematoidea Ivaschkin, 1961	
Habronematidae Ivaschkin, 1961	Jurassic
Hedruridae Railliet, 1916	Permian
Tetrameridae Travassos, 1914	Jurassic
Cystidicolidae Skrjabin, 1964	Triassic

Table 6. *(Continued).*

Nematode taxa	Earliest host fossil record
Superfamily Acuarioidea Railliet, Henry & Siskoff, 1912	
Acuariidae Railliet, Henry & Siskoff, 1912	Triassic
Superfamily Filarioidea Chabaud & Anderson, 1959	
Filariidae Chabaud & Anderson, 1959	Cretaceous
Onchocercidae Leiper, 1911	Carboniferous
Superfamily Aproctoidea Skrjabin & Schikhobalova, 1945	
Aproctidae Skrjabin & Schikhobalova, 1945	Jurassic
Desmidocercidae Cram, 1927	Jurassic
Superfamily Diplotriaenoidea Anderson, 1958	
Diplotriaenidae Anderson, 1958	Permian
Oswaldofilariidae Chabaud & Choquet, 1953	Triassic

Aphelenchoides gynotylurus Timm & Franklin, 1969 (Timm & Franklin, 1969).

All terrestrial lineages of obligate plant parasites use stylets to obtain nourishment from plants. These lineages evolved independently in the Triplonchia (Trichodoridae; Thorne, 1935), Dorylaimia (Longidoridae; Thorne, 1935) and Rhabditida (Tylenchomorpha). Siddiqi (1983) considered that the stylet was a pre-parasitic adaptation first used by microbotrophic nematodes to reach inaccessible microorganisms that was later adapted for feeding on algae, fungi and higher plants. He grouped mycetophagous nematodes together with those feeding on higher plants but considered mycophagy as the primitive trait.

The most primitive stylet-bearing plant parasites would then be mycetophagous members of the Aphelenchidae, Tylenchidae, Paraphelenchidae, Neotylenchidae and Paurodontidae, whose lineages could well extend back to the Silurian with early records of fungi (Table 7). Siddiqi (1983) suggested that both tylenchs and aphelenchoidids originated in the Devonian, but considered that tylenchs originated from algal-feeding lineages. Paramonov (1962) and Maggenti (1971) concluded that tylenchid plant parasites developed from mycetophagous lineages. It is quite likely that both algae and fungi were utilised by these early nematode clades.

Siddiqi (1983) postulated that aphelenchoidid parasites of the aerial parts of plants could have evolved from associations with xylophagous

Table 7. *Estimated dates for the origin of plant-parasitic nematode families inferred from the earliest fossil record of their host group. Plant hosts were obtained from Goodey* et al. *(1965), Hunt (1993) and Siddiqi (2000). Plant fossil records were taken from Collinson* et al. *(1993) and Taylor and Taylor (1993). Nematode classification follows De Ley and Blaxter (2004).*

Nematode taxa	Earliest fossil host record
Class Enoplea Inglis, 1983	
Subclass Enoplia Pearse, 1942	
Order Triplonchida Cobb, 1920	
Suborder Diphtherophorina Micoletzky, 1922	
Superfamily Diphtherophoroidea Micoletzky, 1922	
Family Trichodoridae Thorne, 1935	Triassic
Subclass Dorylaimia Inglis, 1983	
Order Dorylaimida Pearse, 1942	
Suborder Nygolaimina Ahmad & Jairajpuri, 1979	
Superfamily Dorylaimoidea de Man, 1876	
Family Longidoridae Thorne, 1935	Triassic
Class Chromadorea De Ley *et al.*, 2000	
Subclass Chromadorea De Ley *et al.*, 2000	
Order Rhabditida Chitwood, 1933	
Suborder Tylenchina Thorne, 1949	
Infraorder Tylenchomorpha De Ley & Blaxter, 2002	
Superfamily Aphelenchoidea Fuchs, 1937	Silurian
Superfamily Aphelenchoidoidea Skarbilovich, 1947	
Family Aphelenchoididae Skarbilovich, 1947	Triassic
Family Parasitaphelenchidae Rühm, 1956	Triassic
Superfamily Criconematoidea Taylor, 1936	
Family Criconematidae Taylor, 1936	Carboniferous
Family Hemicycliophoridae Skarbilovich, 1959	Triassic
Family Caloosiidae Siddiqi, 1980	Miocene
Family Tylenchulidae Skarbilovich, 1947	Cretaceous
Family Sphaeronematidae Raski & Sher, 1952	Cretaceous
Family Paratylenchidae Thorne, 1949	Triassic
Superfamily Sphaerularioidea Lubbock, 1861	
Family Neotylenchidae Thorne, 1941	Silurian
Family Paurodontidae Thorne, 1941	Silurian
Superfamily Anguinoidea Nicoll, 1935	
Family Anguinidae Nicoll, 1935	Cretaceous

Table 7. *(Continued).*

Nematode taxa	Earliest fossil host record
Superfamily Tylenchoidea Örley, 1880	
Family Hoplolaimidae Filipjev, 1934	Triassic
Family Rotylenchulidae Husain & Khan, 1967	Cretaceous
Family Pratylenchidae Thorne, 1949	Triassic
Family Meloidogynidae Skarbilovich, 1959	Triassic
Family Heteroderidae Filipjev & Schuurmans Stekhoven, 1941	Triassic
Family Ecphyadophoridae Skarbilovich, 1959	Cretaceous
Family Atylenchidae Skarbilovich, 1959	Cretaceous
Family Tylodoridae Paramonov, 1967	Cretaceous
Family Tylenchidae Örley, 1880	Silurian
Family Psilenchidae Paramonov, 1967	Carboniferous
Family Belonolaimidae Whitehead, 1960	Triassic
Family Pratylenchidae Thorne, 1949	Triassic
Family Telotylenchidae Siddiqi, 1960	Triassic
Family Dolichodoridae Chitwood, 1950	Triassic

insects that carried them to those locations. This is quite likely since most of the aphelenchoidids described here, including the Early Cretaceous *Cretaciaphelenchoides burmensis**, could have been carried originally to their aerial locations by xylophagous insects.

Clades of tylenchid and aphelenchoidid parasites that fed on higher plants, as well as the Longidoridae and Trichodoridae, probably occurred by the Carboniferous, especially those lineages that fed on ferns and bryophytes. Those that specialised on gymnosperm groups like Araucariaceae, Cupressaceae and Pinaceae, could have existed by the Triassic (Table 7). All of these early lineages probably shifted to angiosperms in the Early Cretaceous. By the mid-Cretaceous, co-evolution with angiosperms would have resulted in the appearance of many new lineages, some of which became extinct while others exist today.

The oldest fossil stylet-bearing nematode is the Early Cretaceous *Cretaciaphelenchoides burmensis**, followed by the Eocene *Palaeoaphelenchoides balticus**. However, the most complete remains of a plant parasite are eggs, juveniles and adults of Mexican amber *Oligaphelenchoides atrebora* in a single piece of amber. The Dominican amber

*Oligaphelenchoides dominicanus** probably had similar feeding habits since fungal mycelium presumed to be its food source was also preserved. The above species represent free-living mycetophages, which is probably the primitive aphelenchoidid trophic stage. Fossil evidence that aphelenchoidids could have fed in tunnels made by xylophagous beetles was supplied by *Bursaphelenchus similus** in Dominican amber. The presence of a platypodid beetle adjacent to frass material and wood fragments, and the absence of dauer juveniles, indicated that this species was developing under optimum conditions in the beetle's galleries.

Just when aphelenchoidids shifted their diet to higher plants is not known but *Oligaphelenchoides maximus** developing in a monocot rootlet is considered to represent an aerial parasite with a biology similar to some extant *Aphelenchoides* spp. Further evidence of obligate parasitism of aerial portions of plants is provided by *Palaeoanguina dominicana**, which was developing in a seed in Dominican amber. This species probably had a life cycle similar to some extant members of the genus *Anguina* Scopoli, 1777 (Fortuner & Maggenti, 1987).

USING HOST FOSSILS TO ESTIMATE THE ORIGIN OF NEMATODE PARASITES

Since their fossil record is so sparse, determining the evolutionary clocks of parasitic nematodes is difficult. However, phylogenetic time lines can be inferred from the fossil record of their hosts. Previous workers have used this method to reconstruct phylogenies of some vertebrate-parasitic lineages, such as the strongyloids (Dougherty, 1951) and trichostrongyloids (Durette-Desset, 1985). This method of dating assumes that, in general, nematode lineages co-evolved with their hosts, following Farenholz' Rule, "parasite phylogeny reflects host phylogeny". However, there are potential problems with this method of dating nematode lineages. Fossil host dating only provides a minimum age since many extant nematode clades might have originated with host groups now extinct and then switched to those presently surviving. Also, the host range of some nematode lineages could have been quite broad in the past, but became more restricted over time, or *vice versa*. In addition, nematode genera could have switched back and forth between different host groups multiple times.

Since most nematode families have multiple hosts, the earliest fossil of the most primitive host was used to determine the possible date of origin for the nematode group. Thus, if a particular nematode lineage parasitises snakes and lizards today, lizards would be used as the earliest host group since they appeared in the Upper Permian, while the fossil record of snakes dates from the Cretaceous (Carroll, 1988). Knowing these caveats, nematode evolutionary clocks based on fossil host records are presented in Tables 5-7 for invertebrate, vertebrate and plant parasites, respectively.

For determining the origin of nematode parasites of invertebrates (Table 5), invertebrate host records were obtained mainly from Poinar (1975, 1979) and Shepard (1974), with fossil host records taken from Gray (1988), Lehmann and Hillmer (1983) and Rasnitsyn and Quicke (2002). The table also includes the fossil record of intermediate hosts used by vertebrate-parasitic groups. These host records were obtained from Anderson (2000) and Yamaguti (1961). Since so many vertebrate-parasitic lineages have a wide range of intermediate hosts, the earliest fossil record of the most primitive intermediate host is used. A difficulty one encounters by assuming that the vertebrate was the original host in heteroxenous lineages is not knowing when the intermediate host became an essential part of the life cycle. It could have been at the same time as the definitive host or at some later period. Then there is always the problem of the incompleteness of the fossil record. For example, gnathostomatids could have had at least a Devonian origin based on their elasmobranch hosts but the earliest record of their intermediate copepod hosts is the Miocene. In this case, the discrepancy is probably due to the incomplete fossil record of copepods. In some cases, such as with the Physalopteridae, Spiruridae, Gongylonematidae, Cosmocercidae and Ascarididae, the earliest fossil records of the definitive and intermediate host groups are roughly the same.

Utilising the fossil host dating method, the earliest animal-parasitic nematodes would have been members of the marine families Benthimermithidae, Tetradonematidae and Leptolaimidae whose invertebrate host groups extend back to the Cambrian. The earliest terrestrial animal parasites were the Thelastomatoidea, Rhigonematoidea and Ransomnematoidea that parasitised millipedes, which have a fossil record dating back to the Silurian.

The host dating method works best for those parasitic lineages that have restricted host ranges, *e.g.*, parasitise only phylogenetically related hosts. Many vertebrate parasites occur in phylogenetically unrelated hosts that exist in the same environment. Inglis (1965) was of the opinion that, in general, vertebrate-parasitic nematode lineages are not tied to specific host lineages but occur in animals with similar feeding and ecological habits. Chabaud (1971) felt that vertebrates evolving at the same geological time period would be parasitised by nematode lineages that existed at that time. If this is true, it complicates the host dating system. With our present knowledge, the proposed origins of vertebrate-parasitic lineages can only be based on the earliest fossil record of the parasite's range of hosts. With vertebrate-parasitic nematodes (Table 6), vertebrate hosts were taken from Anderson (2000), Yamaguti (1961) and McClelland (2005) and vertebrate host origins are from Carroll (1988). Using this method, the Gnathostomatidae, Anisakidae and Acanthochelidae would represent the earliest vertebrate parasites with appearances by the Devonian (Table 6).

With plant-parasitic nematodes (Table 7), plant hosts were obtained from Hunt (1993), Siddiqi (2000) and Goodey *et al.* (1965) and the fossil records of the plant hosts were taken from Taylor and Taylor (1993) and Collinson *et al.* (1993). The earliest plant parasites were probably mycetophagous lineages of the Aphelenchidae, Paraphelenchidae, Neotylenchidae, Paurodontidae and Tylenchidae that could have appeared by the Silurian. These early lineages may have resembled extant species of *Ditylenchus* and *Aphelenchus* that can obtain nourishment from fungi as well as vascular plants. Another primitive genus is *Psilenchus* de Man, 1921, which Luc *et al.* (1987) considered the closest living form to a prototylenchid. Unfortunately, the feeding habits of this genus are unknown, although Paramonov (1970) felt that *Psilenchus* was probably mycetophagous as well as able to feed on angiosperms.

It is curious why so few nematodes have been reported from gymnosperms, because ever since the Carboniferous there must have been multiple nematode clades feeding on these plants. Perhaps the resinous exudates were a deterrent to many plant nematodes. Extant *Meloidogyne* species feed on *Araucaria*, a plant group extending back to the Triassic, and those nematodes attacking pines could belong to lineages extending back to the Jurassic.

Nematology Monographs & Perspectives, 2011, Vol. 9, 235-246

Functional morphology and the behaviour of fossil nematodes

Behavioural traits of fossil organisms can often be determined by examining the functional morphology of their extant descendants. Functional morphology is defined as the "Interpretation of the function of an organism or organ system by reference to its shape, form and structure" (Lincoln & Boxshall, 1987).

Functional morphology can reveal the feedings habits of fossil nematodes. In plant parasites the stylet is used to penetrate the plant wall and release plant sap into the nematode's gut. Stylets are evident in some of the fossil plant-parasitic nematodes described herein, such as *Cryptaphelenchus dominicus** (Fig. 143) and *Palaeoanguina dominicana**, and also existed or presumably existed in fossil species of *Bursaphelenchus*, *Cretaciaphelenchoides**, *Oligaphelenchoides* and *Palaeoaphelenchoides**. In entomophagous nematodes, stylets are used to enter the host's body cavity. Stylets are evident in the fossil insect parasites, *Howardula helenoschini*, *Palaeoparasitylenchus dominicanus** and *Setonema protera**, and can be assumed to have occurred in *Chalcidonema paradoxa**, *Tripius balticus**, *Paleoiotonchium dominicanum** and *Myrmeconema antiqua**, as well as in species of *Palaeoallantonema**, *Palaeotetradonema**, *Heydenius* and *Cretacimermis*.

Other feeding structures on fossil phytoparasitic nematodes are the buccal teeth of *Palaeonema phyticus*, which were undoubtedly used to disrupt the cortical cells in the stomatal cavities of its Devonian host plants. The function of the long probae or processes on the head of *Nemavermes mackeei* is a mystery (see Fig. 18). They certainly could have served a sensory function and assisted the nematode in locating prey buried in the surrounding sediments. Heavy buccal armature indicating predatory behaviour is shown by the large teeth on the free-living marine nematode, *Palaeoenoploides alessandrellianus* (Fig. 23). Their large size (between 700 and 800 μm in length), heavy sclerotisation and unusual armature (three large teeth, one dorsal with a double hook and two ventral, each with a single hook) suggest that they may have been specialised for particular prey.

Invertebrate-parasitic nematodes use several methods to enter hosts. Free-living infective-stage allantonematids and sphaerulariids use their stylet to direct glandular secretions against the host's cuticle and then puncture an entrance hole in the softened body wall (Poinar & Doncaster, 1965). Such methods were probably employed by all of the fossil insect-parasitic tylenchs, aphelenchoidids and mermithoids presented here. Infective juveniles of the genus *Heterorhabditis* use a small mucron located at the tip of their head to rasp their way through the gut wall or even between the softer intersegmental membranes of potential hosts (Poinar & Georgis, 1990). A similar mucron on the head of an infective juvenile of *Proheterorhabditis** (Fig. 36) shows that this method of entry was established by the Early Cretaceous.

Another example of functional morphology is the character of nematode eggs. A thick outer mammillate shell on the egg of *Ascarites priscus* (Fig. 42) protected the developing embryo. This is similar to the thick shell on eggs of present-day ascarids (Fig. 44), which maintains egg viability for long periods. The shells on other fossil nematode eggs are much thinner, as shown in the transmission electron microscope section through an egg of *Oligaphelenchoides atrebora* in Mexican amber (Figs 3, 5). The eggs of both *O. atrebora* (Fig. 251) and *O. maximus** (Fig. 201) are elliptical, thin-walled and typical of eggs of extant aphelenchoidids.

The eggs of *Myrmeconema antiqua** adjacent to their ant host in Dominican amber are very small but also thick-shelled (Figs 113, 114). It is presumed that the life cycle of this species is similar to that of the extant *Myrmeconema neotropicum*, which also has small eggs (Fig. 117). The thick shell not only protects the eggs during their passage through the alimentary tract of frugivorous birds but also probably ensures viability while they are exposed to the elements in bird droppings.

Behavioural characters also show how fossil nematodes attached themselves to arthropods in phoretic associations. The method used by *Psychodorhabditis attenuatus** for attachment to a mothfly in Dominican amber employed the 'wrap-around' method. The nematode's body is coiled around the abdomen of the carrier (Fig. 169). By contrast, the moth ectoparasitic nematode, *Setonema protera**, in Dominican amber (Figs 202-204) uses a different method of attachment since the females increase in diameter, which keeps them secure under the scales of the host. Members of the extant genus *Noctuidonema* (Figs 205, 206) have a similar body type.

The plant parasite, *Palaeoanguina**, in Dominican amber also has a stout body but that may be due to limited space in the seed. The problem of restricted space for growth also occurs with insect-parasitic nematodes. In order to accommodate large numbers of eggs, as is typical of parasites, nematodes can either become elongated, as is the case of mermithids, which are often three to four times the length of their hosts, or they can increase in diameter. The latter condition is found in many allantonematids and sphaerulariids and occurred in the parasitic female of *Howardula helenoschini* in Baltic amber.

A behavioural characteristic of a few mermithids that parasitise dipteran hosts with short life cycles is to accelerate their own development and complete the final moult to the adult stage while still in the body cavity of the host. Normally, mermithids emerge from their hosts as post-parasitic juveniles and undergo a double moult in the environment before attaining the adult stage. However, extant species of the genus *Heleidomermis* Rubstov can complete their final two moults in the biting midge host and emerge as adults. This same behaviour was noted on individuals of *Cretacimermis protus* that had started to moult inside their hosts in Cretaceous Burmese amber (Poinar & Buckley, 2006; Poinar & Monteys, 2008) (Fig. 39). It is likely that moulting would have been completed within the hosts.

Reproductive behaviour

Fossil evidence of sexual dimorphism in nematodes is rare because both sexes are usually not fossilised. However, based on our knowledge of extant nematodes, we can expect that sexual dimorphism occurred in *Palaeoanguina**, *Setonema** and the fossil allantonematids and sphaerulariids. Female anguinids typically are stouter than and 1.5 to 2.0 times the length of the males (Goodey, 1963). Thus, we can assume that the male of *Palaeoanguina dominicana** was smaller than its mate. Males of the Noctuidonematinae, to which *Setonema** may belong, are also smaller than the females. While the free-living males and females of allantonematids and sphaerulariids are similar in size, the females grow considerably within the host body as they mature.

Most extant nematodes are amphimictic (bisexual) with eggs and sperm produced in separate individuals, although autotoky occurs in some groups (Poinar & Hansen, 1984). Populations of *Palaeorhabditis* *balticus** comprised only juveniles and females, which indicates that they were autotokous and probably hermaphrodites. Since no eggs were detected and young juveniles occurred inside mature females, *P. balticus** could have demonstrated *endotokia matricida* or intrauterine birth resulting in maternal death (Fig. 59).

The presence of the fossil heterorhabditid, *Proheterorhabditis burmanicus**, shows that heterogamy was established by the Early Cretaceous and this species would have been capable of both hermaphroditic and amphimictic life cycles. It cannot be stated that both of these cycles occurred in the fossil rove beetle, since its small size and restricted nutrients could have precluded the amphimictic cycle.

Based on the biology of the extant *Iotonchium californicum* (Poinar, 1991), some inferences can be made about the mating behaviour of *Paleoiotonchium dominicanum**. The males of *I. californicum* rub their expanded head lobes rapidly over the body of the female during mating, with increased intensity during spicular penetration, which may occur through the body wall (traumatic insemination). It is possible that this type of sexual behaviour also occurred in *P. dominicanum**. Also, the fungus-feeding females of *I. californicum* are parthenogenetic and it is likely that this was also the case for *P. dominicanum**.

We can make inferences about the way fossil nematodes used their reproductive organs to obtain nourishment, based on our knowledge of their present descendants at the generic and family level (behavioural fixity). For instance, some lineages of the Sphaerulariidae and Iotonchidae have the unusual character of everting the uterus through the vulva into the body cavity of their host. The outer surfaces of the uterine cells expand and, in *Sphaerularia bombi* (Dufour), are covered with microvillar processes that serve to absorb nutrients (Poinar & Hess, 1972). The extant species *Iotonchium californicum* and *Tripius sciarae* evert their uteri into the body cavity of their fly hosts (Bovien, 1944; Poinar, 1991) and we can assume that the fossil representatives of these lineages, *Paleoiotonchium dominicanum** and *Tripius balticus**, respectively, also had such reproductive modifications.

Fossil evidence of nematode diseases

Since diseases of extant nematodes have been little studied, it is surprising that there are any fossil records of nematode pathogens. However evidence of nematophagous fungi occurred within and around several specimens of the Mexican amber aphelenchoidid, *Oligaphelenchoides atrebora*, and establishes fungal parasitism of nematodes in the mid-Tertiary (Figs 252, 253). Not only were mycelia present inside the nematodes but also spores and mycelium occurred in the surrounding amber (Jansson & Poinar, 1986). These fungi probably belonged to those deuteromycetes (Deuteromycotina) that produce trap devices and invade the nematode's body after it is snared (Barron, 1977).

An Early Cretaceous nematode-trapping fungus was recently described as *Palaeoanellus dimorphus* Schmidt, Dörfelt & Perrichot, 2007 from French amber. This species, considered a member of the Deuteromycotina, possessed unicellular hyphal rings as trapping devices and also had a yeast stage that developed from blastospores (Schmidt *et al.*, 2007). Unfortunately, there were no nematodes associated with *P. dimorphus* in the amber.

Palaeosymbiosis

Phoresis

Dauer juveniles, resistant stages formed by many terrestrial nematodes, are indispensable for species survival. These non-feeding stages are formed under periods of stress and can tolerate low humidity and the absence of food for long periods (Evans & Perry, 1976). Many free-living nematodes depend on insects for dispersal and the dauer juveniles are usually the stages transported. The arthropod carriers introduce the nematodes to new environments suitable for continued development.

The earliest fossil record of dauer juveniles is the Early Cretaceous termite associate, *Cretodiplogaster termitophilus**. The descriptions of many fossil species, including those in the genera *Apifilaria**, *Formicodiplogaster**, *Filotermes**, *Scarabinema**, *Scolytonema**, *Cretodiplogaster**, *Prescafilotermes** and *Panagrolaimus*, are based on dauer juveniles associated with arthropod carriers. The nematodes were captured in resin as they left their phoretic hosts and attempted to locate developmental sites. Most of these dauer juveniles had external (ectophoretic) associations with their carriers, but some may have been carried internally (endophoresis) in the alimentary tract, trachea or pharyngeal glands.

Many fossil dauer stages occur on their carriers in similar locations to their modern descendants. In extant bark and pinhole beetles, dauer juveniles are located in spaces between the elytra and abdomen or in crevices between various body segments (Rühm, 1956), similar to fossil members of the genus *Scolytonema**. The phoretic behaviour indicated by fossil members of the genera *Bursaphelenchus*, *Scolytonema** and *Cryptaphelenchus* with bark and pinhole borers, *Formicodiplogaster myrmenema** with ants, and *Filotermes minutus**, *Prescafilotermes dominicanus**, *P. elongatus** and *Cretodiplogaster termitophilus** with termites, are very similar to the phoretic behaviour of their modern descendants. Even the wrap-around method of attachment shown by *Psychodorhabditis attenuatus** with mothflies is similar to that of some modern rhabditids on mothflies (Bovien, 1937). On the other hand, some phoretic associations, such as *Psocidonema dominicana* on bark lice, *Apifilaria proplebiae** on stingless bees, *Tipulicomis balticus** and *T. dominicanus** on craneflies, *Chilonema filaforma** on centipedes and *Crustafila confina** on isopods, have not been noted previously. It is possible that they exist today but have not yet been discovered and/or reported. Other fossil dauer stages appear to have different attachment sites on their carriers than their modern descendants. Extant diplogastrid dauer juveniles on ants usually occur in the mandibular glands of their carriers; however, dauer stages of the fossil *Formicodiplogaster myrmenema** appear to have been lodged between the abdominal segments of the ants.

The association with diplogastrids and fig wasps probably began with external phoresis, but because female fig wasps have such a smooth surface and the nematodes could be easily injured when the wasp forced its way into the fig fruit, the nematodes entered the body cavity of the wasp, as in *Syconema dominicana**. This species may represent an early stage in the association between fig wasps and diplogastrids since some extant *Parasitodiplogaster* species develop to the adult stage in the body cavity of the wasp (Poinar, 1979).

It is possible that primitive oxyurids also possessed dauer stages that were formed under stressful conditions, such as the death of their hosts. Over the aeons, egg transmission became more effective and the dauer stage was lost. Unfortunately, there is no way of knowing if *Paleothelastoma tipulae** had a dauer stage, was in the process of losing it or never had one. Loss of the ability to form dauer juveniles has occurred in several insect-parasitic rhabditids. One species in the process of losing the ability to form dauer juveniles is *Rhabditis adenobia* Poinar, 1971, which lives in the colleterial glands of dynastid beetles. This species is transmitted from beetle to beetle during mating and has no dauer stage in its life cycle. However, on laboratory cultures seeded with bacteria from the host's colleterial glands, the nematodes will attempt to form dauer juveniles but are never successful. A 'vestigial' dauer with a partially closed mouth is produced but it never moults to the third-stage and soon perishes.

An insect-parasitic rhabditid that completely lost its ability to form dauer juveniles is *Oryctonema genitalis* Poinar from the bursa copulatrix of dynastid beetles (Poinar, 1970). Since the nematodes are transmitted from beetle to beetle during mating, a resistant stage is no longer necessary and the dauer stage has been dropped from the life cycle. This species is incapable of developing on agar plates seeded with bacteria obtained from the bursa copulatrix of the beetle.

Commensalism and mutualism

Some of the fossil cases of phoresis discussed above could be considered examples of commensalism, where one agent (the nematode) benefits and the other (the carrier or transporter) is not harmed. However, if the nematodes were regulating populations of bacteria and fungi that were detrimental to the arthropod carriers, then the association could be considered mutualism. In some instances, mites and pseudoscorpions are also transported by the same carriers that bear nematodes (Plate 17). These arachnids add another element to the equation since they feed on nematodes and thus control their population dynamics. Phoretic mesostigmatid mites of the genus *Proctolaelaps* Berlese remove and devour dauer juveniles of *Cylindrocorpus inevectus* Poinar, Jackson, Bell & Wahid (2003) while both are riding on the weevil, *Elaeidobius kamerunicus* (Faust) (Krantz & Poinar, 2004). If the nematodes were beneficial in some way to the carrier, their reduction by arachnids could upset the balance.

While there are no records of nematodes pollinating higher plants, nematode parasites of insect pollinators such as *Syconema dominicana** could affect the efficiency of pollinating fig wasps. Extant nematodes such as *Sphaerularia bombi* can shorten the life span of bumblebees, thus affecting the pollination of clover (Palm, 1945). Also, the extant allantonematid *Elaeolenchus parthenonema* Poinar, Jackson, Bell &

Wahid, 2003 reduced the efficiency of the palm-pollinating weevil, *Elaeidobius kamerunicus*, which resulted in a decline in fruit set and yield in oil palms (Poinar *et al.*, 2003). If *S. dominicana** fed on bacteria that were detrimental to the fig wasp brood then their activities in the synconium could have benefited the wasps. The above possible scenarios show how difficult it is to interpret associations between nematodes and their carriers.

The bacteria in the beetle parasitised by *Proheterorhabditis burmanicus** are considered to have had a mutualistic association with the nematode, similar to that in extant heterorhabditids. The nematodes protect and introduce the bacteria into the insect's haemocoel, while the multiplying bacteria supply food and produce antibiotics that deter antagonistic microbes (Poinar, 1990b).

Parasitism

The various examples of fossil parasitic associations shown in the present work provide minimum dates for the appearance of certain types of parasitism. Gut parasitism of terrestrial invertebrates had originated by the Early Cretaceous with *Paleothelastoma tipulae** (Thelastomatidae) in an adult cranefly and *Palaeocosmocerca burmanica** (Cosmocercidae) in a snail. Evidence of intestinal parasites of vertebrates also dates from that period with *Ascarites priscus* and *A. gerus* (Ascarididae) in dinosaurs. Evidence of tissue parasites of vertebrates are fossil species of the filariid genus *Cascofilaria** associated with haematophagous insects in Baltic and Dominican amber.

There are many examples of fossil terrestrial nematodes developing in the body cavity of insects. This category includes all species of *Heydenius* and *Cretacimermis* (Mermithidae) attacking insects, *Proheterorhabditis burmanicus** (Heterorhabditidae) in beetles and *Paleoiotonchium dominicanum** (Iotonchidae) parasitising mycetophilid fungus gnats. In addition are the fossil allantonematids such as *Howardula helenoschini* in phorid flies, *Palaeoallantonema apionae** and *Palaeoallantonema zygopinae** in weevils, *Palaeoparasitylenchus dominicanus** from drosophilid flies, *Palaeoallantonema baltica** and *Palaeoallantonema dominicana** in rove beetles, *Palaeoallantonema cephalotae** in ants and *Palaeoallantonema phlebotomae** attacking sand flies in Baltic and Dominican amber. Also in this category are *Tripius balticus** (Sphaerulariidae) parasitising sciarid fungus gnats,

*Syconema dominicana** (Diplogastridae) in fig wasps, *Chalcidonema paradoxa** (Aphelenchoididae) in parasitoid wasps, the tetradonematids, *Palaeotetradonema phlebotomae** in sand flies, *Palaeotetradonema sciarae** attacking sciarid fungus gnats, and *Myrmeconema antiqua** from ants. The only fossil evidence of nematode ectoparasitism is *Setonema protera** on moths in Dominican amber. Fossil nematode parasites of invertebrates in amber are listed under their hosts in Table 8.

Mycetophagous lineages of fossil plant-parasitic nematodes are most often encountered in amber; thus, the aphelenchoidids *Cretaciaphelenchoides burmensis** from Early Cretaceous Burmese amber, *Palaeoaphelenchoides balticus** from Baltic amber, *Oligaphelenchoides atrebora* from Mexican amber and *Bursaphelenchus similus**, *Cryptaphelenchus dominicus**, *Oligaphelenchoides dominicanus** from Dominican amber. Only two aerial parasites of vascular plants have been recovered, *Palaeoanguina dominicana** (Anguinidae) in a monocot seed and *Oligaphelenchoides maximus** (Aphelenchoididae) in a monocot stem, both in Dominican amber. The Devonian *Palaeonema phyticum* is the oldest known plant parasite and, since it lived in the stomatal and substomatal chambers of the primitive vascular plant, *Aglaophyton major*, it can also be regarded as an aerial plant parasite.

Inferences based on behavioural fixity can be made to show how some fossil parasite lineages obtained nutrients from their hosts. In the case of the Baltic amber *Howardula helenoschini*, the parasitic female probably acquired nutrients through microvilli located on the surface of its body wall, similar to that of its modern descendant, *Howardula husseyi* (Riding, 1970). As mentioned previously, the females of *Paleoiotonchium dominicanum** and *Tripius balticus** probably obtained nutrients through their uterine cells that were extruded into the haemocoel of their insect hosts, similar to that of their modern descendants. All of the fossil specimens of *Heydenius* and *Cretacimermis* probably obtained nutrients taken in through minute pores in their cuticles and then absorbed by microvilli on the underlying hypodermal cells, similar to that of modern mermithids (Poinar & Hess, 1977).

Much information about the behaviour of ancient nematodes can be acquired through a careful examination of fossils. Up to the present, most of this information has been acquired through the study of nematodes in amber and, of course, while the habitat of these deposits represented species-rich tropical and subtropical forest types, the nematode populations encountered in the tree bark ecotypes are

Table 8. *Invertebrate hosts and associates of fossil nematodes in amber deposits.*

Taxonomic rank	Nematode	Association	Deposit
Isoptera			
Kalotermitidae	*Cretodiplogaster termitophilus*	Phoretic	Burmese
Hodotermitidae	*Filotermes minutus*	Phoretic	Dominican
Termitidae	*Prescafilotermes dominicanus*	Phoretic	Dominican
Family indet.	*Prescafilotermes elongatus*	Phoretic	Dominican
Psocoptera			
Family indet.	*Psocidonema dominicana*	Phoretic	Dominican
Hemiptera			
Achiliidae	*Heydenius brownii*	Parasitic	Baltic
Coleoptera			
Brentidae	*Palaeodiplogaster brentiphila*	Phoretic	Dominican
Curculionidae	*Palaeoallantonema apionae*	Parasitic	Dominican
	Palaeoallantonema zygopinae	Parasitic	Dominican
Lampyridae	*Heydenius lamprophilus*	Parasitic	Dominican
Scarabaeidae	*Scarabinema minuta*	Phoretic	Dominican
Scolytidae	*Bursaphelenchus similus*	Phoretic	Dominican
	Cryptaphelenchus dominicus	Phoretic	Dominican
	Panagrolaimus neotropicus	Phoretic	Dominican
	Scolytonema dominicana	Phoretic	Dominican
	Scolytonema hispaniolae	Phoretic	Dominican
	Scolytonema mexicana	Phoretic	Mexican
Staphylinidae	*Palaeoallantonema baltica*	Parasitic	Baltic
	Palaeoallantonema dominicana	Parasitic	Dominican
Staphylinidae?	*Proheterorhabditis burmanicus*	Parasitic	Burmese
Lepidoptera			
Family indet.	*Setonema protera*	Parasitic	Dominican
Diptera			
Anisopodidae	*Heydenius saprophilus*	Parasitic	Dominican
Cecidomyiidae	*Heydenius cecidomyae*	Parasitic	Baltic
Ceratopogonidae	*Cretacimermis protus*	Parasitic	Burmese
Chironomidae	*Cretacimermis chironomae*	Parasitic	Burmese
	Cretacimermis libani	Parasitic	Lebanese
	Heydenius neotropicus	Parasitic	Dominican
	Heydenius matutinus	Parasitic	Baltic
Culicidae	*Cascofilaria dominicana*	Parasitic	Dominican
	Cascofilaria parva	Parasitic	Dominican
	Heydenius dominicus	Parasitic	Dominican
Drosophilidae	*Palaeoparasitylenchus dominicanus*	Parasitic	Dominican
Milichiidae?	*Heydenius dipterophilus*	Parasitic	Dominican

Table 8. *(Continued).*

Taxonomic rank	Nematode	Association	Deposit
Mycetophilidae	*Paleoiotonchium dominicanum*	Parasitic	Dominican
Phoridae	*Howardula helenoschini*	Parasitic	Baltic
Phlebotomidae	*Palaeoallantonema phlebotomae*	Parasitic	Dominican
	Palaeotetradonema phlebotomae	Parasitic	Mexican
Psychodidae	*Heydenius psychodae*	Parasitic	Dominican
	Psychodorhabditis attenuatus	Phoretic	Dominican
	Psychodorhabditis balticus	Phoretic	Baltic
	Psychodorhabditis dominicanus	Phoretic	Dominican
Scatopsidae	*Heydenius scatophilus*	Parasitic	Dominican
Sciaridae	*Heydenius sciarophilus*	Parasitic	Baltic
	Palaeotetradonema sciarae	Parasitic	Baltic
	Tripius balticus	Parasitic	Baltic
Simuliidae	*Cascofilaria baltica*	Parasitic	Baltic
Tipulidae	*Paleothelastoma tipulae*	Parasitic	Burmese
	Tipulacomis balticus	Phoretic	Baltic
	Tipulacomis dominicanus	Phoretic	Dominican
Hymenoptera			
Agaonidae	*Syconema dominicana*	Parasitic	Dominican
Apidae	*Apifilaria proplebiae*	Phoretic	Dominican
Chalcidoidea	*Chalcidonema paradoxa*	Parasitic	Dominican
Formicidae	*Formicodiplogaster myrmenema*	Phoretic	Dominican
	Heydenius formicinus	Parasitic	Baltic
	Heydenius myrmecophila	Parasitic	Dominican
	Myrmeconema antiqua	Parasitic	Dominican
	Palaeoallantonema cephalotae	Parasitic	Dominican
Arachnida			
Thomisidae	*Heydenius araneus*	Parasitic	Baltic
Trechaleidae	*Palaeoarachnus balticus*	Phoretic	Baltic
Chilopoda			
Scutigeridae	*Chilonema filaforma*	Phoretic	Dominican
Diplopoda			
Polydesmidae	*Diplonema minuta*	Phoretic	Dominican
Isopoda			
Philosciidae	*Crustafila confina*	Phoretic	Dominican
Mollusca			
Punctidae	*Palaeocosmocerca burmanica*	Parasitic	Burmese
Tardigrada			
Family indet.	*Vetus microcosmus*	Associate	Dominican

restricted mainly to microbotrophic, mycetophagous and entomogenous forms. Only in rare instances are vertebrate and higher plant parasites encountered. Nevertheless, it is still possible to construct fossil time lines and evolutionary clocks with the nematode lineages that do exist, obtain minimum dates when specific hosts were acquired, learn when certain types of behaviour appeared and study the co-evolution of nematode associates and parasites with their hosts.

Nematology Monographs & Perspectives, 2011, Vol. 9, 247-357

Appendix I

Taxonomic descriptions of previously known and new fossil nematode taxa *

Phylum Nematoda Potts, 1932
Class Enoplea Inglis, 1983
Subclass Enoplia Pearse, 1942
Order Enoplida Schuurmans Stekhoven & De Coninck, 1933
Family Palaeonematidae Poinar, Kerp & Hass, 2008

Genus *Palaeonema* Poinar, Kerp & Hass, 2008

Palaeonema phyticum Poinar, Kerp & Hass, 2008
(Figs 13-16; Plate 1)

DESCRIPTION (modified from Poinar *et al.*, 2008)

Medium-sized, slender nematodes up to 890 μm in length and 28 μm in diam.; cuticle with transverse striations; six slightly offset lips, each bearing a seta; amphids spiral-shaped; buccal cavity tubular, composed of a strongly sclerotised anterior portion bearing basal protuberances and non-sclerotised unarmed posterior portion; pharynx cylindrical, lacking isthmus and median bulb; terminal portion slightly expanded, with gland nuclei located at extreme base; nerve ring positioned at mid-pharynx; ovaries didelphic, reflexed; vulva located at *ca* 40% of body length; tail elongate, with swollen tip.

COMMENTS

Developing nematodes occurred in the stomatal and substomatal chambers of the host plant, *Aglaophyton major*. *Palaeonema* represented an early stage in plant parasitism and probably possessed a behaviour

* In this section, the lists of species from extant genera include only those known from fossils.

similar to present-day dyssaprobes that enter plants through stomata and use their buccal armature to tear open the cells. It represents the earliest known terrestrial herbivore.

TYPE MATERIAL

Deposited in the Forschungsstelle für Paläobotanik, Westfälische Wilhelms-Universität, Münster, Germany.

TYPE HOST

Early Devonian (396 mya) land plant, *Aglaophyton major* (Kidson & Lang).

TYPE LOCALITY

Rhynie Chert, Aberdeen, Scotland.

Family Palaeoenoploididae n. fam.

DIAGNOSIS

Medium sized nematodes with three large hooked mandibles; dorsal mandible doubly hooked; single-hooked ventral mandibles connected by two bar-like apophyses.

TYPE GENUS

Palaeoenoploides Mattavelli & Bracchi, 2008

TYPE SPECIES

P. alessandrellianus Mattavelli & Bracchi, 2008

Palaeoenoploides alessandrellianus **Mattavelli & Bracchi, 2008**
(Figs 22, 23)

DESCRIPTION

The generic and specific description is based on 172 specimens. Length 19 (8-46) mm, diam. 900 (300-2300) μm, chitinised stomatal apparatus composed of three hooked teeth, dorsal tooth bifid at tip, ventral teeth simple but with apophyses, spicules small, thin, rod-shaped.

COMMENTS

This type of buccal armature indicates a carnivorous habit for this species. All 172 specimens were dorso-ventrally or laterally compressed on slabs of yellow-greyish limestone. Both part and counterpart are preserved in 21 specimens. Occasionally, nematodes, polychaete annelids, decapod crustaceans and/or fish were preserved on the same slab.

TYPE MATERIAL

Holotype: MSNM (Museo Civico di Storia Naturale di Milano, Sezione di Paleontologia, C.so Venezia) i13285; paratype: MSNM i26242ab (part and counterpart, MSNM i24941 and MSNM i24908abI (part and counterpart).

TYPE HABITAT AND LOCALITY

Free-living marine predator. Haqel, Lebanon (Late Cretaceous).

Subclass Dorylaimia Inglis, 1983
Order Mermithida Hyman, 1951
Superfamily Mermithoidea Braun, 1883
Family Mermithidae Braun, 1883

Genus *Heydenius* Taylor, 1935

This is a collective group genus established for fossil mermithids that could not be placed in extant genera due to the absence or obstruction of pertinent characters. The genus *Heydenius* is now limited to fossil mermithids from the Tertiary period. The collective group genus *Cretacimermis* Poinar, 2003 was erected for Cretaceous mermithids. Since the majority of extant mermithid parasites are limited to hosts of a single arthropod family or group, the various species listed below are appropriately limited to fossil members of a particular family or group.

TYPE SPECIES

H. antiquus (von Heyden, 1860) Taylor, 1935*
= *Mermis antiqua* von Heyden, 1860

* Type by subsequent designation.

OTHER SPECIES

H. araneus Poinar, 2000
H. brownii Poinar, 2001
H. cecidomyae **n. sp.**
H. dipterophilus **n. sp.**
H. dominicus Poinar, 1984
H. formicinus Poinar, 2002
H. lamprophilus **n. sp.**
H. matutinus (Menge, 1866) Taylor, 1935
= *Mermis matutina* Menge, 1866
H. myrmecophila Poinar, Lachaud, Castillo & Infante, 2006
H. neotropicus **n. sp.**
H. psychodae **n. sp.**
H. quadristriatus (Menge, 1872) Taylor, 1935
= *Mermis quadristriata* Menge, 1872
H. saprophilus **n. sp.**
H. scatophilus **n. sp.**
H. sciarophilus **n. sp.**
H. tabanae **n. sp.**

Heydenius antiquus **(von Heyden, 1860) Taylor, 1935**
= *Mermis antiqua* von Heyden, 1860
(Fig. 60)

The single specimen was in the process of emerging from the rectum of a cerambycid beetle (Coleoptera: Cerambycidae).

DESCRIPTION (modified from von Heyden, 1860, 1862)

Length about 25 mm although a portion of the body still resides inside the beetle host.

COMMENTS

Mermithid parasitism of cerambycid beetles occurs today, although none of the nematodes appears to have been described (Poinar, 1975).

TYPE MATERIAL

Whereabouts unknown.

TYPE HOST

Hesthesis immortua Heyden (Coleoptera: Cerambycidae).

TYPE LOCALITY

Rhine lignite (Eocene brown coal).

Heydenius araneus Poinar, 2000
(Fig. 97)

A completely emerged mermithid adjacent to a spider in Baltic amber.

DESCRIPTION (modified from Poinar, 2000)

Specimen in a loose coil; head attenuated with rounded tip; tail bluntly rounded, slightly asymmetrical; tail appendage not visible (white deposit covering extremity); cuticle smooth, with some folding from fossilisation process. Length 24 mm; greatest body diam. 350 μm; a = 69.

COMMENTS

This is the only fossil mermithid parasite of an arachnid. Extant spiders are parasitised by mermithids throughout the world today and the life cycle is often indirect with larvae of aquatic insects serving as paratenic hosts (Poinar, 1987; Poinar & Early, 1990) (Fig. 98).

TYPE MATERIAL

Deposited in the Andrew Cholewinski amber collection, Redondo Beach, California.

TYPE HOST

A juvenile crab spider of the family Thomisidae (Arachnida).

TYPE LOCALITY

Northern Europe between 40 and 50 mya.

Heydenius brownii Poinar, 2001
(Fig. 77; Plate 12)

This mermithid was in the process of emerging from a Baltic amber planthopper. The tip of the head is missing and the posterior portion of the body is still inside the planthopper.

DESCRIPTION (modified from Poinar, 2001a)

White, cuticle smooth, exposed body portion 12 mm in length with the greatest diam. 400 μm.

COMMENTS

Soon after this description was published, the first extant mermithid-infected planthoppers were recovered from Europe (Helden, 2008) (Fig. 80).

TYPE MATERIAL

Deposited in the Poinar amber collection (accession no. N-3-51) maintained at Oregon State University.

TYPE HOST

A planthopper of the family Achiliidae (Hemiptera: Fulgoroidea).

TYPE LOCALITY

Baltic area in northern Europe between 40 and 50 mya.

Heydenius cecidomyae n. sp.
(Figs 91, 92)

Two specimens, one completely emerged and positioned adjacent to a gall gnat and the second with part of its body still in the haemocoel of the same host in Baltic amber.

Description

Head and tail rounded; completely emerged specimen 1.3 mm in length and 80 μm in diam. a = 16; specimen partly in host 1.2 mm in length and 63 μm in diam.

Comments

The specimen emerging from the fly leaves no doubt about the parasitic association. As far as could be determined, there are no extant reports of mermithids parasitising gall gnats.

Type material

Specimen (accession no. N-3-53) deposited in the Poinar amber collection maintained at Oregon State University.

Type host

A female gall gnat (length 917 μm) (Diptera: Cecidomyiidae).

Type locality

Baltic region some 40-50 mya.

Etymology

The specific name reflects the host family (Cecidomyiidae).

***Heydenius dipterophilus* n. sp.**
(Fig. 180)

The species is based on a single mermithid that completely emerged from and is still adjacent to an acalyptrate fly in Dominican amber.

Description

Head and tail rounded, length 4.4 mm; diam. 100 μm; a = 44.

COMMENTS

The fly host is tentatively identified as a member of the family Milichidae. There are no extant records of mermithids from this family of flies.

TYPE MATERIAL

Deposited in the Poinar amber collection (accession no. N-3-66) maintained at Oregon State University.

TYPE HOST

A tentative member (length 2.5 mm) of the family Milichidae.

TYPE LOCALITY

Amber mines located between Santiago and Puerto Plata in the Cordillera Septentrional of the northern portion of the Dominican Republic.

ETYMOLOGY

The specific name is composed from the host order (Diptera) and the Greek word 'philia' for love.

Heydenius dominicus Poinar, 1984
(Figs 213-215; Plates 26, 27)

Two mermithids adjacent to an adult mosquito in Dominican amber.

DESCRIPTION (modified from Poinar, 1984a)

Head blunt, truncated. Specimen 1 (with single coil): length 3.35 mm, diam. 50 μm, a = 68; length terminal tail appendage, 69 μm. Specimen 2 with double coil: length 3.41 mm; diam. 50 μm, a = 68; tail appendage composed of a proximal acute portion and a terminal filiform portion.

COMMENTS

A third specimen of *H. dominicus* was coiled in the abdomen of a mosquito (*Culex* sp., length 4.3 mm) in a second piece of Dominican amber (Fig. 213; Plate 27). Mermithid parasitism of mosquitoes occurs throughout the world today and several parasite and host genera are involved (Poinar, 1975, 1977b) (Fig. 216).

TYPE MATERIAL

Specimen 1 (accession no. DO-3201-W-1) and specimen 2 (accession no. DO-3201-W-2) deposited in the Museum Staatliches für Naturkunde in Stuttgart, Germany. Specimen 3 inside the mosquito deposited in the Poinar amber collection (accession no. N-3-42) maintained at Oregon State University.

TYPE HOST

Mosquitoes of the genus *Culex* (Diptera: Culicidae).

TYPE LOCALITY

Amber mines located between Santiago and Puerto Plata in the Cordillera Septentrional of the northern portion of the Dominican Republic.

Heydenius formicinus Poinar, 2002
(Figs 73, 74; Plate 11)

DESCRIPTION (modified from Poinar, 2002)

Specimen partially emerged from host ant (Fig. 73); body colour white in reflected light; cuticle smooth but body surface wrinkled and folded (preservation artefact); length of body outside host 8 mm; diam. 207 μm; head rounded, semi-hemispherical; mouth opening subterminal. Another specimen of *H. formicinus* (accession no. N-3-69) in Baltic amber is positioned adjacent to a worker ant (Fig. 74). This specimen is 13.7 mm in length and 142 μm in diam. The ant host is 2.5 mm in length.

COMMENTS

While the total length of *H. formicinus* cannot be established since part of the nematode still resides within the ant, it can be estimated on the basis of body thickness that approximately half of the nematode has emerged. Thus, the total length would be about 16 mm. A comparison of *H. formicinus* with extant ant mermithids shows a resemblance to the European species, *Allomermis myrmecophila* Baylis, 1921. Mermithid parasites of ants are widespread today (Poinar *et al.*, 2006b, 2007b) (Figs 75, 76).

TYPE SPECIMEN

Deposited in the Poinar amber collection (accession nos N-3-52 and N-3-69) maintained at Oregon State University.

TYPE HOST

Alate male of *Prenolepis henschei* Mayr, 1868.

TYPE LOCALITY

Baltic region some 40-50 mya.

Heydenius lamprophilus n. sp.

(Figs 209, 210; Plate 24B)

This species is based on a single specimen adjacent to an adult beetle of the family Lampyridae in Dominican amber. The tip of the beetle's abdomen is partially detached from the remainder of the body and it is the presumed exit point of the mermithid.

DESCRIPTION

The specimen is highly oxidised, which makes it difficult to see many details. Length 14.2 mm; diam. 320 μm; a = 44.

COMMENTS

The large size of the mermithid indicates development in a large host like the beetle. There are no reports of extant mermithid parasites of Lampyridae (Poinar, 1975).

TYPE MATERIAL

Deposited in the Poinar amber collection (accession no. N-3-17) maintained at Oregon State University.

TYPE HOST

A beetle (length 5.8 mm) of the family Lampyridae (Coleoptera).

TYPE LOCALITY

Amber mines located between Santiago and Puerto Plata in the Cordillera Septentrional of the northern portion of the Dominican Republic.

ETYMOLOGY

The specific name reflects the family name of the host beetle (Lampyridae) and the Greek word 'philia' for love.

***Heydenius matutinus* (Menge, 1866) Taylor, 1935**
= *Mermis matutina* Menge, 1866
(Figs 61-70; Plates 8-10)

This species was originally based on three mermithids associated with a chironomid in Baltic amber. One specimen was partly emerged from the host while the other two had completely emerged. This taxon is considered a collective species for all mermithids parasitising midges of the family Chironomidae in Baltic amber. While there are probably several different genera involved, it is impossible to distinguish between them when only parasitic and post-parasitic juveniles are available.

DESCRIPTION (modified from Menge, 1866)

Body cylindrical, yellowish, head and tail bluntly pointed. The measurements were provided for only one specimen, which was 3.5 mm in length and 100 μm in diam.

TYPE SPECIMEN

Whereabouts unknown.

TYPE HOST

Adult members of the family Chironomidae (Diptera) in Baltic amber.

TYPE LOCALITY

Baltic region some 40-50 mya.

ADDITIONAL VOUCHER SPECIMENS (from the Poinar amber collection)

N-3-15A. A single mermithid (length 3.5 mm, diam. 113 μm, a = 31) partly emerged from its chironomid host (length 2.3 mm).
N-3-15C. A single mermithid (length 3.4 mm, diam. 123 μm, a = 28) partly emerged from its midge host (length 1.2 mm) (Fig. 65).
N-3-15E. Two mermithids, one (length 1.3 mm, diam. 77 μm, a = 17) partly emerged from the midge host (length 980 μm) and the second (length 420 μm, diam. 56 μm, a = 7.5) completely emerged from the host (Fig. 66).
N-3-35A. A single mermithid (length 5.4 mm, diam. 106 μm, a = 51) completely emerged from its midge host (length 2.4 mm). Two parasitic mites are also present. One (length 356 μm) (arrow) is adjacent to the fly while the second (length 320 μm) is attached to the tip of the abdomen of the midge (Fig. 64).
N-3-35B. Two mermithids (lengths 8.6 mm and 6.4 mm) completely emerged from their adjacent midge host (length 2.0). A parasitic mite (length 288 μm) is also adjacent to the chironomid (Fig. 69).
N-3-54A. A single mermithid (length 2.5 mm) partly emerged from its chironomid host (length 945 μm) (Fig. 67).
N-3-54B. A single mermithid (length 4.1 mm) partly emerged from its midge host (length 1.6 mm).
N-3-56A. A single mermithid (length 2.5 mm, diam. 63 μm, a = 40) completely emerged from its midge host (length 770 μm) and a second specimen still inside the host (Fig. 68).
N-3-56B. A single mermithid (1.6 mm) partly emerged from its host (length 1.2 mm).
N-3-76. A single mermithid (length 1.4 mm, diam. 63 μm, a = 22) behind its chironomid host (length 1.4 mm). The nematode is still connected by a strand of congealed haemolymph to its host (Fig. 70; Plate 10). This association was noted previously (Poinar & Malgarin, 2007).

There are additional specimens of *H. matutinus* in various private and public collections around the world since this is probably the most common fossil mermithid. Two specimens are in the Copenhagen

Zoological Museum amber collection (Fig. 63) and another, illustrated by Schlee and Glockner (1978; see their Table 11), is in the Stuttgart amber collection.

Heydenius myrmecophila **Poinar, Lachaud, Castillo & Infante, 2006**
(Figs 110, 111)

This species is based on a parasitic juvenile adjacent to a worker ant in Dominican amber.

DESCRIPTION (modified from Poinar *et al.*, 2006b)

Body completely opaque, internal structures not visible; length 1.9 mm; diam. 56 μm; a = 34; head truncate; tail attenuated, pointed at tip.

COMMENTS

The nematode probably emerged from a wound on the gaster of the ant. Some 18 neotropical ant species are known hosts of mermithids and certainly many more exist (Poinar *et al.*, 2006b).

TYPE MATERIAL

Parasitic juvenile in Dominican amber deposited in the Poinar amber collection (accession no. N-3-65) maintained at Oregon State University.

TYPE HOST

Linepithema sp. (Hymenoptera: Formicidae).

TYPE LOCALITY

Amber mines located between Santiago and Puerto Plata in the Cordillera Septentrional of the northern portion of the Dominican Republic.

Heydenius neotropicus **n. sp.**
(Figs 211, 212; Plate 25)

This species is erected for mermithid parasites of chironomid midges in Dominican amber.

The description is based on mermithids parasitising midges in two pieces of Dominican amber.

DESCRIPTION

Amber piece N-3-13A contains two mermithids partly emerged from the fly host. These are fairly mature parasitic juveniles that were probably ready to emerge naturally from the host. Head and tail bluntly pointed; length 2.3 mm, diam. 60 μm, a = 38 for specimen 1 and length 2.8 mm, diam. 60 μm, a = 47 for specimen 2. Amber piece N-3-13B contains a midge parasitised by four mermithids, three of which exited the host and are adjacent to it in the amber; the fourth specimen is still in the body cavity of the midge. Head rounded to truncate; head appears to be slightly swollen in two specimens; tail rounded; body length, diam. and ratio a; 1.4 mm, 50 μm, a = 28; 1.6 mm, 44 μm, a = 36; 1.2 mm, 41 μm, a = 29; 0.9 mm, 38 μm, a = 24, respectively.

COMMENTS

Midge parasitism by mermithids in Dominican amber appears to be not as common as in Baltic amber and thus far no records occur in Mexican amber. However, it is obvious from the variety of mermithids in extant chironomid midges (Poinar, 1975) (Fig. 38) that these parasitic associations have been maintained over millions of years.

TYPE MATERIAL

Parasitic juveniles in Dominican amber deposited in the Poinar amber collection (accession no. N-3-13A, N-3-13B) maintained at Oregon State University.

TYPE HOST

Members of the family Chironomidae (Diptera).

TYPE LOCALITY

Amber mines located between Santiago and Puerto Plata in the Cordillera Septentrional of the northern portion of the Dominican Republic.

ETYMOLOGY

The specific name reflects the geographical region of the fossil.

Heydenius psychodae n. sp.
(Figs 172, 173)

This species is erected for mermithid parasites of members of the family Psychodidae (mothflies). It is based on two completely emerged mermithids adjacent to a mothfly in Dominican amber.

DESCRIPTION

Head attenuated but rounded at tip; tail rounded; length, diam. and ratio a; 1.6 mm, 57 μm, a = 28, and 1.1 mm, 63 μm, a = 17, respectively.

COMMENTS

The size, shape and apparent juvenile state of the parasites are all indications of mermithid nematodes. These were apparently close to maturity and ready to leave the host. As far as could be determined, there are no records of extant mermithids parasitising mothflies.

TYPE MATERIAL

Two parasitic juveniles in Dominican amber deposited in the Poinar amber collection (accession no. N-3-57B) maintained at Oregon State University.

TYPE HOST

An adult female (length 1.3 mm) of the family Psychodidae (Diptera).

TYPE LOCALITY

Amber mines located between Santiago and Puerto Plata in the Cordillera Septentrional of the northern portion of the Dominican Republic.

ETYMOLOGY

The specific name is based on the host family (Psychodidae).

Heydenius quadristriatus **(Menge, 1872) Taylor, 1935**
= *Mermis quadristriata* Menge, 1872
(Figs 71, 72)

This specimen was described by Menge (1872) in Baltic amber without any association with an invertebrate host.

DESCRIPTION (modified from Menge, 1872)

Large, white nematode, 60 mm in length and 1 mm in diam.; a = 60; cuticle completely smooth except for four thread-like black lines occurring over surface of body from one end to the other.

COMMENTS

In the same piece of amber with the mermithid were the remains of an ant and a phalangid although, due to the large size of the nematode, Menge (1872) felt that the host must have been a large orthopteran or caterpillar. This specific name is used here for all mermithids in Baltic amber that have no associated insect host.

TYPE MATERIAL

Whereabouts unknown. Three additional specimens: N-3-15A, B and C (A and B shown in Fig. 74) deposited in the Poinar amber collection. Measurements: specimen A, length 6.1 mm; diam. 80 μm; a = 76; specimen B, length 5.8 mm; diam. 71 μm; a = 81; specimen C, length 5.4 mm; diam. 67 μm; a = 81. Two additional specimens (lengths, 7.0 and 7.3 mm) are illustrated in Weitschat and Wichard (2002; their Plate 7d, f).

HOST

None evident.

TYPE LOCALITY

Baltic region between 40 and 50 mya.

Heydenius saprophilus n. sp.
(Figs 223, 224)

This species is erected for mermithid parasites of members of the family Anisopodidae. It is based on a single mature parasitic juvenile adjacent to a wood gnat in Dominican amber.

Description

Body covered with brownish deposit. Head and tail attenuated; length 5.1 mm; diam. 99 μm; a = 52; length of tail spike 27 μm.

Comments

There are no reports of extant mermithids parasitising anisopodids. This species is also represented by two parasitic juveniles adjacent to a *Mycetobia* anisopodid in Dominican amber deposited in the American Museum of Natural History collection (Grimaldi, 2003). This species probably had a direct cycle as is typical of mermithid parasites of Diptera. It indicates that female anisopodids in the Dominican amber forest returned to preferred habitats for oviposition and such habitats contained infective juveniles of *H. saprophilus* **n. sp.**

Type material

Deposited in the Poinar amber collection (accession no. N-3-102) maintained at Oregon State University.

Type host

Representative (length 5.1 mm) of the family Anisopodidae (Diptera).

Type locality

Amber mines located between Santiago and Puerto Plata in the Cordillera Septentrional of the northern portion of the Dominican Republic.

Etymology

The specific name is taken from the Greek 'sapros' for decayed, indicating the developmental habitat of the host larvae.

***Heydenius scatophilus* n. sp.**
(Figs 174, 175)

This species is erected for mermithid parasites of members of the family Scatopsidae. It is based on five specimens near a large scatopsid fly in Dominican amber.

DESCRIPTION

Head and tail rounded, lengths 4.7, 4.6, 3.6, 4.2 and 2.7 mm; diam. 95, 104, 45, 45 and 45 μm, respectively.

COMMENTS

Although the nematodes are not adjacent to the fly, the body cavity of the fly is hollow and the scatopsid is presumed to have been the host. No extant mermithids have been reported from members of this host family.

TYPE MATERIAL

Deposited in the Poinar amber collection (accession no. N-3-8) maintained at Oregon State University.

TYPE HOST

Adult (length 2.6 mm) of the family Scatopsidae (Diptera).

TYPE LOCALITY

Amber mines located between Santiago and Puerto Plata in the Cordillera Septentrional of the northern portion of the Dominican Republic.

ETYMOLOGY

The specific name is based on the name of the host family (Scatopsidae) and the Greek work ‘philia’ for love.

Heydenius sciarophilus n. sp.
(Figs 79-83)

This species is erected for mermithid parasites of members of the family Sciaridae (Diptera) in Baltic amber. It is based on 14 specimens associated with adult sciarids in five separate pieces of Baltic amber.

DESCRIPTION

Amber piece no. N-3-46 (Fig. 79), with one large and one smaller specimen that had just emerged from the fly host. Head rounded in both specimens, tail rounded in larger specimen, tail pointed in smaller specimen; length larger specimen 2.8 mm, diam. larger specimen 88 μm; a = 32: length smaller specimen 1.4 mm; diam. smaller specimen 63 μm; a = 22: length of fly host 1.7 mm.

Amber piece no. N-3-59 (Fig. 80). One mermithid emerging from host; head pointed; emerged portion, length 1.9 mm; diam. 63 μm; a = 30; length of fly host 2.0 mm.

Amber piece N-3-15B (Fig. 81). A single mermithid; length 2.4 mm; diam. 114 μm; a = 21; completely emerged from its sciarid host (length 1.6 mm).

Amber piece N-3-84 (Fig. 82; Plate 13). Four mermithids, three free in the amber and one partially emerged; head and tail rounded except for the smallest one, which has a pointed head; length and diam. of partially emerged specimen 1, 1.1 mm and 41 μm; specimen 2, length 3.0 mm; diam. 58 μm; a = 52; specimen 3, length 2.9 mm; diam. 54 μm, a = 54; specimen 4, length 1.1 mm, diam. 41 μm; a = 27. Length of fly host 2.1 mm.

Amber piece N-3-85 (Fig. 83). Five mermithids, three adjacent to the fly, one of which has the anterior portion polished away and two smaller mermithids in the amber but some distance away from the fly. Head and tail rounded on all specimens. Length, diam. and ratio a of those adjacent to the fly, 2.6 mm, 55 μm, 47; 2.4 mm, 63 μm, 38; 1.6 (partial specimen), 63 μm; lengths, diam. and ratio a of other two specimens, 2.1 mm, 95 μm, 22 and 1.9 mm, 110 μm, 17; length of fly host 1.8 mm.

COMMENTS

While it is impossible to know if all these specimens belong to the same species, they are treated so here since diagnostic characters are obscured. They establish the presence of parasitism of sciarids by mermithoid nematodes in the Early Tertiary. Consideration was given to the possibility that some of these are tetradonematids since they appear to be more common in sciarids than mermithids today. However, in none of the specimens was there any evidence of egg development, which is characteristic of tetradonematids. Only a single described mermithid, *Orthomermis oedobranchus* Poinar, 1964, is known from sciarids today (Fig. 84) and this mermithid also parasitises a chironomid midge and a beetle in the same habitat (Poinar, 1964). An additional specimen of *H. sciarophilus* in Baltic amber is illustrated in Weitschat and Wichard (2002: their Plate 7c).

TYPE HOST

Sciarid fungus gnats (Diptera: Sciaridae).

TYPE LOCALITY

Baltic area some 40-50 mya.

ETYMOLOGY

The specific name reflects the host family (Sciaridae) and the Greek word 'philia' for love.

Heydenius tabanae n. sp.

This species is based on one specimen reported by Grabenhorst (1985) emerging from the body of an adult horsefly in Pliocene sedimentary deposits dated at approximately 2.5 mya.

DESCRIPTION (modified from Grabenhorst, 1985)

While only 13 mm of the body of the nematode is protruding from the host, the total length was estimated at *ca* 28 mm. Body diam. is 0.5 mm. The colour is dark brown but becomes paler near the extremities.

COMMENTS

There are a number of extant mermithid parasites of horseflies and deerflies worldwide (Poinar, 1985) (Fig. 256). The interesting feature of this fossil is that it is emerging from an adult horsefly, while all extant cases are larval parasitism. Of course, many mermithids that initially infect the larvae of holometabolous insects pass through the pupal stage into the adult. Mermithid parasites of horseflies can have either direct or indirect developmental cycles.

TYPE MATERIAL

Deposited in the Grabenhorst fossil collection (accession no. 2-271), Paul-Gerhard-Schule, D-3354 Dassel, Germany.

TYPE HOST

The extant horsefly species, *Tabanus sudeticus* Zeller (Diptera: Tabanidae).

TYPE LOCALITY

Late Pliocene of Willershausen, Kreis Osterode, Germany (2.5 mya).

ETYMOLOGY

The specific name reflects the name of the host family (Tabanidae).

Genus *Cretacimermis* Poinar, 2001

Cretacimermis was established as a collective genus for mermithids from the Cretaceous that cannot be identified with an existing genus or completely described due to the obscurity of diagnostic characters (Poinar, 2001b).

TYPE SPECIES

C. libani (Poinar, Acra & Acra, 1994) Poinar, 2001
= *Heleidomermis libani* Poinar, Acra & Acra, 1994

OTHER SPECIES

C. chironomae **n. sp.**
C. protus Poinar & Buckley, 2006

Cretacimermis chironomae n. sp.

This species is based on two mermithids associated with an adult chironomid in Burmese amber. One of the nematodes has completely exited from the host while the other has a short section of its body inside the insect's abdomen.

DESCRIPTION (modified from Grimaldi *et al.*, 2002)

Length 3.3 and 3.4 mm; tail pointed.

COMMENTS

The original description of these fossils (Grimaldi *et al.*, 2002) contains many errors. Not only did the authors confuse the head with the tail but they also described an 'array of spicules' emerging from the anterior portion of the nematode. These 'spicules' appear to be artefacts in the amber. Grimaldi *et al.* (2002) also state that the diam. is 50 mm, which is clearly incorrect – it is likely that they meant 50 μm.

TYPE SPECIMEN

American Museum of Natural History (accession no. Bu 320).

TYPE HOST

An adult Chironomidae (Diptera).

TYPE LOCALITY

Amber mine in the Hukawng Valley, southwest of Maingkhwan in the state of Kachin (26°20′N, 96°36′E), northern Myanmar (Burma).

ETYMOLOGY

The specific name is based on the host family (Chironomidae).

Cretacimermis libani (Poinar, Acra & Acra, 1994) Poinar, 2001

= *Heleidomermis libani* Poinar, Acra & Acra, 1994
(Fig. 25; Plate 3)

A mermithid coiled in the abdominal cavity of a chironomid midge in Lebanese amber.

DESCRIPTION (modified from Poinar *et al.*, 1994)

Length 6.0 mm; diam. 120 μm; a = 60; tail appendage absent.

COMMENTS

The mermithid was moulting to the adult stage while still in the body cavity of the adult midge. This behaviour is rare but does occur in *Hydromermis contorta* (Kohn), an extant parasite of chironomid midges (Poinar & Tourenq, 1972) and in extant species of *Heleidomermis* Rubstov, parasites of biting midges.

TYPE HOST

Adult midge (Diptera: Chironomidae).

TYPE MATERIAL

Deposited in the Acra collection of Lebanese amber, Beirut.

TYPE LOCALITY

Between Jezzine and Dar-al-Baidhar in southern Lebanon. Beds originate from the Neocomian division of the Early Cretaceous some 130-135 mya.

Cretacimermis protus Poinar & Buckley, 2006

(Fig. 39; Plate 6)

This species was described from a moulting post-parasitic juvenile emerging from a biting midge in Burmese amber.

DESCRIPTION (modified from Poinar & Buckley, 2006)

White, partially transparent; length 4.3 mm; diam. 46 μm; a = 93; head narrowed; shed third-stage cuticle smooth, lacking cross fibres, extending out over head; mouth opening on third-stage cuticle slightly subventral; tail pointed; tail appendage not observed.

COMMENTS

The partially shed cuticle, slightly subventral mouth and absence of a tail appendage suggest that this species belongs to the *Heleidomermis* clade. Species of *Heleidomermis* moult, mate and initiate egg development in the body cavity of their biting midge hosts (Poinar & Monteys, 2008).

TYPE HOST

A member of the family Ceratopogonidae (Diptera).

TYPE MATERIAL

Deposited in the amber collection of Ron Buckley, Florence, Kentucky.

Locality: Amber mine in the Hukawng Valley, southwest of Maingkhwan in the state of Kachin (26°20′N, 96°36′E), northern Myanmar (Burma).

NOTE

Two additional specimens adjacent to a biting midge in another piece of Burmese amber are now considered to belong to this species. They were described as *Cretacimermis* sp. in Poinar and Monteys (2008).

DESCRIPTION (modified from Poinar & Monteys, 2008)

Both specimens are darkened from the fossilisation process; lengths 2.7 and 2.5 mm; heads and tails tapered; larger specimen with faint spicule 43 μm from tail tip; spicule length 12 μm; cloacal body diam. 16 μm; male tail surrounded by loose cuticle.

COMMENTS

The observed characters and ceratopogonid hosts suggest that these specimens are also basal members of the *Heleidomermis* clade.

TYPE MATERIAL

Deposited in the amber collection of Ron Buckley, Florence, Kentucky.

TYPE HOST

Biting midge (Diptera: Ceratopogonidae).

TYPE LOCALITY

Amber mine in the Hukawng Valley, southwest of Maingkhwan in the state of Kachin (26°20′N, 96°36′E), northern Myanmar (Burma).

Another two specimens of *C. protus* that were emerging from a biting midge in Burmese amber are shown in Figure 40. The exposed portions of these specimens are 492 μm and 738 μm in length and both are 54 μm in diam. Unfortunately, their poor state of preservation obscures other characters.

TYPE MATERIAL

Deposited in the Poinar amber collection (accession no. B-N-3) maintained at Oregon State University.

TYPE HOST

Biting midge (Diptera: Ceratopogonidae).

TYPE LOCALITY

Amber mine in the Hukawng Valley, southwest of Maingkhwan in the state of Kachin (26°20′N, 96°36′E), northern Myanmar (Burma).

Family Tetradonematidae Cobb, 1919

Genus *Myrmeconema* Poinar & Yanoviak, 2008

TYPE SPECIES

Myrmeconema neotropicum Poinar & Yanoviak, 2008

OTHER SPECIES

Myrmeconema antiqua **n. sp.**

Myrmeconema antiqua n. sp.
(Figs 113-115)

This species is represented by hundreds of eggs and a few hatched juveniles in a piece of Dominican amber containing an injured ant of the genus *Cephalotes* Latreille.

DESCRIPTION

Eggs spherical to slightly oval, smooth, max. diam. (n = 12) 36 (30-56) μm; juveniles emerging from or adjacent to egg (n = 4); head bluntly rounded; tail pointed, length 132 (122-149) μm; diam. 10 (9-11) μm; a = 13.

COMMENTS

The identification and systematic placement of this species is based on the extant *Cephalotes* parasite, *Myrmeconema neotropicum* Poinar & Yanoviak, 2008 (Yanoviak *et al.*, 2008) (Figs 116, 117). The parasitised *Cephalotes* ant in the amber has a lighter coloured area on its abdomen and a hole in its gaster through which the nematode eggs were released into the resin. The hole in the ant's abdomen could have been caused by a bird that mistook the red gaster for a fruit and while attempting to swallow it dropped the ant in a pool of resin. The smaller eggs and juveniles of *M. antiqua* **n. sp.** distinguish this species from the extant *M. neotropicum*.

TYPE MATERIAL

Deposited in the Poinar amber collection (accession no. N-3-75) maintained at Oregon State University.

TYPE HOST

Cephalotes serratus (Vierberger & Scheven) (Hymenoptera: Formicidae).

TYPE LOCALITY

Amber mines located between Santiago and Puerto Plata in the Cordillera Septentrional of the northern portion of the Dominican Republic.

Palaeotetradonema **n. gen.**

This genus is erected as a collective genus for fossil members of the family Tetradonematidae that cannot be further identified.

TYPE SPECIES

P. phlebotomae **n. gen., n. sp.**

Palaeotetradonema phlebotomae **n. gen., n. sp.**

(Figs 254, 255)

This species is represented by three immature females adjacent to a sand fly (Phlebotomidae) in Mexican amber.

DESCRIPTION

Head attenuated but rounded at tip; tail rounded; length, diam. and ratio a, respectively; specimen 1: 720 μm, 27 μm, 27; specimen 2: 652 μm, 27 μm, 24; specimen 3: 720 μm, 32 μm, 23; developing ova in specimen 2 range from 8 to 12 μm in diam.

COMMENTS

The large diameter and developing ova place the nematodes in the Tetradonematidae. Developing ova in the extant sand fly tetradonematid, *Didilia ooglypta* (Tang *et al.*, 1993), range from 9 to 17 μm, which is within the size range noted in the female fossil.

TYPE MATERIAL

Three parasitic juveniles in Mexican amber deposited in the Poinar amber collection (accession no. N-3-49) maintained at Oregon State University.

TYPE HOST

A female sand fly (length 1.1 mm) of the family Phlebotomidae (Diptera).

TYPE LOCALITY

Chiapas, Mexico. La Quinta Formation, Late Oligocene-Early Miocene.

ETYMOLOGY

The specific name is based on the name of the host family (Phlebotomidae).

Palaeotetradonema sciarae n. sp.
(Fig. 85)

The description is based on one specimen that is almost completely emerged from a sciarid fungus gnat in Baltic amber.

DESCRIPTION

Tail and head rounded, length 4.6 mm; diam. 92 μm; a = 50; body cavity containing developing ova ranging from 12 to 15 μm in diam.

COMMENTS

This specimen is similar in appearance to the mermithoid, *Heydenius sciarophilus* **n. sp.**; however, because it contains developing ova, it is described in the family Tetradonematidae. At least two extant species of tetradonematids attack sciarids, *Tetradonema plicans* Cobb (Ferris & Ferris, 1966) and *Aproctonema entomophagum* Keilin (Keilin & Robinson, 1933).

TYPE MATERIAL

Deposited in the Poinar amber collection (accession no. N-3-62) maintained at Oregon State University.

TYPE HOST

Adult sciarid fungus gnat (Sciaridae) in Baltic amber; length of host 1.7 mm.

TYPE LOCALITY

Baltic region some 40-50 mya ago.

ETYMOLOGY

The specific name is based on the name of the host family (Sciaridae).

Order Chromadorida Chitwood, 1933
Suborder Chromadorina Filipjev, 1917
Superfamily Chromadorina Filipjev, 1929

Family Nemavermidae n. fam.

DIAGNOSIS

Large, stout nematodes with setae covering body; head bearing fleshy protuberances.

TYPE GENUS

Nemavermes Schram, 1973

TYPE SPECIES

N. mackeei Schram, 1973

Nemavermes mackeei Schram, 1973
(Figs 17, 18)

The description is based on several specimens from several private and public collections.

DESCRIPTION (modified from Schram, 1973, 1979)

Length (n = 4) 118 (100-140) mm; diam. (n = 4) 5.5 (5.0-6.0) mm; a = 21; tail pointed to somewhat rounded; head rounded, bearing elongate labial papillae; fairly long setae (bristles) widely scattered over entire body surface.

COMMENTS

The fleshy, elongate structures protruding from the head of some specimens are a diagnostic character of the family. Their purpose is unknown, but it is possible that they are sensory organs used to probe the bottom in search of food.

TYPE MATERIAL

Field Museum of Natural History, Chicago (accession no. PE 21551).

TYPE HABITAT

Offshore marine environment with possible freshwater inflows. Detritus-filled guts in several specimens suggest that they were deposit feeders.

TYPE LOCALITY

Peabody Coal Company, Pit II, Will, Greenly and Kankakee counties, Illinois. Middle Pennsylvanian Carbondale formation, Francis Creek Shale, Essex biota (320 mya). In a later study, Schram (1979) found nine additional specimens in Bear Gulch Limestone, Uppermost Mississippian (Namurian) in central Montana (310 mya).

Order Desmoscolecida Filipjev, 1929
Superfamily Desmoscolecoidea Shipley, 1896

Family Eophasmidae n. fam.

DIAGNOSIS

Elongate forms with coarsely annulated cuticle; body bearing short setae.

TYPE GENUS

Eophasma Arduini, Pinna & Teruzzi, 1983

TYPE AND ONLY SPECIES

E. jurasicum Arduini, Pinna & Teruzzi, 1983

Eophasma jurasicum Arduini, Pinna & Teruzzi, 1983
(Fig. 19; Plate 2A)

DESCRIPTION (modified from Arduini *et al.*, 1983)

Based on six compression fossil specimens in sedimentary deposits. Elongate, vermiform body; cuticle coarsely annulated; some short setae in anterior and posterior regions; head truncate; buccal region with possible denticles; pharynx short; intestine faint, represented by dark stain; possible valvular area between pharynx and intestine; dimensions of holotype: length *ca* 55 mm; max. diam. *ca* 2 mm; a = 28; posterior portion of body tapering to an acute, flexible terminus.

COMMENTS

Eophasma is placed in the Desmoscolecoidea on the basis of its coarsely annulated cuticle.

TYPE MATERIAL

Holotype (accession no. i784) and paratypes (accession nos i783, i785, i786, i790, i1685) deposited in the Museo di Storia Naturale, Milan.

TYPE HABITAT

Free-living marine nematode that probably functioned as a detritivore.

TYPE LOCALITY

Como, Italy, in the Lower Sinemurian of the Early Jurassic.

Order Plectida Malakhov, 1982

Family Captivonemidae n. fam.

DIAGNOSIS

Small forms with finely annulated cuticle and irregularly positioned long somatic setae.

TYPE GENUS

Captivonema Boström *in* Manun, Bose, Sayer & Boström, 1994

TYPE AND ONLY SPECIES

C. cretacea Boström *in* Manun, Bose, Sayer & Boström, 1994

Captivonema cretacea Boström *in* Manun, Bose, Sayer & Boström, 1994

(Figs 20, 21; Plate 2B)

DESCRIPTION (modified from Manun *et al.*, 1994)

Vermiform fossils with unsegmented but finely annulated cuticle (7-8 annules every 10 μm) and prominent, irregularly distributed setae ranging up to 25 μm in mid-body region: setae absent from anterior and posterior ends: length 0.5 mm; diam. 20 μm; a = 25; body tapering posteriorly to pointed terminus; anterior end slightly tapering; gut about one-third body diam. and extending for about half body length.

COMMENTS

The small body size, fine cuticular annulations and dense covering of long, somatic setae are the main diagnostic characters for this family. Its manner of preservation is unique since the nematode was embedded in the wall of an Early Cretaceous clitellate annelid cocoon.

Extant nematodes with an annulated cuticle and dense coating of setae are *Greeffiella* Cobb, 1922 and *Epsilonema* Steiner, 1927. However, there are not enough characters present to align *Captivonema* with either of these lineages.

TYPE HABITAT

Free-living in freshwater wetlands (bogs, swamps).

TYPE MATERIAL

Palaeontological Museum, Oslo (PMO no. 120.864).

TYPE LOCALITY

Bohemanflya, Spitsbergen, Norway: Helvetiafjellet Formation, Glitrefjellet Member (Barremian – 127 mya).

Family Plectidae Örley, 1880

Genus *Oligoplectus* Taylor, 1935

Oligoplectus succini (von Duisburg, 1862) Taylor, 1935
=*Anguillula succini* von Duisburg, 1862
(Fig. 47)

The collective genus *Oligoplectus* Taylor, 1935 was erected for fossil members of the Plectidae that could not be identified to genus. This species, the type and only nominal member of the genus, was based on a single specimen first described by von Duisburg in 1862. It was later transferred to *Oligoplectus* by Taylor (1935).

DESCRIPTION (modified from von Duisburg, 1862 and Taylor, 1935)

Head knob-like; tail curved ventrally, pointed at tip; cuticular annulations 2 μm apart; length 855 μm; diam. 16 μm; a = 53.

TYPE SPECIMEN

Whereabouts unknown.

HABIT

Unknown.

TYPE LOCALITY

Baltic region some 40-50 mya.

Order Rhabditida Chitwood, 1933
Suborder Spirurina Railliet & Henry, 1915
Infraorder Oxyuridomorpha De Ley & Blaxter, 2002
Superfamily Thelastomatoidea Travassos, 1929
Family Thelastomatidae Travassos, 1929

Genus *Paleothelastoma* n. gen.

This is a collective group genus for fossil members of the family Thelastomatidae.

TYPE AND ONLY SPECIES

P. tipulae **n. gen., n. sp.**

Paleothelastoma tipulae n. gen., n. sp.
(Figs 33, 34; Plate 5)

The description is based on a single specimen adjacent to an adult limoniid cranefly in Burmese amber.

DESCRIPTION

Outstretched, with slender, elongate tail and truncate head; length 376 μm; diam. 25 μm; a = 15; cuticular annulations *ca* 10 μm apart in anterior half of body; cylindrical pharynx 123 μm long; slight constriction at base of pharynx possibly representing separation of isthmus from basal bulb; tail length 61 μm; anal body diam. 11 μm.

COMMENTS

The fossil is probably a pre-adult male related to the extant thelastomatid genus *Cephalobellus* Cobb, 1920, although representatives of three other extant genera of thelastomatids also parasitise tipulids (Leibersperger, 1960; Dale, 1965; Skrjabin *et al.*, 1966; Jarry, 1969; Vojtková, 1971). The males of the tipulid parasite, *Cephalobellus tipulae*

tipulae Leibersperger, 1960 range from 370 to 1330 μm in length, which overlaps with that of the fossil. While there is a suggestion of a single straight spicule in the genital chamber of *P. tipulae* **n. gen., n. sp.**, there is no evidence of genital papillae or lateral alae, which suggests that the specimen is an immature male rather than a juvenile female. The head is actually more slender than it appears in the photo since the specimen obviously moved its head in the resin, resulting in a wider head diameter.

While four genera of thelastomatids occur in tipulid larvae, none has been reported from adult flies. However, thelastomatids do occur in adult Hydrophilidae and Passalidae (Coleoptera) (Jarry, 1964).

TYPE SPECIMEN

Deposited in the Poinar amber collection (accession no. B-N-4) maintained at Oregon State University.

TYPE HOST

An adult female of the genus *Gonomyia* Meigen, 1818 (Diptera: Limoniidae). The larvae develop in sandy or rich organic soil near water sources and feed on detritus.

TYPE LOCALITY

Amber mine in the Hukawng Valley, southwest of Maingkhwan in the state of Kachin (26°20′N, 96°36′E), northern Myanmar (Burma).

ETYMOLOGY

The specific name is based on the superfamily of the host (Tipuloidea), which includes members of the families Cylindrotomidae, Limoniidae, Pediciidae and Tipulidae.

Infraorder Spiruromorpha De Ley & Blaxter, 2002
Superfamily Filarioidea Chabaud & Anderson, 1959
Family Filariidae Chabaud & Anderson, 1959

Genus *Cascofilaria* n. gen.

This genus is erected as a collective group genus for fossil nematodes of the family Filariidae.

TYPE SPECIES

C. baltica **n. gen., n. sp.**

OTHER SPECIES

C. dominicana **n. gen., n. sp.**
C. parva **n. gen., n. sp.**

***Cascofilaria baltica* n. gen., n. sp.**
(Fig. 96)

This species is based on a single specimen adjacent to a female blackfly in Baltic amber.

DESCRIPTION

Bluntly rounded head; pointed tail with short terminal spine; anterior portion of body coiled; length 294 μm; diam. 12 μm; a = 25; specimen covered with sheath or air bubble; coiled head region hampers detailed observations.

COMMENTS

It is extremely rare to find nematodes associated with blackflies in Baltic amber, even though extant members of this group are hosts to mermithid and filarial nematodes. Free-living nematodes have never been reported from blackflies (Poinar, 1975). Based on the shape and size of this fossil, it is considered to represent a microfilaria. Extant blackflies vector filarial nematodes of the genera *Onchocerca* Diesing, 1841 and *Splendidofilaria* Skrjabin, 1923. Microfilaria of the 27 species of *Onchocerca*, a worldwide genus infecting mammals, range from 190 to 395 μm in length and have rounded heads and pointed tails (Anderson, 2000). It is likely that the blackfly, whose abdomen is extended, fed on a mammal just before flying into the resin and the microfilaria was attached to the insect's mouthparts.

TYPE MATERIAL

Deposited in the Poinar amber collection (accession no. Sy-1-172) maintained at Oregon State University.

TYPE HOST

A female blackfly (Diptera: Simuliidae) (length 2.3 mm).

TYPE LOCALITY

Baltic region some 40-50 mya.

ETYMOLOGY

The generic name is from the Latin 'cascus' for old. The specific name is from the Latin 'filum' for thread, which is used as a suffix for certain extant nematode genera.

Cascofilaria dominicana **n. gen., n. sp.**
(Figs 217, 218)

Two nematodes representing different stages of a filarial parasite are adjacent to the head of a female *Culex* mosquito in Dominican amber.

DESCRIPTION

Nematode A (Fig. 217). Short, wide body, length 238 μm; diam. 23 μm; a = 10; head rounded, body of uniform thickness except for narrow tail spur, 41 μm in length.

COMMENTS

This is considered to be a pre-infective stage of a filarial nematode close to the genus *Foleyella* or *Waltonia*. Extant members of these genera develop in the haemocoel of mosquitoes and infect frogs and toads. The pre-infectives of *F. duboisi* range from 230 to 650 μm in length and have a tail stub similar to that of the fossil (Witenberg & Gerichter, 1944). Pre-infectives are active and it is presumed that the nematode emerged prematurely from the adjacent mosquito. A cell attached to the body of the nematode could represent a mosquito haemocyte.

DESCRIPTION

Nematode B (Fig. 218). Head and tail attenuated but rounded at tip; length 540 μm; diam. 32 μm; a = 17.

COMMENTS

This specimen is considered to be an infective stage of *C. dominicana* **n. gen., n. sp.** The size is close to that of the infective stages of *F. duboisi*, which range from 640 to 840 μm in length and 15 to 22 μm in diam. (Witenberg & Gerichter, 1944). Members of *Foleyella* infect frogs in both the Old and New World, including Brazilian tree frogs (*Leptodactylus* spp.). A related genus of tree frogs, *Eleutherodactylus*, has been recovered in Dominican amber (Poinar, 1992).

TYPE MATERIAL

Deposited in the Poinar amber collection (accession no. N-3-83) maintained at Oregon State University.

TYPE HOST

Culex sp. (Diptera: Culicidae).

TYPE LOCALITY

Amber mines located between Santiago and Puerto Plata in the Cordillera Septentrional of the northern portion of the Dominican Republic.

Cascofilaria parva n. gen., n. sp.
(Fig. 219)

This species is based on a single infective stage filariid positioned adjacent to a female *Culex* mosquito in Dominican amber.

DESCRIPTION

Head slightly narrowed and round at tip; body uniformly thick; length 209 μm; diam. 11 μm; a = 19; tail tip bearing a small knob.

COMMENTS

This specimen is unique since there are no known filariids with such short infective stages. However, ornamentation at the tail tip is a common feature of infective stage filariids (Bain & Chabaud, 1986).

TYPE MATERIAL

Deposited in the Poinar amber collection (accession no. Sy-1-165) maintained at Oregon State University.

TYPE HOST

Culex sp. (Diptera: Culicidae).

TYPE LOCALITY

Amber mines located between Santiago and Puerto Plata in the Cordillera Septentrional of the northern portion of the Dominican Republic.

Infraorder Ascaridomorpha De Ley & Baxter, 2002
Superfamily Ascaridoidea Baird, 1853
Family Ascarididae Baird, 1853

Genus *Ascarites* Poinar & Boucot, 2006

This collective group genus was erected for fossil members of the family Ascarididae that cannot be placed in extant genera.

Ascarites priscus Poinar & Boucot, 2006
(Fig. 42)

This species is based on two eggs extracted from a dinosaur coprolite.

DESCRIPTION (modified from Poinar & Boucot, 2006)

Eggs spherical, 42 and 46 μm in diam.; eggshell mammillated, 4.2 μm in thickness; diam. of developing juveniles in egg, 8.1 and 8.4 μm.

COMMENTS

The fossil eggs are similar to those of the extant *Ascaris lumbricoides* (Fig. 44) in size and in having a mammillated outer surface.

TYPE SPECIMEN

Deposited in the Department of Zoology, Oregon State University, Corvallis, OR (slide no. Y2 852 (2)).

TYPE HOST

The predatory dinosaur, *Megalosaurus* sp.

TYPE LOCALITY

Early Cretaceous Bernissart Wealden Iguanodon shaft, Belgium.

Ascarites gerus Poinar & Boucot, 2006
(Fig. 43)

This species is based on a single egg extracted from a dinosaur coprolite.

DESCRIPTION (modified from Poinar & Boucot, 2006)

Egg spherical, 49-52 μm in diam.; eggshell smooth, 3.4-3.6 μm in thickness; enclosed juvenile nematode narrow, 5.4 μm in diam.; coiled at least twice inside egg.

TYPE SPECIMEN

Deposited in the Department of Zoology, Oregon State University, Corvallis, OR (slide no. Y2 852 (2)).

TYPE HOST

The predatory dinosaur, *Megalosaurus* sp.

TYPE LOCALITY

Early Cretaceous Bernissart Wealden Iguanodon shaft, Belgium.

Superfamily Cosmocercoidea Travassos, 1925
Family Cosmocercidae Travassos, 1925

Genus *Palaeocosmocerca* n. gen.

This collective group genus is for fossil nematodes of the family Cosmocercidae.

TYPE AND ONLY SPECIES

P. burmanica **n. gen., n. sp.**

Palaeocosmocerca burmanica n. gen., n. sp.

(Figs 30-32; Plate 4)

DESCRIPTION

Two morphological types present; Type 1 presumably represents second-stage juveniles. Type 2 presumably represents pre-infective stages.

Type 1: (n = 4): lips offset; anterior portion of pharynx constricted; tail attenuated, pointed; length 455 (410-473) μm; diam. 23 (21-26) μm; a = 19.

Type 2: (n = 2): slender body; lips not offset; with attenuated head and tail; length 526 (473-578) μm; diam. 16 (13-18) μm; a = 33.

COMMENTS

Based on the morphological features of the second-stage juveniles and pre-infectives adjacent to and emerging from the mantle cavity of the fossil snail, the fossils most closely resemble *Cosmocercoides dukae* (Holl, 1928). The latter species is a parasite of the snails, *Discus cronkhitei* (Newcombe) and *Zonitoides arborea* (Say), in North America (Anderson, 1960). The lengths of the second-stage juveniles of *C. dukae* (410-480 μm) overlap those of the fossils (410-473 μm) as do the body lengths of the infective stages of *C. dukae* (578-614 μm). Also, the offset lips, flat head and constricted anterior portion of the pharynx are other features shared by the two species.

The life cycle of *Palaeocosmocerca* n. gen. could have been similar to that of the extant *C. dukae*. The adults lived in the intestinal tract of the snail and the second-stage juveniles and third-stage infectives occurred in the mantle cavity. When the snail made contact with the resin, the juveniles and pre-infective stages emerged from the mantle cavity. The extant sibling species of *C. dukae*, *Cosmocercoides variabilis* (Harwood, 1930), occurs in the rectum of frogs and toads today. It is possible that this species evolved from snail parasites that were acquired when the amphibians ingested infected hosts, although molluscs could have been secondary hosts (Vanderburgh & Anderson, 1987).

TYPE MATERIAL

Deposited in the Poinar amber collection (accession no. B-N-6) maintained at Oregon State University.

TYPE HOST

An arboreal juvenile snail of the family Punctidae.

TYPE LOCALITY

Amber mine in the Hukawng Valley, southwest of Maingkhwan in the state of Kachin (26°20′N, 96°36′E), northern Myanmar (Burma).

ETYMOLOGY

The specific name refers to the geographical locality of the fossil.

Suborder Tylenchina Thorne, 1949
Infraorder Panagrolaimomorpha De Ley & Blaxter, 2002
Superfamily Panagrolaimoidea Thorne, 1937
Family Panagrolaimidae Thorne, 1937

Genus *Panagrolaimus* Fuchs, 1930

***Panagrolaimus neotropicus* n. sp.**
(Figs 144, 145)

This species, the only known fossil member of the genus, is represented by 12 specimens located near the posterior end of an adult platypodid beetle in Dominican amber. The beetle has its elytra partly opened with the membranous wings extended.

Description

Small, slender species with long pointed tail; head rounded or truncate with deposit extruded from the mouth in form of a small droplet or elongate plug; length (n = 8) 229 (208-258) μm; diam. (n = 8) 12 (10-13) μm; a = 19.

Comments

The small size, long pointed tail and mouth deposit are characters of dauer juveniles of *Panagrolaimus*. According to Rühm (1956), the dauer of *Panagrolaimus* species usually possess evidence of a secreted deposit in the mouth area. This plug partly extrudes from the mouth of the extant *P. scheucherae* Rühm, 1956. In *P. neotropicus* **n. sp.**, the secretion appears to have been ejected from the mouth of several individuals but remains as a small droplet attached to the head. The dauer of the extant *P. verrucosus* Fuchs, 1930 have a slender tail and range from 254-288 μm in length, which overlaps with that of *P. neotropicus* **n. sp.** The former species occurs under the elytra and in dorsal folds of bark beetles with 20-30 individuals per beetle (Rühm, 1956). It is apparent that *P. neotropicus* **n. sp.** also occurred under the elytra and was associated with the membranous pair of wings. A second platypodid in Dominican amber contained eight dauer juveniles of *P. neotropicus* **n. sp.** under its elytra. It is interesting that along with this nematode were seven larger dauer stages of another species. Unfortunately, these latter nematodes were partly masked by the body of the beetle and could not be described.

Type material

Deposited in the Poinar amber collection (accession no. N-3-21) maintained at Oregon State University.

Type host

Phoretically associated with an adult platypodid beetle (length 3.0 mm) of the genus *Cenocephalus* (Coleoptera: Platypodidae).

Type locality

Amber mines located between Santiago and Puerto Plata in the Cordillera Septentrional of the northern portion of the Dominican Republic.

Infraorder Tylenchomorpha De Ley & Blaxter, 2002
Superfamily Aphelenchoidea Fuchs, 1937
Family Aphelenchoididae Skarbilovich, 1947

Genus *Bursaphelenchus* Fuchs, 1937

Bursaphelenchus similus n. sp.

(Figs 136-138)

This species, the only known fossil member of the genus, is represented by a population of developing aphelenchoidid nematodes in Dominican amber associated with a platypodid beetle and diffuse frass material (wood fragments) that probably originated from the beetle's gallery.

DESCRIPTION

Long, slender nematodes with lips offset; head truncate; stylet present; male tail curved, female tail bearing a terminal spike (digitate); length (n = 12) 464 (126-888) μm; diam. (n = 7) 13 (6-9) μm; a = 36.

COMMENTS

The size of the adults, presence of a stylet, curved male tail and digitate female tail align this species with members of the genus *Bursaphelenchus* (Fig. 139). A similar size range occurs in extant members of the genus (Hunt, 1993). Also the females of *B. fraudulentus* Rühm (1956) and *B. chitwoodi* Rühm (1956) have digitate tails similar to those of the fossils. It appears that a population of *B. similus* **n. sp.** was developing in the gallery of the platypodid beetle and both the beetle and the nematodes, together with some of the frass, were carried into the resin. There is no evidence of dauer juveniles. Extant species of this genus are associated with a wilting disease of pines (Mamiya, 1972).

TYPE MATERIAL

Deposited in the Poinar amber collection (accession no. N-3-88) maintained at Oregon State University.

TYPE HOST

Fungi growing in the gallery of the platypodid beetle.

TYPE LOCALITY

Amber mines located between Santiago and Puerto Plata in the Cordillera Septentrional of the northern portion of the Dominican Republic.

ETYMOLOGY

The specific name is taken from the Latin 'similis' = like.

Genus *Chalcidonema* n. gen.

TYPE AND ONLY SPECIES

C. paradoxa **n. gen., n. sp.**

This is a collective genus for fossil nematodes from the Tertiary associated with members of the parasitic wasp superfamily Chalcidoidea.

Chalcidonema paradoxa n. gen., n. sp.

(Figs 237-239; Plate 30)

This species is based on seven specimens of various sizes associated with a parasitic wasp of the superfamily Chalcidoidea in Dominican amber.

DESCRIPTION

Body narrow; head attenuated with offset lips; tail acute with short pointed tip; length (n = 6) 309 (72-518) μm; diam. (n = 7) 11 (5-18) μm; a = 28.

COMMENTS

It is difficult to determine the stages of the larger nematodes adjacent to the wasp. The largest individual, which is 518 μm in length (Fig. 237), could represent an adult, although the specimen protruding from the rear of the wasp appears to be even larger and may represent a parasitic female. Based on the size and location of the specimens (inside the body cavity), *C. paradoxa* **n. gen., n. sp.** is considered to have functioned as a parasite.

Other aphelenchoidid-insect associations involve *Bursaphelenchus cocophilus* (Cobb, 1919), which has an endophoretic association with coconut weevils (Hunt, 1993) and *B. kevini* Giblin, Swan & Kaya (1984), which is associated with halictid bees (Hymenoptera: Halictidae). Some features of *C. paradoxa* **n. gen., n. sp.** (offset lips, pointed juvenile tail) resemble those of *B. kevini*, but in the latter only the dauer juveniles are associated with the bees and they are located in the reproductive system, with further development occurring on fungi in the nest of the bee. This latter species is much larger than *C. paradoxa* **n. gen., n. sp.** and clearly has a different biology. *Chalcidonema paradoxa* **n. gen., n. sp.** is only tentatively assigned to the Aphelenchoididae as a stylet was not observed.

TYPE MATERIAL

Deposited in the Poinar amber collection (accession no. N-3-99) maintained at Oregon State University.

TYPE HOST

A member of the Chalcidoidea (Hymenoptera) tentatively assigned to the family Pteromalidae.

TYPE LOCALITY

Amber mines located between Santiago and Puerto Plata in the Cordillera Septentrional of the northern portion of the Dominican Republic.

ETYMOLOGY

The specific name is from the Latin 'paradoxus' meaning contrary to expectations.

Genus *Cretaciaphelenchoides* n. gen.

This is a collective genus for members of the Aphelenchoidea from Cretaceous deposits.

TYPE AND ONLY SPECIES

C. burmensis **n. gen., n. sp.**

Cretaciaphelenchoides burmensis n. gen., n. sp.
(Figs 28, 29)

This species is based on 11 nematodes together with fungal mycelium in a piece of Burmese amber.

DESCRIPTION

Body narrow and elongate; narrowly rounded head; lips not offset; tail pointed, tightly coiled in males; length males (n = 4) 524 (330-475) μm; diam. males 16 (11-17) μm; a = 33: females large, gently curved; length (n = 3) 392 (363-426) μm; diam. 17 (15-19) μm; a = 23. Shorter forms with moderate diameter are considered to represent juveniles; length juveniles (n = 2) 267-252 μm; diam. 12-15 μm. Short, narrow forms are considered to represent dauer juveniles; length dauers (n = 2) 174 μm for both specimens, diam. 8 and 10 μm; a = 19.

COMMENTS

These are very small aphelenchoidids and their size separates them from extant species. These are the earliest known nematodes feeding in an aerial habitat and they probably were originally carried to this site by xylophagous beetles.

TYPE MATERIAL

Deposited in the Poinar amber collection (accession no. B-N-7) maintained at Oregon State University.

TYPE HOST

Unidentified fungus with extensive mycelium and spores.

TYPE LOCALITY

Amber mine in the Hukawng Valley, southwest of Maingkhwan in the state of Kachin (26°20′N, 96°36′E), northern Myanmar (Burma).

ETYMOLOGY

The specific name refers to the country where the fossil originated.

Genus *Cryptaphelenchus* Fuchs, 1937

Cryptaphelenchus dominicus n. sp.

(Figs 140-143; Plate 16)

This species, the only known fossil member of the genus, is based on 57 dauer juveniles associated with a platypodid beetle in Dominican amber. Most of the nematodes are located near the head of the beetle, with some emerging between the head and thorax.

DESCRIPTION

Small, slender forms with truncate head; lips not offset but neck region slightly constricted in some specimens; tail greatly attenuated and filiform; stylet (only visible in some specimens) length 5 μm; nematode length (n = 14) 286 (248-326) μm; diam. (n = 10) 9 (6-10) μm; a = 32.

COMMENTS

The small size, presence of a stylet and location on the beetle (near the head) is why this species is placed in the genus *Cryptaphelenchus* (see Hunt, 1993). Bark beetles often transport dauer and juvenile females in the fold between the head and thorax, which is the situation with *C. dominicus* **n. sp.** The greatly attenuated tail with a filiform tip is similar to that of the extant *C. leptocaudus* Rühm (1956). The latter species is small, with females ranging from 325 to 382 μm (Rühm, 1956). Rühm gave no measurements for the dauer stages of *C. leptocaudus*, but his measurements of juvenile females are in the size range of *C. dominicus* **n. sp.** Rühm (1956) reported that the average number of dauer juveniles of *C. leptocaudus* on bark beetles was 30, with a maximum of 127.

TYPE MATERIAL

Deposited in the Poinar amber collection (accession no. N-3-19) maintained at Oregon State University.

TYPE HOST

A bark beetle belonging to the family Platypodidae (Coleoptera).

TYPE LOCALITY

Amber mines located between Santiago and Puerto Plata in the Cordillera Septentrional of the northern portion of the Dominican Republic.

ETYMOLOGY

The specific name refers to the country where the fossil originated.

Genus *Oligaphelenchoides* Poinar, 1977

Erected as a collective genus for Tertiary members of the family Aphelenchoididae. It is herein restricted to Aphelenchoididae from New World Tertiary deposits, the newly established collective genus *Palaeoaphelenchoides* (see below) being for Aphelenchoididae from Old World Tertiary deposits.

TYPE SPECIES

O. atrebora Poinar, 1977

OTHER SPECIES

O. dominicanus **n. sp.**
O. maximus **n. sp.**

Oligaphelenchoides atrebora Poinar, 1977
(Figs 248-251; Plate 31)

DESCRIPTION (modified from Poinar, 1977a)

Elongate nematodes with smooth cuticle; females straight or slightly curved ventrally; lip region slightly offset; males with tail tip reflexed ventrally; females (n = 5): length 1.1 (0.9-1.3) mm; diam. 22 (20-24) μm; a = 50; egg length 62 μm; egg diam. 23 μm: newly hatched juveniles (n = 5); length 184 (154-216) μm; diam. 9 (8-9) μm: males (n = 5); length 1.1 (1.0-1.2) mm; diam. 24 (20-26) μm; a = 46.

COMMENTS

It is rare to find a complete population of mycetophagous nematodes in a single piece of amber, together with their food source. Nematophagous fungi attacking *O. atrebora* were also present (Figs 252, 253).

TYPE MATERIAL

Deposited in the University of California Museum of Paleontology (Holotype: no. 14504); (Paratype male and female: nos 14505 and 14506, respectively).

TYPE HOST

Mycelium of unknown fungus.

TYPE LOCALITY

Palo-Blanco Mine about 22 km north east of Tapilula, Chiapas, Mexico. La Quinta Formation, Late Oligocene-Early Miocene.

Oligaphelenchoides dominicanus n. sp.

(Figs 154-156)

This species is represented by two adults and three pre-adults adjacent to a mat of fungal hyphae in Dominican amber.

DESCRIPTION

Long, slender nematodes with tail drawn out to a fine point; head narrowed with lips offset; stylet with small knobs, 19 μm long ($n = 1$); cuticle apparently with three lateral lines; length of two adult specimens attached to each other, 1.04 and 1.06 mm, diam. 27 μm; lengths of three juveniles, 693-850 μm; diam. 19-22 μm; $a = 36$.

COMMENTS

In the amber just below these nematodes is a layer of mycelium, which is presumed to have been the food source. Two smaller developing rhabditiform nematodes are also adjacent to the fungal mat. A pair of adults that were possibly mating was also present (Fig. 155).

TYPE MATERIAL

Deposited in the Poinar amber collection (accession no. N-3-91) maintained at Oregon State University.

TYPE HOST

Mat of fungal hyphae.

TYPE LOCALITY

Amber mines located between Santiago and Puerto Plata in the Cordillera Septentrional of the northern portion of the Dominican Republic.

ETYMOLOGY

The specific name refers to the country where the fossil originated.

Oligaphelenchoides maximus n. sp.
(Figs 197-201)

This species is represented by numerous specimens of all stages in and adjacent to a core of higher plant tissue in Dominican amber.

DESCRIPTION

Eggs, juveniles and adults are present. Head rounded, lips not offset; tail not long but wedge-shaped with pointed tip; stylet present; male tail C-shaped; length of males ($n = 2$), 1.2-1.3 mm; diam. of males, 28-32 μm; $a = 43$; length of females ($n = 5$), 1.2 (0.97-1.40) mm; diam. 28 (22-35) μm; $a = 43$; eggs elliptical, length ($n = 2$), 70-72 μm; diam. 23-25 μm.

COMMENTS

This species is quite long and slender. The shape, presence of a stylet and C-shaped male tail are all characters found in the Aphelenchoididae. The pointed tail, lips not offset and large size are characters that do not comfortably match any extant genera in that family. The presence of eggs, juveniles and adults indicates that *O. maximus* **n. sp.** was developing in the plant tissue. The section of fossil plant tissue, which is 3 mm in diam. and 14 mm in length, is probably that of a monocot since the vascular bundles are scattered. The nematodes are throughout the core and especially along its outer edge. There is no evidence of fungi associated with the plant tissue so *O. maximus* **n. sp.** is considered to represent a migratory root parasite with a biology similar to members of the genus *Aphelenchoides*. Species of *Aphelenchoides* are known to develop in orchids (Feder, 1951) and these epiphytes, as well as various grasses, were present in the Dominican amber forest (Poinar & Poinar, 1999).

TYPE MATERIAL

Deposited in the Poinar amber collection (accession no. N-3-14) maintained at Oregon State University.

TYPE HOST

Rootlet or root hair of a monocot.

TYPE LOCALITY

Amber mines located between Santiago and Puerto Plata in the Cordillera Septentrional of the northern portion of the Dominican Republic.

ETYMOLOGY

The specific epithet is derived from the Latin 'maximus' = greatest.

Genus *Palaeoaphelenchoides* n. gen.

This is a collective genus for Tertiary nematodes from Old World deposits that can be assigned to the Aphelenchoididae.

TYPE AND ONLY SPECIES

P. balticus **n. gen., n. sp.**

Palaeoaphelenchoides balticus n. gen., n. sp.
(Figs 56, 57)

This species is based on several hundred specimens in a block of Baltic amber together with seven female braconid wasps (Hymenoptera: Braconidae), a bark beetle (Coleoptera: Scolytidae) and a fungus gnat (Diptera: Sciaridae). There is no apparent association between the nematodes and any of the insects.

DESCRIPTION

Measurements of juvenile and adult stages were taken from a random sample. Length (n = 11) 607 (126-1040) μm; diam. 11 (6-25) μm; a = 55; body slender; head narrowed; faint stylet visible in some larger specimens; tail of female rounded: tail of male pointed and strongly curved; spicules curved; some extremely slender forms ranged from 140 to 170 μm in length and 8 to 12 μm in diam.

COMMENTS

The medium sized females and slightly smaller males with a tightly coiled tail strongly suggest that this species is a member of the Aphelenchoididae. Females of the genera *Aphelenchoides* Fischer, 1894 and *Bursaphelenchus* Fuchs, 1937 are similar in length (Goodey, 1963) and occur in rotting wood, presumably the habitat of the fossil species. It is likely that the braconid wasps recently emerged from a wood-boring insect and somehow became entrapped in the resin. The mixed stages with mature females indicate that this was a breeding population. The very slender smaller specimens may represent resistant (dauer) stages.

TYPE MATERIAL

Deposited in the Poinar amber collection (accession no. N-3-80) maintained at Oregon State University.

TYPE HOST

Free living but probably mycetophagous in insect galleries or on rotting portions of the bark.

TYPE LOCALITY

Baltic region some 40-50 mya.

ETYMOLOGY

The generic epithet is derived from the Greek 'palaios' = ancient and Aphelenchoides. The generic epithet is derived from the source of the fossil.

Family Acugutturidae Hunt, 1980

Genus *Setonema* n. gen.

This collective genus is erected for fossil members of the family Acugutturidae Hunt, 1980.

TYPE AND ONLY SPECIES

S. protera **n. gen., n. sp.**

Setonema protera n. gen., n. sp.

(Figs 202-204; Plate 23A)

This species is based on several specimens in a single piece of amber from the Dominican Republic. One specimen is adjacent to a moth while a second smaller individual is some distance from the moth. Additional specimens are on the body of the moth.

DESCRIPTION

Specimen positioned adjacent to moth: head not offset, cuticular annulations only present in short area near anterior end; faint stylet present; excretory pore located 29 μm from head tip; anal opening posterior, 14 μm from terminus; several fine attached droplets protruding from anus; length 453 μm; diam. 50 μm; a = 9. Specimen positioned some distance from moth: head rounded, tail drawn out, faint stylet; length 126 μm; diam. 10 μm; a = 13; no details noted on specimens still attached to body of moth.

COMMENTS

This species is assigned to the extant subfamily Noctuidonematinae Hunt, 1993 and shows that the ectoparasitic life habits of aphelenchs have been in existence since the mid-Tertiary. The larger specimen near the moth is considered to be a late stage juvenile female while the smaller specimen an early juvenile stage. The anterior cuticular annulations, faint stylet, anterior excretory pore and short tail indicate that this fossil lineage is related to the extant genera *Noctuidonema* Remillet & Silvain, 1988 and *Vampyronema* Hunt, 1993 in the Noctuidonematinae Hunt, 1993, which are ectoparasites of adult moths (Remillet & Silvain, 1988; Anderson & Laumond, 1992; Marti & Rogers, 1995; Marti *et al.*, 2000). Females of *N. guyanense* Remillet & Silvain range from 530 to 770 μm in length and 70 to 150 μm in diam. (Figs 205, 206; Plate 23B). Additional specimens of *Setonema* **n. gen.** occur on the posterior portion of the abdomen of the fossil moth, although overlapping scales hindered measurements.

TYPE MATERIAL

Deposited in the Poinar amber collection (accession no. N-3-31) maintained at Oregon State University.

TYPE HOST

Unidentified moth (Lepidoptera).

TYPE LOCALITY

Amber mines located between Santiago and Puerto Plata in the Cordillera Septentrional of the northern portion of the Dominican Republic.

ETYMOLOGY

The generic epithet is derived from the Greek 'seto' = moth and the Greek 'nema' = thread. The specific epithet is derived from the Greek 'protos' = first.

Superfamily Sphaerularioidea Lubbock, 1861
Family Anguinidae Nicoll, 1935

Genus *Palaeoanguina* n. gen.

This is a collective group genus for fossil members of the family Anguinidae (after Fortuner & Maggenti, 1987) or subfamily Anguininae Nicoll, 1935 (after Siddiqi, 2000).

TYPE AND ONLY SPECIES

P. dominicana **n. gen., n. sp.**

Palaeoanguina dominicana n. gen., n. sp.
(Figs 207, 208; Plate 24A)

This species is based on a juvenile female adjacent to a monocot seed in Dominican amber.

DESCRIPTION

Obese immature female with body curled in C-shape; cuticle finely annulated; head pointed, stylet slightly protruding; tail conoid, with peg-like tip; length 1.6 mm; diam. 200 μm; a = 8.

COMMENTS

The specimen is adjacent to an immature seed (Fig. 207). Apparently, *P. dominicana* **n. gen., n. sp.** was inside the seed when it encountered the resin. Females of *Anguina* can reach nearly 4 mm in length and 1-3 females and an equal number of males can occur in the same seed gall (Thorne, 1961). Based on the structure of the seed it appears to be that of a monocot, possibly that of a grass since they have been described from Dominican amber (Poinar & Judziewicz, 2005).

TYPE MATERIAL

Deposited in the Poinar amber collection (accession no. N-3-6A) maintained at Oregon State University.

TYPE HOST

Immature monocot seed.

TYPE LOCALITY

Amber mines located between Santiago and Puerto Plata in the Cordillera Septentrional of the northern portion of the Dominican Republic.

ETYMOLOGY

The generic epithet is derived from the Greek 'palaios' = ancient and *Anguina*. The specific epithet is derived from the source of the fossil.

Family Iotonchidae Goodey, 1935

Genus *Paleoiotonchium* n. gen.

This genus is erected for fossil members of the family Iotonchidae.

TYPE AND ONLY SPECIES

P. dominicanum **n. gen., n. sp.**

Paleoiotonchium dominicanum n. gen., n. sp.
(Figs 181-183; Plate 20)

This species is based on 15 juvenile nematodes adjacent to a fossil fungus gnat in Dominican amber.

DESCRIPTION

Short, moderately narrow nematodes; lips not offset; length (n = 12), 233 (208-296) μm; diam. 13 (12-16) μm; a = 18.

COMMENTS

Besides the nematodes that surrounded the fly there were a number of individuals within the body cavity of the fungus gnat. It is presumed that the life cycle of *P. dominicanum* **n. gen., n. sp.** was similar to that of its present-day descendant, *Iotonchium californicum* (Poinar, 1991b) (Figs 184, 185). This latter species has two separate life cycles, one in a mushroom and the other in a fungus gnat. Mycetophagous juveniles emerge from the fly and undergo several life cycles in the mushroom. However, when the latter begins to decay, entomophagous stages are formed that mate in the mushroom. The females then enter the body cavity of fly larvae and produce young that develop into mycetophagous juveniles. It was this stage that was captured in amber.

TYPE MATERIAL

Deposited in the Poinar amber collection (accession no. N-3-23) maintained at Oregon State University.

TYPE HOST

Female fungus gnat of the tribe Xylophilini (Diptera: Mycetophilidae).

TYPE LOCALITY

Amber mines located between Santiago and Puerto Plata in the Cordillera Septentrional of the northern portion of the Dominican Republic.

ETYMOLOGY

The generic name is from the Greek 'palaios' = ancient and the genus *Iotonchium*. The specific name refers to the locality of the fossil.

Family Allantonematidae (Pereira, 1932) Chitwood & Chitwood, 1937

Genus *Howardula* Cobb, 1921

Howardula helenoschini Poinar, 2003

(Fig. 90; Plate 15)

This species, the only known fossil member of the genus, is based on five nematodes clustered at the tip of a phorid fly in Baltic amber. Only two could be well studied because of their orientation. Both nematodes appear to have been in the process of moulting. One was considered to be a male and the other a female.

DESCRIPTION (modified from Poinar, 2003c)

Anterior end bluntly rounded to truncate; lip region not offset; tail narrowly pointed: Male: length 772 μm; diam. 36 μm; a = 21; tail length 36 μm; length spicules 10 μm; length testis 203 μm: Female: length 780 μm; diam. 32 μm; a = 24; length stylet (partial), 8 μm; length uterus and ovary, 256 μm; vulva positioned on small protuberance; % vulva 77; length lemon-shaped parasitic female in body of host 650 μm.

COMMENTS

The life cycle of *H. helenoschini* was probably similar to that of the extant *H. husseyi* Richardson, Hesling & Riding, 1977 that parasitises the phorid fly, *Megaselia halterata* Wood. However, it appears that the development in the host is more rapid in the former species since while the pre-adults of *H. helenoschini* were exiting the host, the second-stage juveniles of *H. husseyi* leave the host.

TYPE SPECIMEN

Deposited in the Poinar amber collection (accession no. N-3-64) maintained at Oregon State University.

TYPE HOST

A female of the genus *Triphleba* Rondani (Diptera: Phoridae).

TYPE LOCALITY

Baltic region some 40-50 mya.

Genus *Palaeoallantonema* n. gen.

This is a collective genus for fossil Allantonematidae that cannot be assigned to an extant genus.

TYPE SPECIES

P. apionae **n. gen., n. sp.**

OTHER SPECIES

P. baltica **n. gen., n. sp.**
P. cephalotae **n. gen., n. sp.**
P. dominicana **n. gen., n. sp.**
P. phlebotomae **n. gen., n. sp.**
P. zygopinae **n. gen., n. sp.**

Palaeoallantonema apionae n. gen., n. sp.
(Figs 189, 190)

This species is based on five nematodes associated with an adult apionid weevil in Dominican amber.

DESCRIPTION

Pre-adult female adjacent to weevil: head and tail rounded, length 356 μm; diam. 14 μm; a = 25. Pre-adult male: tail tightly curved, pointed; spicule tip protruding; length 270 μm; diam. 14 μm; a = 19; three additional juveniles with lengths of 226, 135 and 113 μm.

COMMENTS

The shape and size of the pre-adult female and male are similar to members of the genus *Howardula* Cobb. While the fourth-stage female is smaller in length than those of the extant *Apion* parasite, *H. apioni* Poinar *et al.*, 1980, it is close to the range of females of the beetle parasite, *H. madecassa* Remillet & Van Waerebeke, 1975 (370-396 μm).

TYPE MATERIAL

Deposited in the Poinar amber collection (accession no. N-3-41) maintained at Oregon State University.

TYPE HOST

An adult weevil (length 1.4 mm) of the family Apionidae (Coleoptera).

TYPE LOCALITY

Amber mines located between Santiago and Puerto Plata in the Cordillera Septentrional of the northern portion of the Dominican Republic.

ETYMOLOGY

The generic epithet is derived from the Greek 'palaios' = ancient and *Allantonema*. The specific epithet is derived from the host of the fossil.

***Palaeoallantonema baltica* n. gen., n. sp.**
(Figs 93-95)

This species is represented by 92 specimens surrounding an adult rove beetle in Baltic amber.

DESCRIPTION

The nematodes fall into two size categories. Category A (n = 7); stout forms with heads narrowly rounded and tail bluntly rounded; length 253 (226-293) μm; diam. 29 (27-36) μm; a = 9. Category B (n = 7): slender forms with rounded heads; tail narrowly rounded; length 181 (153-216) μm; diam. 22 (18-23) μm; a = 8.

COMMENTS

The size range of *P. baltica* **n. gen., n. sp.** is significantly shorter than all extant allantonematid and entaphelenchid parasites of staphylinids (Wachek, 1955; Hunt, 1993). The life cycle could have been similar to that of extant *Parasitylenchus* spp., which has two parasitic cycles in the host. The stout forms could represent adults of the first parasitic cycle while the more slender forms could be free-living pre-adults of the second generation.

TYPE SPECIMEN

Deposited in the Poinar amber collection (accession no. N-3-37) maintained at Oregon State University.

TYPE HOST

An adult female rove beetle (Coleoptera: Staphylinidae).

TYPE LOCALITY

Baltic region some 40-50 mya.

ETYMOLOGY

The specific epithet is derived from the locality of the fossil.

Palaeoallantonema cephalotae n. gen., n. sp.
(Figs 118, 119)

This species is represented by seven specimens adjacent to a worker ant of the genus *Cephalotes* Latreille, 1802, in Dominican amber.

DESCRIPTION

(n = 7): Head rounded, tail attenuated but rounded at tip; length 298 (261-339) μm; diam.; 18 (16-20) μm; a = 17. A specimen considered to be a pre-adult female contained a faint stylet, 18 μm in length, as well as a vulval opening located 226 μm from the anterior end.

COMMENTS

The shape of the nematodes resembles pre-adults of extant allantonematids recovered from *Cephalotes christophersen*i (Forel) in Peru (Fig. 120). These extant forms have not yet been described but the larger pre-adults separate them from the fossil specimens.

TYPE MATERIAL

Deposited in the Poinar amber collection (accession no. N-3-77) maintained at Oregon State University.

TYPE HOST

Worker (length 4.2 mm) of *Cephalotes serratus* Vierbergen & Scheven (Hymenoptera: Formicidae).

TYPE LOCALITY

Amber mines located between Santiago and Puerto Plata in the Cordillera Septentrional of the northern portion of the Dominican Republic.

ETYMOLOGY

The specific epithet is derived from the host of the fossil.

***Palaeoallantonema dominicana* n. gen., n. sp.**
(Figs 225-227; Plate 29)

This species is represented by 44 juvenile allantonematids of uniform length, together with some smaller stages, associated with an adult staphylinid beetle in Dominican amber.

DESCRIPTION

Body wall smooth; lips indistinct; tail blunt; length 268 (235-300) μm; greatest diam. 10 (7-13) μm; a = 27; smaller forms adjacent to a bubble in the amber were 100-200 μm in length.

COMMENTS

As noted by Poinar and Brodzinsky (1986), who reported these fossils earlier, the size of the juveniles is significantly smaller than extant allantonematid and entaphelenchid staphylinid parasites (Wachek, 1955; Hunt, 1993). The closest extant species of similar size is *Proparasitylenchus platystethis* Wachek, 1955, whose males range from 330 to 385 μm in length and females from 415 to 468 μm in length. However, there are not enough characters visible in the fossils to place them in that genus.

TYPE MATERIAL

Deposited in the Poinar amber collection (accession no. N-3-9) maintained at Oregon State University.

TYPE HOST

A member of the genus *Neoxantholinus* Cameron, 1944 (Coleoptera: Staphylinidae) (length 4.0 mm).

TYPE LOCALITY

Amber mines located between Santiago and Puerto Plata in the Cordillera Septentrional of the northern portion of the Dominican Republic.

ETYMOLOGY

The specific epithet is derived from the locality of the fossil.

***Palaeoallantonema phlebotomae* n. gen., n. sp.**
(Figs 220-222)

This species is based on three juveniles that exited from a sand fly in Dominican amber.

DESCRIPTION

Head rounded; tail short, but with a pointed tip; length (n = 3), 326 (317-344) μm; diam. 15 (14-16) μm; a = 22; body wall with cuticular annulations; one specimen is ensheathed; some nematodes occur in the coelom of the sand fly but their characters are unclear.

COMMENTS

Extant sand flies are parasitised by allantonematid nematodes in Africa, South America and the Middle East. One species parasitising sand flies in Colombia, SA, is *Anadranema phlebotophaga* Poinar *et al.*, 1993. This species is hermaphroditic with stages emerging from the host ranging from 277 to 364 μm in length and 16 to 23 μm in diam. The cuticle of *A. phlebotophaga* is annulated, similar to that of the fossil. The infective fourth-stage hermaphrodites of *A. phlebotophaga* often retain their last-stage cuticle, which appears to be the case in at least one of the fossil specimens. However, due to the absence of additional characters, it is not possible to assign the fossil species to *Anadranema* at this time.

TYPE MATERIAL

Deposited in the Poinar amber collection (accession no. N-3-93) maintained at Oregon State University.

TYPE HOST

An adult female sand fly of the genus *Lutzomyia* (Diptera: Phlebotomidae).

TYPE LOCALITY

Amber mines located between Santiago and Puerto Plata in the Cordillera Septentrional of the northern portion of the Dominican Republic.

ETYMOLOGY

The specific epithet is derived from the host of the fossil.

***Palaeoallantonema zygopinae* n. gen., n. sp.**
(Figs 191, 192)

This species is based on eight nematodes associated with an adult zygopine weevil in Dominican amber. A ninth specimen (nema B) differs from the others and is probably a dauer juvenile of a separate species.

DESCRIPTION

Third-stage juvenile (n = 3); body straight, slender, head rounded; lips not offset; stylet faint, length 9 μm; cephalic constriction absent; tail wedge-shaped, rounded at tip; length 340 (337-344) μm; diam. 15 (14-15) μm; a = 23. Nema B (n = 1): head and tail pointed, length 371 μm; diam. 11 μm; a = 34.

COMMENTS

Third-stage juveniles are the stage of allantonematids that usually exit from the gut or reproductive system of the host. The rounded tail and head of *P. zygopinae* **n. gen., n. sp.** are typical characters of many allantonematids (Rühm, 1956), including the extant weevil parasite, *Howardula apioni* Poinar, Laumond & Bonifassi, 1980. The length of the third-stage juveniles falls within the range of the free-living juveniles and females of the bark beetle nematode parasites, *Parasitylenchus curvidentis* (Fuchs, 1914) (Rühm, 1956) and *Sulphuretylenchus kleinei* (Rühm, 1956), and the drosophilid parasite, *Parasitylenchus nearcticus* Poinar, Jaenke & Dombeck, 1997.

TYPE MATERIAL

Deposited in the Poinar amber collection (accession no. N-3-106) maintained at Oregon State University.

TYPE HOST

An adult weevil (length 4.7 mm) of the subfamily Zygopinae (Coleoptera: Curculionidae).

TYPE LOCALITY

Amber mines located between Santiago and Puerto Plata in the Cordillera Septentrional of the northern portion of the Dominican Republic.

ETYMOLOGY

The specific epithet is derived from the host of the fossil.

Genus *Palaeoparasitylenchus* n. gen.

This is a collective genus for fossil allantonematids belonging to the *Parasitylenchus* clade. As with extant species of the genus *Parasitylenchus*, *Palaeoparasitylenchus* **n. gen.** had two amphimictic cycles in the host and the juveniles of the second parasitic cycle mature to adult stages inside the host.

TYPE AND ONLY SPECIES

P. dominicanus **n. gen., n. sp.**

Palaeoparasitylenchus dominicanus n. gen., n. sp.
(Figs 159-165; Plate 18)

The preservation of the specimens was excellent so it is possible to see more characters in these fossils than in previous ones. The description is based on 120 nematodes that were in the process of emerging from a drosophilid fly in Dominican amber. Additional specimens were still in the body cavity of the host. This association was noted previously (Poinar, 1984b, c).

DESCRIPTION

Vermiform females and pre-adult females with stylet, vulva and penetration glands; length 317 (271-351) μm; diam. 11 (9-14) μm; a = 29; head truncate; distance from head to middle of nerve ring 22 (18-25) μm; length of stylet 6 (6-7) μm; distance from head to terminus of first pharyngeal gland 101 (67-124) μm; distance from head to terminus of second pharyngeal gland 137 (117-156) μm; distance from head to terminus of third pharyngeal gland 178 (174-181) μm; tail length 39 (29-50) μm; % vulva 70 (61-75); tail pointed, with tip often reflexed.

Developing second generation parasitic females; length 155 (136-226) μm; diam. 15 (12-18) μm; a = 10; head rounded, body dense, tail rounded, sometimes with a ventral protuberance or terminal swelling.

COMMENTS

The closest extant species to *P. dominicanus* **n. gen., n. sp.** is *Parasitylenchus nearcticus* Poinar *et al.*, 1997 from eastern North America (Fig. 166). Free-living females of *P. dominicanus* **n. gen., n. sp.** are larger than those of *P. nearcticus*, which are 239 (228-254 μm) in length and possess a smaller stylet, longer pharyngeal glands and a shorter distance from the head to the nerve ring. The parasitic second generation females of *P. nearcticus* are only slightly larger than the free-living vermiform females. The second generation parasitic females of *P. dominicanus* **n. sp.** are smaller than the free-living females, possibly because they had not reached full size.

The host genera of the two species differs since *P. nearcticus* has been reported only in members of the genus *Drosophila* Fallén, while the fossil species parasitised an extinct species of the genus *Chymomyza* Czerny, 1903.

TYPE MATERIAL

Deposited in the Poinar amber collection (accession no. N-3-1) maintained at Oregon State University.

TYPE HOST

An extinct species of drosophilid fly, *Chymomyza primavera* Grimaldi (Diptera: Drosophilidae).

TYPE LOCALITY

Amber mines located between Santiago and Puerto Plata in the Cordillera Septentrional of the northern portion of the Dominican Republic.

ETYMOLOGY

The generic epithet is derived from the Greek 'palaios' = ancient and *Parasitylenchus*. The specific epithet is derived from the locality of the fossil.

Family Sphaerulariidae Lubbock, 1865

Genus *Tripius* Chitwood, 1935

Tripius balticus n. sp.

(Figs 86-88; Plate 14)

This species, the only known fossil member of the genus, is represented by 18 fourth-stage juveniles that had emerged or were in the process of emerging from an adult sciarid fungus gnat in Baltic amber.

DESCRIPTION

Head rounded, tail attenuated to an acute tip; length (n = 9) 268 (235-317) μm; diam. 16 (14-18) μm; a = 17.

COMMENTS

The stages emerging from the adult sciarid are very similar in shape and diameter to those of the extant species, *T. sciarae* (Bovien, 1944)

(Poinar, 1965) (Fig. 89); however, the parasitic adult and length of the fourth-stage juveniles are smaller than those of *T. sciarae*.

TYPE SPECIMEN

Deposited in the Poinar amber collection (accession no. N-3-67) maintained at Oregon State University.

TYPE HOST

Female fungus gnat of the family Sciaridae (Diptera).

TYPE LOCALITY

Baltic region some 40-50 mya.

ETYMOLOGY

The specific epithet is derived from the locality of the fossil.

Suborder Rhabditina Chitwood, 1933
Infraorder Diplogastromorpha De Ley & Blaxter, 2002
Superfamily Diplogastroidea Micoletzky, 1922

Genus *Apifilaria* n. gen.

This is a collective genus for fossil Diplogastroidea associated with bees.

TYPE AND ONLY SPECIES

A. proplebiae **n. gen., n. sp.**

***Apifilaria proplebiae* n. gen., n. sp.**
(Figs 121-123)

This species is represented by five dauer nematodes in one piece of Dominican amber (accession no. N-3-25) and three nematodes (one dauer and two developmental forms) in a second piece of Dominican amber (accession no. N-3-47). In both amber pieces the nematodes are adjacent to a stingless bee. No other insect inclusions are present in either piece.

Description

(Accession no. N-3-25): Head attenuated with blunt tip; lips offset; tail attenuated, pointed; length (n = 5) 392 (378-405) μm; diam. (n = 5) 16 (15-18) μm; a = 25; (accession no. N-3-47): one dauer 405 μm long and 13 μm wide and two developmental juveniles with rounded heads and short, rounded tails; length 292 and 563 μm; diam. 27 and 32 μm.

Comments

The present species is tentatively assigned to the Diplogastridae based on its general shape. Dauer nematodes have not been recovered from extant stingless bees but several types are associated with solitary bees. Two found in halictids are the aphelenchoidid *Bursaphelenchus* (=*Huntaphelenchoides*) *kevini* (Giblin *et al.*, 1984) (Roubik, 1989) and the diplogastrid *Acrostichus* (=*Aduncospiculum*) *halicti* (Giblin & Kaya, 1984). The dauer stages of the former species range from 674 to 820 μm in length and 17 to 25 μm in diam. while those of *A. halicti* range from 337 to 443 μm in length and 13 to 28 μm in diam. (Giblin & Kaya, 1984; Giblin *et al.*, 1984). An undescribed diplogastrid is associated with bees of the genus *Andrena* (Erteld, 1994). *Apifilaria proplebiae* **n. gen., n. sp.** probably colonised the bee's nest and the nematodes crawled on worker bees as they left the hive.

Type material

Both specimens (accession nos N-3-25 and N-3-47) deposited in the Poinar amber collection maintained at Oregon State University.

Type host

Workers of the social bee, *Proplebeia dominicana* (Hymenoptera: Apidae): length of bees (accession no. N-3-24) 1.6 mm; (accession no. N-3-47) 3.2 mm.

Type locality

Amber mines located between Santiago and Puerto Plata in the Cordillera Septentrional of the northern portion of the Dominican Republic.

ETYMOLOGY

The generic epithet is derived from the Latin 'apis' = bee and the Latin 'filum' = thread. The specific epithet is derived from the host genus.

Genus *Chilonema* n. gen.

This is a collective genus for fossil Diplogastroidea associated with members of the Chilopoda (centipedes).

TYPE AND ONLY SPECIES

C. filaforma **n. gen., n. sp.**

Chilonema filaforma n. gen., n. sp.
(Figs 232, 233)

This species is based on a dauer juvenile associated with a member of the centipede family Scutigeridae in Dominican amber.

DESCRIPTION

Head bluntly rounded; lips not offset; tail attenuated, ensheathing cuticle; length 235 μm; diam. 16 μm; a = 15.

COMMENTS

The small size and shape of the dauer is why this species is assigned to the Diplogastridae. No reports of dauer nematodes associated with extant centipedes could be located.

TYPE MATERIAL

Specimens (accession no. N-3-96) deposited in the Poinar amber collection maintained at Oregon State University.

TYPE HOST

A member of the family Scutigeridae (Scutigeromorpha: Chilopoda).

TYPE LOCALITY

Amber mines located between Santiago and Puerto Plata in the Cordillera Septentrional of the northern portion of the Dominican Republic.

ETYMOLOGY

The generic epithet is derived from the order of the centipedes = Chilopoda. The specific epithet is derived from the Latin 'filum' = thread and the Latin 'forma' = shape.

Genus *Cretodiplogaster* n. gen.

This is a collective genus for Diplogastroidea associated with Cretaceous termites.

TYPE AND ONLY SPECIES

C. termitophilus **n. gen., n. sp.**

Cretodiplogaster termitophilus n. gen., n. sp.
(Fig. 26)

This species is based on two dauer juveniles associated with a termite in Burmese amber.

DESCRIPTION

Both specimens with ensheathing cuticle; specimen A, length 195 μm; diam. 11 μm; a = 18; specimen B, length 215 μm; diam. 9 μm; a = 24.

COMMENTS

While it would be desirable to have additional characters, on the basis of their size the fossils are placed in the Diplogastroidea and are probably members of the Diplogastridae. All extant nematodes with dauers associated with termites belong to the families Rhabditidae and Diplogastridae (Fig. 27). In general most dauer stages of the diplogastrids are smaller than those of the Rhabditidae (Table 2). None of the dauer stages of extant termitophilous Rhabditidae is less than 300 μm in length.

TYPE MATERIAL

Specimen (accession no. B-N-5) deposited in the Poinar amber collection, Oregon State University.

TYPE HOST

Kalotermes burmensis Poinar, 2009 (Isoptera: Kalotermitidae).

TYPE LOCALITY

Amber mine in the Hukawng Valley, southwest of Maingkhwan in the state of Kachin (26°20′N, 96°36′E), northern Myanmar (Burma).

ETYMOLOGY

The generic epithet is derived from the Latin 'creta' = chalk and the genus *Diplogaster*. The specific epithet is derived from termite and the Greek 'philia' = love.

Genus *Diplonema* n. gen.

This is a collective genus for fossil Diplogastroidea associated with members of the Diplopoda (millipedes).

TYPE AND ONLY SPECIES

D. minuta **n. gen., n. sp.**

Diplonema minuta n. gen., n. sp.
(Figs 228-230)

This species is based on two dauer juveniles associated with a polydesmid millipede in Dominican amber.

DESCRIPTION

Dauer small, with acutely rounded head; tail pointed; lengths 190 and 195 μm; diam. 9 μm; a = 21; ensheathing cuticle not observed.

COMMENTS

The small size of the dauer is why the nematodes are placed in the Diplogastroidea. They are probably members of the Diplogastridae since the two extant species of *Rhabditis* associated with millipedes (Poinar, 1986; Sudhaus & Schulte, 1989) have much larger dauer juveniles (504-611 μm for *R. myriophora* Poinar (Fig. 231) and 535-670 μm for *R. necromena* Sudhaus & Schulte). No records of diplogastroids associated with extant millipedes could be found.

TYPE MATERIAL

Specimens (accession no. N-3-97) deposited in the Poinar amber collection maintained at Oregon State University.

TYPE HOST

Millipede of the family Polydesmidae (Diplopoda).

TYPE LOCALITY

Amber mines located between Santiago and Puerto Plata in the Cordillera Septentrional of the northern portion of the Dominican Republic.

ETYMOLOGY

The generic epithet is derived from the Diplopoda and the Greek 'nema' = thread. The specific epithet is derived from the Latin 'minuta' = small.

Genus *Filotermes* n. gen.

This is a collective genus for fossil Diplogastroidea associated with Tertiary termites.

TYPE AND ONLY SPECIES

F. minutus **n. gen., n. sp.**

Filotermes minutus n. gen., n. sp.
(Fig. 124)

This species is based on numerous dauer juveniles associated with termites in Dominican amber.

Description

Head rounded, lips not offset; tail pointed, body surface oily; length 145 (137-153) μm; greatest diam. 8 (6-12) μm; a = 18.

Comments

The small size of the dauer stages and the oily body surface (some dauers appear to be attached to each other) is why this species is placed in the Diplogastroidea. They probably are members of the Diplogastridae. The nematodes were probably reproducing in the termite nest, and, when conditions deteriorated, produced dauer juveniles that attached themselves to the termites.

Type material

Specimens (accession no. N-3-26F) deposited in the Poinar amber collection maintained at Oregon State University.

Type host

Worker termites belonging to the family Hodotermitidae.

Type locality

Amber mines located between Santiago and Puerto Plata in the Cordillera Septentrional of the northern portion of the Dominican Republic.

Etymology

The generic epithet is derived from the Latin 'filum' = thread and termite. The specific epithet is derived from the Latin 'minuta' = small.

Oligodiplogaster Poinar, 1977

A collective genus for free-living nematodes from Tertiary deposits that can be assigned to the family Diplogastridae.

TYPE AND ONLY SPECIES

O. antiqua Poinar, 1977

Oligodiplogaster antiqua Poinar, 1977
(Fig. 245)

This species is represented by two free-living specimens in Mexican amber.

DESCRIPTION (modified from Poinar, 1977a)

Lips indistinct; posterior portion of body bent ventrally; length 500 and 540 μm; diam. 20 μm; $a = 26$; tail bearing a long filament; length of tail filament, 40-80 μm; specimens surrounded by a thin black deposit; swollen areas are considered preservation artefacts.

COMMENTS

The shape of the tail with the extended tail filament is characteristic of male Diplogastridae.

TYPE MATERIAL

Holotype male (accession no. 14507) and paratype male (accession no. 14508) deposited in the University of California Museum of Paleontology (block no. B-7051-5, 6), Berkeley, CA.

HABITAT

Free in amber.

TYPE LOCALITY

Vega de La Compana, Rio San Pedro Valley about 7 km north of Simojovel, Chiapas, Mexico, La Quinta formation, Late Oligocene-Early Miocene.

Genus *Palaeodiplogaster* n. gen.

This is a collective genus for Diplogastroidea associated with Tertiary beetles (Coleoptera).

TYPE AND ONLY SPECIES

P. brentiphila **n. gen., n. sp.**

Palaeodiplogaster brentiphila n. gen., n. sp.
(Figs 157, 158)

This species is based on a single adult adjacent to a straight-snouted weevil in Dominican amber.

DESCRIPTION

Head truncate, lips not offset, tail drawn out to a fine filament; length 1.2 mm; diam. 58 μm; a = 21; length of tail (approximate) 105 μm; length of tail filament 68 μm.

COMMENTS

The measurements suggest that this is an adult, probably a female and the drawn-out tail suggests it is a diplogastrid. Wood-boring insects, especially Scolytidae and Platypodidae, are known to cohabit their galleries with diplogastrids, which then form phoretic associations with the beetles (Rühm, 1956; Massey, 1974).

Measurements of the fossil are very close to some extant members of the genus *Rhabdontolaimus* Fuchs, 1931 (length 1.1; diam. 55 μm; a = 20) and *Micoletzkya* Weingärtner, 1955 (length 0.85-1.51 mm; diam. 33 μm; a = 22). Species of both genera are associated with bark beetles in North America (Massey, 1974).

TYPE MATERIAL

Deposited in the amber collection of Ron Cauble (accession no. D 010), Berkeley, CA.

TYPE HOST

Dominibrentus leptus Poinar, 2009 (Coleoptera: Brentidae).

TYPE LOCALITY

Amber mines located between Santiago and Puerto Plata in the Cordillera Septentrional of the northern portion of the Dominican Republic.

ETYMOLOGY

The generic epithet is derived the Greek from 'palaios' = old and the nematode genus *Diplogaster*. The specific epithet is derived from the host family and the Greek 'philia' = love.

Genus *Formicodiplogaster* n. gen.

This is a collective genus for Diplogastroidea associated with Tertiary ants (Hymenoptera: Formicidae).

TYPE AND ONLY SPECIES

F. myrmenema **n. gen., n. sp.**

Formicodiplogaster myrmenema n. gen., n. sp.
(Figs 105-109)

This species is based on dauer juveniles adjacent to worker *Azteca* ants in several separate pieces of Dominican amber and reproducing forms associated with portions of *Azteca* nests in two pieces of Dominican amber. This particular association was noted previously (Poinar, 1982).

DESCRIPTION

Reproducing stages: head rounded, lips not offset; tail pointed: (accession no. N-3-18 F) (n = 10): length 274 (194-325) μm; diam. 27 (12-39) μm; a = 10: (accession no. N-3-18 G) (n = 5): length 218 (145-252) μm; diam. 29 (19-39) μm; a = 8. Dauer stages: head rounded; lips not offset; tail pointed: (accession nos N-3-18, A-E) (n = 8): length 172 (120-208) μm; diam. 14 (10-15) μm; a = 12.

COMMENTS

It is surmised that *F. myrmenema* **n. gen., n. sp.** reproduced in the nests of the ants and when conditions deteriorated, dauer stages were produced. Undescribed diplogastrids occur phoretically on extant *Azteca* worker ants in Costa Rica (Paul E. Hanson, pers. comm., 12 September 2004).

TYPE MATERIAL

All specimens (accession nos N-3-18 A-G) deposited in the Poinar amber collection maintained at Oregon State University.

TYPE HOST

Ants of the genus *Azteca* (Hymenoptera: Formicidae).

TYPE LOCALITY

Amber mines located between Santiago and Puerto Plata in the Cordillera Septentrional of the northern portion of the Dominican Republic.

ETYMOLOGY

The generic epithet is derived from the Latin 'formica' = ant and the nematode genus *Diplogaster*. The specific epithet is derived from the Greek 'myrmex' = ant and the Greek 'nema' = thread.

Genus *Scarabinema* n. gen.

This is a collective genus for Diplogastroidea associated with fossil dung beetles.

TYPE AND ONLY SPECIES

S. minuta **n. gen., n. sp.**

Scarabinema minuta n. gen., n. sp.
(Figs 176, 177)

This species is based on two dauer juveniles adjacent to a dung beetle in Dominican amber.

DESCRIPTION

Short, stout forms with narrowed, blunt-tipped head and short, pointed tail; lengths 109 and 111 μm; diam. 8 and 9 μm: a = 14.

COMMENTS

The small size and shape of the dauer are the reason they are placed with the diplogastrids. Dauer juveniles of rhabditids and diplogastrids commonly occur under the elytra, in the intersegmental membranes and in the genital tract of extant dung beetles (Sachs, 1950). Some diplogastrids also occur in the body cavity of dung beetles (Bovien, 1937; Poinar *et al.*, 1976), and at least one rhabditid forms cysts on the surface of adult beetles (Triffit & Oldham, 1927) (Figs 178, 179). Nematodes associated with neotropical dung beetles have been little investigated. Species of *Parasitorhabditis* and *Cruznema* occur in larval guts of the scarabaeid, *Cyclocephala signaticollis* (Burmeister) in Argentina, but apparently no dauer stages are formed (Reboredo & Camino, 2000).

TYPE MATERIAL

Specimens (accession no. N-3-94) deposited in the Poinar amber collection maintained at Oregon State University.

TYPE HOST

Adult scarab beetle (Coleoptera: Scarabaeidae) in Dominican amber.

TYPE LOCALITY

Amber mines located between Santiago and Puerto Plata in the Cordillera Septentrional of the northern portion of the Dominican Republic.

ETYMOLOGY

The generic epithet is derived from the family of host beetles and the Greek 'nema' = thread. The specific epithet is derived from the Latin 'minuta' = small.

Genus *Scolytonema* n. gen.

This is a collective genus for fossil Diplogastroidea associated with members of the wood-boring families Scolytidae and Platypodidae (Coleoptera). The species described below are based on dauer juveniles on or adjacent to their beetle carriers. In some cases, there may be more than one species of nematode present, since multiple nematode species are known to be associated with individual bark beetles today (Rühm, 1956). Dauer nematodes associated with extant bark beetles include members of the families Diplogastridae, Panagrolaimidae, Rhabditidae and Aphelenchoididae (Rühm, 1956; Shepard, 1974; Poinar, 1975).

TYPE SPECIES

S. dominicana **n. gen., n. sp.**

OTHER SPECIES

S. hispaniolae **n. gen., n. sp.**
S. mexicana **n. gen., n. sp.**

Scolytonema dominicana **n. gen., n. sp.**
(Figs 146-150, 153; Plate 17)

This species is based on dauer juveniles associated with several beetles in separate pieces of Dominican amber. Amber piece N-3-60 contained ten dauer stages associated with a scolytid bark beetle. Amber piece N-3-95 contained four dauer stages associated with a platypodid pinhole borer. Amber piece N-3-90 contained three dauer stages associated with another platypodid and a platypodid in a piece of amber from the Ru Smith collection contained ten dauer stages.

DESCRIPTION

For specimens in N-3-60 (Figs 146, 147): head attenuated, rounded at tip; tail long, slender, pointed; length (n = 10) 229 (207-275) μm; diam. (n = 10) 12 (10-15) μm; a = 19; some individuals surrounded by enclosing cuticle of previous stage.

For specimens in Ru Smith collection (RS.P1030) (Figs 148, 149): head attenuated, rounded at tip, tail long, slender, pointed at tip; length (n = 10) 240 (216-293) μm; diam. (n = 10) 11 (10-14) μm; a = 22.

For specimens in amber specimen N-3-90, head attenuated, rounded to truncate at tip; tail long, pointed; length (n = 3) 243 (226-310) μm; diam. (n = 3) 12 (11-14) μm; a = 20. The nematodes were positioned under the elytra and in the surrounding amber (Fig. 150).

For nematodes in amber piece N-3-95; head attenuated; rounded at tip; tail long, pointed; length (n = 4) 223 (204-238) μm; diam. (n = 4) 14 (11-16) μm; a = 16. Some of these dauer were still in the intersegmental folds of the beetle. This platypodid was particularly interesting since it was also carrying a pseudoscorpion, thus providing an example of double phoresis in Dominican amber (Fig. 153; Plate 17).

COMMENTS

A diagnostic feature of these fossils is their relatively large size; however, not enough details are present to definitely assign the species to an extant genus or family. Unfortunately, few workers provide quantitative information about the dauer stages of bark beetle nematodes but Rühm (1956) provided measurements of some species. Genera that have dauer stages overlapping in length and diameter with *S. dominicana* **n. gen., n. sp.** are *Cryptaphelenchus* (272-318 μm) (Aphelenchoididae), *Panagrolaimus* (254-288 μm) (Panagrolaimidae) and *Micoletzkya* (264-413 μm) (Diplogastridae). Based on the size and general shape, these nematodes are considered to belong to the Diplogastridae.

TYPE MATERIAL

Specimens (accession nos N-3-60, N-3-90 and N-3-95) deposited in the Poinar amber collection maintained at Oregon State University: R.S.P1030 deposited in the Ru Smith amber collection.

TYPE HOST

Nematodes in N-3-60 are associated with *Scolytus poinari* Bright (length 4.8 mm) (Coleoptera: Scolytidae) (Bright & Poinar, 1994). Nematodes in N-3-90 and the Ru Smith specimen are associated with unidentified platypodid beetles (Coleoptera: Platypodidae). The nematodes in N-3-95 are associated with a male of the platypodid beetle, *Cenocephalus rhinoceroide* Schwaller (length = 2.3 mm).

TYPE LOCALITY

Amber mines located between Santiago and Puerto Plata in the Cordillera Septentrional of the northern portion of the Dominican Republic.

ETYMOLOGY

The generic epithet is derived from the family of host beetles and the Greek 'nema' = thread. The specific epithet is derived from the locality of the fossil.

***Scolytonema hispaniolae* n. gen., n. sp.**

(Figs 151-153; Plate 17)

This species is represented by 21 specimens resting on or adjacent to the dorsum of the abdomen (the elytra and membranous wings are outstretched, thus exposing the abdomen) of a platypodid beetle in Dominican amber (N-3-89). An additional six specimens occur adjacent to a platypodid beetle in a separate piece of amber (N-3-95) that also contains juveniles of *S. dominicana* **n. gen., n. sp.**

DESCRIPTION

For all specimens in both N-3-89 and N-3-95; head narrowed but rounded at tip; lips not offset; tail short, awl-shaped; length specimens in N-3-89; (n = 9) 163 (136-199) μm; diam. (n = 10) 11 (9-14) μm; a = 15. Length specimens in N-3-95: (n = 6) 177 (159-192) μm; diam. (n = 6) 13 (11-16) μm; a = 14.

COMMENTS

This species is shorter and also has a shorter tail than *S. dominicana* **n. gen., n. sp.** It is clear that the fossil nematodes were being carried under the elytra of the beetle in N-3-89 since most still occur in that position. There is a 'plug' of frass adjacent to the beetle but a careful search did not reveal any nematodes in the frass pellet.

TYPE MATERIAL

Specimens (accession nos N-3-89 and N-3-95) deposited in the Poinar amber collection maintained at Oregon State University.

TYPE HOST

Adult platypodid beetle (lengths 3.9 and 3.6 mm) (Coleoptera: Platypodidae) in Dominican amber.

TYPE LOCALITY

Amber mines located between Santiago and Puerto Plata in the Cordillera Septentrional of the northern portion of the Dominican Republic.

ETYMOLOGY

The specific epithet is derived from the region where the amber was found.

Scolytonema mexicana n. gen., n. sp.

(Figs 246, 247)

This species is represented by 14 specimens adjacent to a platypodid beetle in Mexican amber.

DESCRIPTION

Head attenuated; tail long, pointed; length (n = 8) 185 (154-211) μm; diam. (n = 8) 8 (7-9) μm; a = 23.

COMMENTS

This species is characterised by its small size.

TYPE MATERIAL

Specimens (accession no. N-3-39) deposited in the Poinar amber collection maintained at Oregon State University.

TYPE HOST

An adult platypodid beetle (Coleoptera: Platypodidae) (length 3.3 mm).

TYPE LOCALITY

Semijovel, Chiapas, Mexico. La Quinta Formation, Late Oligocene-Early Miocene.

ETYMOLOGY

The specific epithet is derived from the locality of the amber.

***Syconema* n. gen.**

This is a collective genus for nematodes associated with fossil fig wasps.

TYPE AND ONLY SPECIES

S. dominicana **n. gen., n. sp.**

***Syconema dominicana* n. gen., n. sp.**
(Figs 193-195; Plate 21)

This species is based on 59 nematodes emerging from the body cavity of an adult female fig wasp in Dominican amber.

DESCRIPTION

N-3-24A: head not offset; strongly narrowed at apex; stylet absent; length (n = 18) 292 (159-419) μm; diam. (n = 18) 12 (11-16) μm; a = 24.

Two other pieces of Dominican amber with fig wasps and associated nematodes are N-3-24B, with a fig wasp (length 1.14 mm) surrounded by four nematodes; length 145 (125-172) and diam. 11 (9-14) μm; a = 13; and amber piece N-3-24C, with a fig wasp (length 2.2 mm) with five associated nematodes. Only three of the nematodes are clearly visible and they range from 193 (170-211) μm in length and 13 (11-14) μm in diam.; a = 15. While these other specimens are not as clear or well preserved as those in N-3-24A, they appear to be conspecific.

COMMENTS

It is clear that several stages are represented in this sample and the range in size indicates that the nematodes were developing inside the body cavity of the fig wasps. No adults were detected so it is likely that the nematodes represent third- and fourth-stage juveniles with possibly some second-stage juveniles as well. This species is placed in the *Parasitodiplogaster* clade based on the attenuated head with the rounded (not offset) lip region and greatly attenuated tail. A comparison with the extant *Parasitodiplogaster sycophilon* Poinar, 1979 reveals a similar attenuated head and tail (Fig. 196). Juveniles of *P. nymphanema* Poinar & Herre, 1991 removed from the genital system of extant female fig wasps ranged in length from 222 to 288 μm and in diam. from 10 to 13 μm, which overlaps with the measurements of the fossil forms.

It should be noted, however, that size alone cannot be used to distinguish members of the *Parasitodiplogaster* clade from those of the genus *Schistonchus* Cobb, 1927, which also occur in the body cavity of fig wasps. Second-stage juveniles of *S. macrophylla* Lloyd & Davies, 1997 and *S. altermacrophylla* Lloyd & Davies, 1997 ranged from 220 to 317 μm in length, which also overlaps with the lengths of the fossils. However, *Schistonchus* has a well developed stylet and an offset blunt-tipped, lip region, which are lacking on the fossils. The fig wasp, which is 2.2 mm in length, is surrounded by pollen grains resembling those of extant *Ficus* (Fig. 193).

TYPE MATERIAL

Specimens (accession nos N-3-24A, B and C) deposited in the Poinar amber collection maintained at Oregon State University.

TYPE HOST

Fig wasp belonging to the family Agaonidae (Hymenoptera).

TYPE LOCALITY

Amber mines located between Santiago and Puerto Plata in the Cordillera Septentrional of the northern portion of the Dominican Republic.

ETYMOLOGY

The generic epithet is derived from the Greek 'sykon' = fig and the Greek 'nema' = thread. The specific epithet is derived from the location of the amber.

Infraorder Rhabditomorpha De Ley & Blaxter, 2002
Superfamily Rhabditoidea Örley, 1880
Family Rhabditidae Örley, 1880

Genus *Palaeoarachnus* n. gen.

This collective genus is for dauer nematodes associated with fossil spiders in Tertiary amber.

TYPE AND ONLY SPECIES

P. balticus **n. gen., n. sp.**

***Palaeoarachnus balticus* n. gen., n. sp.**

This species is based on two dauer rhabditids adjacent to a spider in Baltic amber.

DESCRIPTION (based on drawing and photos in Wunderlich, 2004)

Body broad; tail pointed; head rounded; lengths and diam., respectively, specimen 1; 461 μm, 23 μm, a = 20: specimen 2; 292 μm, 15 μm, a = 19.

COMMENTS

The nematodes are adjacent to the opisthosoma and right femur of the spider host. The measurements were calculated from Figure 5, p. 182 of Wunderlich (2004) and the description was based on Figure 5 and photo 337 on p. 465 of Wunderlich (2004). Wunderlich (2004) included these nematodes in the Rhabditida and this placement has been followed here. Phoretic nematodes associated with spiders have been little studied, but Noordam *et al.* (1998) reported rhabditid dauer stages on extant dune spiders in Holland.

TYPE SPECIMEN

Deposited in the Wunderlich amber collection, Hirschberg, Germany.

TYPE HOST

Holotype male spider, *Eotrechalea annulata* Wunderlich, 2004 (Arachnida: Trechaleidae).

TYPE LOCALITY

Baltic region some 40-50 mya.

ETYMOLOGY

The generic epithet is derived from the Greek 'palaois' = old and the Greek 'arachne' = spider. The specific epithet is derived from the locality of the amber.

Genus *Crustafila* n. gen.

This is a collective group genus for fossil nematodes associated with members of the Isopoda and Amphipoda.

TYPE AND ONLY SPECIES

C. confina **n. gen., n. sp.**

Crustafila confina n. gen., n. sp.
(Figs 234-236)

This species is based on several specimens associated with an adult and juvenile pillbug in Dominican amber. The size and shape of the specimens indicate that a developing colony occurred in the microhabitat of the pillbug. One specimen, considered to represent a dauer juvenile, was adjacent to the pillbug.

DESCRIPTION

Body characters variable, depending on stage; head bluntly rounded; tail attenuated; lengths range from 176 to 399 μm; diam. range from 7 to 17 μm; a = 25; dauer juvenile, 305 μm in length and 12 μm in diam. with three cuticular striations in tail region.

COMMENTS

This species is tentatively placed in the Rhabditidae based on the shape and size of the dauer stage. Dauer juveniles of the extant rhabditid *Caenorhabditis vulgaris* Baird, Fitch & Emmons, 1994 occur on the inner surface of the dorsal carapace of pill bugs. After the host dies the nematodes initiate development on the cadaver within 1-2 days (Baird *et al.*, 1994). It is possible that a similar life pattern occurred with *C. confina* **n. gen., n. sp.** and the fossils represent stages that were able to initiate development after the pillbug perished in the resin. The dauer stages of *C. vulgaris* range from 620 to 680 μm in length and 22.5 to 25.0 μm in diam. and are much larger than those of *C. confina* **n. gen., n. sp.**

TYPE MATERIAL

Specimen (accession no. N-3-98) deposited in the Poinar amber collection maintained at Oregon State University.

TYPE HOST

A pillbug of the family Philosciidae (Isopoda) in Dominican amber.

TYPE LOCALITY

Amber mines located between Santiago and Puerto Plata in the Cordillera Septentrional of the northern portion of the Dominican Republic.

ETYMOLOGY

The generic epithet is derived from the Latin 'crusta' = hard body surface and the Latin 'filum' for thread. The specific epithet is derived from the Latin 'confinis' = adjoining.

Genus *Palaeorhabditis* n. gen.

This is a collective group genus for fossil rhabditids associated with frass from wood-boring insects.

TYPE AND ONLY SPECIES

P. balticus **n. gen., n. sp.**

Palaeorhabditis balticus n. gen., n. sp.
(Figs 58, 59)

This species is described from several hundred specimens associated with frass from a wood-boring insect in Baltic amber.

DESCRIPTION

Head attenuated, rounded at tip; tail elongate-conoid; length (n = 12) 742 (248-1400) μm; diam. 34 (18-50); a = 22; V = 56: intrauterine juvenile: length 126 μm; diam. 8 μm.

COMMENTS

The insect frass is probably from a member of the bark beetle family Scolytidae, which occurs in Baltic amber (Poinar, 1992). No males could be detected in this large population of mixed stages. Also no eggs could be detected either in the amber or between the wood filaments and particles forming the frass. Based on these observations, *P.*

balticus **n. gen., n. sp.** is presumed to be an autotokous, self-fertilising hermaphrodite, although parthenogenesis and pseudogamy also occur in members of the Rhabditidae (Poinar & Hansen, 1983). The only other free-living nematode groups whose females are autotokous are the Monhysteridae and Oncholaimidae, both of which are associated with an aquatic or moist habitat. Rühm (1956) did find two monhysterids (*Monhystera vulgaris* de Man, 1880 and *Prismatolaimus dolichurus* de Man, 1880) in bark beetle frass, but these were of rare occurrence and he concluded that they entered the galleries from the soil during periods of high moisture.

A comparison was made with the free-living nematodes in the genus *Vetus* described previously from Baltic amber. While these species are not associated with insect frass, they could have been at one time. Specimens of *Vetus capillaceus* (Menge) are thinner and longer (up to 2 mm in length) than *P. balticus* **n. gen., n. sp.**; those of *V. duisburgi* Taylor are smaller (650-850 μm in length and 17-18 μm in diam.) than *P. balticus* **n. gen., n. sp.** and males occur in the population. The stages of *V. pristinus* (Menge) (139 μm in length) are minute and have pointed tails.

Type specimen

Deposited in the Poinar amber collection (accession no. N-3-81) maintained at Oregon State University.

Type habitat

Developing in frass of a wood-boring beetle.

Type locality

Baltic region some 40-50 mya.

Etymology

The generic epithet is derived from the Greek 'palaios' = old. The specific epithet is derived from the location of the amber.

Genus *Prescafilotermes* n. gen.

This collective genus is established for fossil nematodes of the family Rhabditidae associated with termites.

TYPE SPECIES

P. dominicanus **n. gen., n. sp.**

OTHER SPECIES

P. elongatus **n. gen., n. sp.**

Prescafilotermes dominicanus n. gen., n. sp.
(Figs 125-128)

This species occurs in five pieces of Dominican amber and is represented by both reproducing and dauer individuals associated with termites of the families Hodotermitidae, Kalotermitidae and Rhinotermitidae.

DESCRIPTION

Reproducing stages: Lips not offset, head rounded, tail pointed. Specimens (n = 6): In accession no. N-3-26B; length 376 (311-495) μm; diam. 22 (14-27) μm; a = 17. Specimens (n = 9): In accession no. N-3-26 C; length 508 (315-888) μm; diam. 48 (23-74) μm; a = 11. Specimens (n = 5): In accession no. N-3-26D; length 310 (203-486) μm; diam. 26 (19-39) μm; a = 12. Dauer stages: Lips not offset, body slender; tail elongate, filiform. Specimens (n = 2): In accession no. N-3-26A; length 428 (420-436) μm; diam. 17 (16-18) μm; a = 25. Specimens (n = 4): In accession no. N-3-26G; length 443 (378-513) μm; diam. 20 (17-22) μm; a = 22.

COMMENTS

The size and general body shape are the reason these nematodes are placed in the family Rhabditidae. Present-day damp wood termites *Zootermopsis angusticollis* (Hagen) in Oregon carry dauer stages of a rhabditid similar to *P. dominicanus* **n. gen., n. sp.** (Fig. 133). These extant dauer stages average 450 μm in length and 16 μm in diam.

TYPE MATERIAL

Specimens (accession nos N-3-26A, B, C, D and G) deposited in the Poinar amber collection maintained at Oregon State University.

TYPE HOSTS

All specimens associated with worker termites of the families Hodotermitidae, Kalotermitidae and Rhinotermitidae (Isoptera).

TYPE LOCALITY

Amber mines located between Santiago and Puerto Plata in the Cordillera Septentrional of the northern portion of the Dominican Republic.

ETYMOLOGY

The generic epithet is derived from the Latin 'prescus' = ancient, the Latin 'filum' = thread and the termite genus, *Termes*. The specific epithet is derived from the location of the amber.

***Prescafilotermes elongatus* n. gen., n. sp.**
(Figs 129, 130)

This species is represented by a single dauer adjacent to a worker termite of the family Termitidae.

DESCRIPTION

Body slender, elongate, head narrowed at tip; tail filiform; length 654 μm; diam. 29 μm; a = 23.

COMMENTS

The large size of the dauer stage and its association with a member of the Termitidae distinguishes this species differs from *P. dominicanus* **n. gen., n. sp.**

TYPE MATERIAL

Specimen (accession no. N-3-26E) deposited in the Poinar amber collection maintained at Oregon State University.

TYPE HOST

A worker termite (*Nausutitermes* sp.) of the family Termitidae (Isoptera).

TYPE LOCALITY

Amber mines located between Santiago and Puerto Plata in the Cordillera Septentrional of the northern portion of the Dominican Republic.

ETYMOLOGY

The specific epithet is derived from the Latin 'elongatus' = elongated.

Genus *Psocidonema* n. gen.

This is a collective genus for fossil rhabditoid nematodes associated with members of the order Psocida (bark lice).

TYPE AND ONLY SPECIES

P. dominicana **n. gen., n. sp.**

Psocidonema dominicana n. gen., n. sp.
(Figs 134, 135)

This species is based on two specimens associated with a bark louse in Dominican amber. One specimen is adjacent to the insect while a second specimen is attached to the abdomen of the bark louse.

DESCRIPTION

Specimen adjacent to the bark louse: head rounded, tail truncate; outer cuticle with at least two lateral lines; length 430 μm; diam. 23 μm; a = 19. Due to its position on the insect, the second specimen could not be measured.

Comments

These are placed in the family Rhabditidae since the size of the juvenile falls within the size range of most rhabditid dauer juveniles. A phoretic association is indicated with the specimen still attached to the bark louse. No records of extant nematode associates or parasites of bark lice could be found.

Type material

Specimens (accession no. N-3-92) deposited in the Poinar amber collection maintained at Oregon State University.

Type host

An adult bark louse (Psocoptera).

Type locality

Amber mines located between Santiago and Puerto Plata in the Cordillera Septentrional of the northern portion of the Dominican Republic.

Etymology

The generic epithet is derived from the order of the host and the Greek 'nema' = thread. The specific epithet is derived from the location of the amber.

Genus *Psychodorhabditis* n. gen.

This is a collective genus for fossil rhabditoid nematodes associated with members of the dipterous families Psychodidae and Phlebotomidae.

Type species

P. attenuatus **n. gen., n. sp.**

Other species

P. balticus **n. gen., n. sp.**
P. dominicanus **n. gen., n. sp.**

***Psychodorhabditis attenuatus* n. gen., n. sp.**
(Figs 167-169)

This species is represented by two dauer nematodes associated with a female mothfly in Dominican amber. One dauer is wrapped around the abdomen of the mothfly and could not be measured.

DESCRIPTION

Body long and slender; length 450 μm; diam. 11 μm; a = 41; tail attenuated.

COMMENTS

The habit of extant rhabditid nematodes wrapping their bodies around the abdomen of mothflies was first reported for *Rhabditis dubia* Bovien, 1937. The fossil dauer stage is shorter and thinner than those of *R. dubia* and the closely related *Rhabditis berolina* Sudhaus & Kühne, 1989 which, along with several other species of *Rhabditis* and a *Hexatylus*, have similar phoretic habits.

TYPE MATERIAL

Deposited in the Poinar amber collection (accession no. N-3-57A) maintained at Oregon State University.

TYPE HOST

An adult female (length 1.3 mm) of the family Psychodidae (Diptera).

TYPE LOCALITY

Amber mines located between Santiago and Puerto Plata in the Cordillera Septentrional of the northern portion of the Dominican Republic.

ETYMOLOGY

The generic epithet is derived from the name of the host family and the nematode genus *Rhabditis*. The specific epithet is derived from the Latin ‘attenuatus’ = drawn out.

Psychodorhabditis balticus n. gen., n. sp.
(Figs 54, 55)

This species is based on two dauer nematodes adjacent to a mothfly in Baltic amber.

Description

Head rounded; tail attenuated: lengths 1.1 mm and 1.2 mm; diam. 56 μm; a = 20.

Comments

While many details are obscured, the size and attenuated tails are similar to those of the dauer stages of the extant *R. dubia* (length 800-1210 μm; diam. 36-54 μm) and *R. berolina* Sudhaus & Kühne (length 800-950 μm; diam. 26-40 μm). The dauer stages of both these species are phoretic on mothflies (Bovien, 1937; Sudhaus & Kühne, 1989).

Type specimen

Deposited in the Poinar amber collection (accession no. N-3-74) maintained at Oregon State University.

Type host

An adult mothfly (Diptera: Psychodidae).

Type locality

Baltic region some 40-50 mya.

Etymology

The specific epithet is derived from the location of the amber.

Psychodorhabditis dominicanus n. gen., n. sp.
(Figs 170, 171)

This species is represented by a single dauer juvenile adjacent to an adult mothfly in Dominican amber.

DESCRIPTION

Head rounded at tip; tail rounded; length 848 μm; diam. 17 μm; a = 50.

COMMENTS

This species is almost twice as long as *P. attenuatus* **n. gen., n. sp.** from the same amber source and has a rounded tail. Its size is within the range of the extant species, *R. dubia* (800-1210 μm) and *R. berolina* (800-950 μm). The dauer of both these species are phoretic on mothflies (Bovien, 1937; Sudhaus & Kühne, 1989) but have not been reported outside of Europe.

TYPE MATERIAL

Deposited in the Poinar amber collection (accession no. N-3-57C) maintained at Oregon State University.

TYPE HOST

An adult mothfly (length 1.2 mm) of the family Psychodidae (Diptera).

TYPE LOCALITY

Amber mines located between Santiago and Puerto Plata in the Cordillera Septentrional of the northern portion of the Dominican Republic.

ETYMOLOGY

The specific epithet is derived from the location of the amber.

Genus *Tipulacomis* n. gen.

Tipulacomis is erected as a collective genus for fossil rhabditoid nematodes associated with members of the Tipuloidea (including the families Cylindrotomidae, Limoniidae, Pediciidae and Tipulidae).

TYPE SPECIES

T. balticus **n. gen., n. sp.**

OTHER SPECIES

T. dominicanus **n. gen., n. sp.**

Tipulacomis balticus n. gen., n. sp.

(Figs 51-53; Plate 7)

This species is based on dauer and developmental forms associated with an adult cranefly in Baltic amber.

DESCRIPTION

Dauer juveniles (n = 9): slender with narrowed, rounded head, long, narrow stoma and elongate straight filiform tail; length 287 (266-312) μm; diam. 13 (11-15) μm; a = 22. Developmental stages (n = 2). Short, stout forms with head bluntly rounded; tail abruptly attenuated (spiked at the tip); length 328-365 μm; diam. 40 μm; a = 9.

COMMENTS

Large numbers of dauers, as well as the two developmental stages, are clustered around the abdominal segments of the cranefly. The two developmental stages show that both dauer juveniles as well as non-dauer stages can have phoretic associations with adult craneflies. A separate piece of Baltic amber with an adult Limoniidae associated with numerous specimens of *T. balticus* **n. gen., n. sp.** ranging in length from 180 to 280 μm is illustrated in Weitschat and Wichard (2002: their Plate 7a, b).

Lam and Webster (1971) described two rhabditoids associated with the larvae of *Tipula paludosa* Meigen in North America and Sudhaus (1993) reported one of these species, *Rhabditis* (*Oscheius*) *tipulae* Lam & Webster, 1971, from European *T. paludosa* larvae. Since none of the authors gave any measurements of the dauer stages of these cranefly-associated rhabditids, comparisons with the dauer of *T. balticus* **n. gen., n. sp.** are not possible. It is likely that *T. balticus* **n. gen., n. sp.** developed in the same habitat as the cranefly and climbed on the adult fly when it emerged from the pupal cell in the soil. These associations are considered to be strictly phoretic. No reports of extant nematodes associated with adult craneflies could be located.

TYPE SPECIMEN

Deposited in the Poinar amber collection (accession no. N-3-70) maintained at Oregon State University.

TYPE HOST

An adult of the family Limoniidae (Diptera).

TYPE LOCALITY

Baltic region some 40-50 mya.

ETYMOLOGY

The generic epithet is derived from the host family and the Latin 'comis' = friendly. The specific epithet is derived from the location of the amber.

***Tipulacomis dominicanus* n. gen., n. sp.**
(Figs 186-188)

This species is based on dauer juveniles associated with an adult cranefly in Dominican amber.

DESCRIPTION

(n = 5): Long, slender forms with rounded head and pointed tail; tail in at least half of population with characteristic bend; length 395 (346-432) μm; diam. 16 (14-17) μm; a = 25.

COMMENTS

The length of the dauer stages and a characteristic bend in their tail region separate this species from *T. balticus* **n. gen., n. sp.** in Baltic amber. Lam and Webster (1971) described two rhabditoids associated with the larvae of *Tipula paludosa* Meigen in North America and Sudhaus (1993) reported one of these species, *Rhabditis* (*Oscheius*) *tipulae*, from European *T. paludosa* larvae. Since neither of these publications gave any measurements of the dauer stages of these cranefly-associated rhabditids, comparisons with the dauer of *T. dominicanus* **n. gen., n. sp.** cannot be made. It is likely that *T. dominicanus* **n. gen., n. sp.** developed in the same habitat as the cranefly and climbed on the adult fly when it emerged from the pupal cell in the soil. These associations are considered to be strictly phoretic. No reports of extant nematodes associated with adult craneflies could be located.

TYPE MATERIAL

Specimen (accession no. N-3-72) deposited in the Poinar amber collection maintained at Oregon State University.

TYPE HOST

An adult female of the genus *Dicranomyia* (Diptera: Limoniidae).

TYPE LOCALITY

Amber mines located between Santiago and Puerto Plata in the Cordillera Septentrional of the northern portion of the Dominican Republic.

ETYMOLOGY

The specific epithet is derived from the location of the amber.

Superfamily Heterorhabditoidea n. grad.
Family Heterorhabditidae Poinar, 1976

Genus *Proheterorhabditis* n. gen.

This genus is erected for entomopathogenic nematodes with characteristics of the family Heterorhabditidae.

TYPE AND ONLY SPECIES

Proheterorhabditis burmanicus **n. gen., n. sp.**

Proheterorhabditis burmanicus n. gen., n. sp.
(Figs 35-37)

This species is represented by three infective stage juveniles adjacent to a partially decomposed beetle in Burmese amber.

Description

Third-stage juveniles surrounded by second-stage cuticle with longitudinal striations; head bluntly rounded, tip of head bearing minute tooth; tail attenuated to a fine point; length (n = 3) 517 (485-533) μm; diam. (n = 3) 26 (24-29) μm; a = 20. Associated bacteria: A detailed examination of the haemocoel and adipose tissue in the body cavity of the rove beetle revealed rod-shaped bacteria ranging from 5.1 to 6.6 μm in length (Fig. 37).

Comments

Characters that relate these fossils to extant heterorhabditids are their lengths (485-533 μm), which compares with that of *Heterorhabditis indicus* Poinar *et al.*, 1992 (479-573 μm), the ensheathing cuticle with longitudinal striations and the small terminal tooth. The presence in the body cavity of the beetle host of bacterial cells that are similar in size and shape to those of *Photorhabdus luminescens* associated with extant *Heterorhabditis* infections is another feature connecting the fossils to the heterorhabditid clade.

Type material

Deposited in the Poinar amber collection (accession no. B-N-8) maintained at Oregon State University.

Type host

Length 1.3 mm; body oval, not flattened; antennae threadlike, with rounded segments, antennal club indistinct; tarsi, 5-5-5; elytra abbreviated, exposing last five abdominal segments, elytral striations indistinct. While the specimen is somewhat decomposed, which makes an exact determination difficult, the above characters are similar to those of some Staphylinidae (Coleoptera). This assignment is only tentative since many diagnostic characters are masked. However, rove beetle adults and larvae are susceptible to infection by extant species of *Heterorhabditis* (Georgis *et al.*, 1991).

Type locality

Amber mine in the Hukawng Valley, southwest of Maingkhwan in the state of Kachin (26°20′N, 96°36′E), northern Myanmar (Burma).

ETYMOLOGY

The generic epithet is derived from the Greek 'pro' = before and the nematode genus *Heterorhabditis*. The specific epithet is derived from the location of the amber.

Genera without family status

Copalonema n. gen.

This collective genus is erected for sub-recent (sub-fossil) nematodes in copal deposits around the world.

TYPE AND ONLY SPECIES

Copalonema colombiana **n. gen., n. sp.**

Copalonema colombiana n. gen., n. sp.
(Figs 257, 258)

This species is based on 50 juvenile and adult nematodes in a piece of copal from Colombia, South America.

DESCRIPTION

Head evenly rounded; tail conoid, awl-shaped; length of adults (n = 2) 1.8 and 1.9 mm; diam. of adults (n = 2) 69 and 91 μm; a = 26; length of juveniles (n = 3) 182, 297 and 411 μm.

COMMENTS

These fossils are in Colombian copal, which originates from at least two extant species of *Hymenaea* trees. Using carbon dating techniques, Colombian copal was found to range from 10 to 500 years of age (Poinar, 1996). While *C. colombiana* **n. gen., n. sp.** is most certainly a contemporary nematode still living in South America, it has probably not been described since neotropical free-living nematodes have been little investigated. Inclusions in copal allow biologists to study the effect of short-term climatic or ecological changes on extant biota. Colombian copal, which is usually lighter in colour than amber, is sometimes sold as amber with ages given in millions of years.

TYPE MATERIAL

Deposited in the Poinar amber collection (accession no. N-3-11) maintained at Oregon State University.

TYPE HABITAT

Free living on the bark of trees.

TYPE LOCALITY

Magdalena River valley in Colombia, South America.

ETYMOLOGY

The generic epithet is derived from the aboriginal American term 'copalli' = resin from leguminous trees. The specific epithet is derived from the location of the amber.

Genus *Vetus* Taylor, 1935

This collective genus was erected by Taylor (1935) for fossil free-living nematodes that cannot be placed in an extant family.

TYPE SPECIES

V. capillaceus (Menge, 1866) Taylor, 1935
=*Anguillula capillacea* Menge, 1866

OTHER SPECIES

V. duisburgi Taylor, 1935
V. elongatus **n. sp.**
V. libani **n. sp.**
V. maximus **n. sp.**
V. microcosmus **n. sp.**
V. pristinus (Menge, 1866) Taylor, 1935
=*Anguillula pristina* Menge, 1866
V. robustus Poinar, 1977

Vetus capillaceus **(Menge, 1866) Taylor, 1935**
=*Anguillula capillacea* Menge, 1866
(Fig. 49)

This species was based on some 50 individuals in a single piece of amber.

DESCRIPTION (modified from Menge, 1866)

Body cylindrical, very slender, head rounded, tail pointed; cuticle without annulations; length 0.7-2.0 mm; greatest diam. 100 μm; a = 10.

TYPE SPECIMEN

Location unknown.

HABITS

Free in amber.

TYPE LOCALITY

Baltic region some 40-50 mya.

Vetus duisburgi **Taylor, 1935**
(Fig. 50)

This species was described by Taylor (1935) based on two specimens that were reported by von Duisburg (1862).

DESCRIPTION (modified from Taylor, 1935)

Body fusiform; head rounded, tail bluntly pointed; cuticle bearing annules 1 μm wide; lengths, diam. and ratio a, respectively; specimen A, 850 μm, 18 μm, 47; specimen B, 650 μm, 17 μm, 38.

TYPE SPECIMEN

Location unknown.

HABITAT

Free in amber.

TYPE LOCALITY

Baltic region some 40-50 mya.

Vetus elongatus n. sp.
(Figs 103, 104)

Represented by eight nematodes in a piece of Dominican amber.

DESCRIPTION

Body elongate, narrow, length (n = 5) 1.128 (0.945-1.260) mm; diam. 22 (10-25) μm; a = 51; head attenuated but rounded at tip; stylet length *ca* 20 μm; tail short, rounded; anus located near tail tip.

COMMENTS

This species has characteristics of free-living, entomogenous females of *Deladenus* spp. The entomogenous infective females of *Deladenus siricidicola* Bedding, 1972b range from 0.80 to 1.60 mm in length, are 20 μm in diam. and the style length varies from 19 to 25 μm. All of these measurements fall within those of *V. elongatus* **n. sp.** There are two distinct generations of *D. siricidicola* in wood infested by *Sirex* woodwasps. One is composed of mycetophagous females that feed on fungi symbiotically associated with the woodwasps. Under conditions of stress, insect-infective stages are formed and, after mating, the females move through the plant host until they locate *Sirex* larvae (Bedding, 1972b; Poinar, 1979). The fossil species could be infective stage entomogenous females of other neotylenchs, since neotropical members of this family have been little investigated. The nematodes could have been in an insect gallery in the wood and pushed out by a resin flow.

TYPE SPECIMEN

Deposited in the Poinar amber collection (accession no. N-3-33) maintained at Oregon State University.

HABITAT

Free living.

TYPE LOCALITY

Amber mines located between Santiago and Puerto Plata in the Cordillera Septentrional of the northern portion of the Dominican Republic.

ETYMOLOGY

The specific epithet is derived from the Latin 'elongatus' = elongated.

***Vetus libani* n. sp.**
(Fig. 24)

This species is represented by two free-living nematodes in Lebanese amber.

DESCRIPTION

Small forms with head sharply truncate; tail pointed, but not filamentous; length (n = 2) 239 and 248 μm; diam. (n = 2) 14 and 18 μm; a = 17.

COMMENTS

These are considered to be microbotrophic, free-living forms that were living under the bark or in fissures in the bark on the amber tree. There are no other animal inclusions in the amber. They were previously noted in a survey of life forms in Lebanese amber (Poinar & Milki, 2001).

TYPE MATERIAL

Deposited in the Milki amber collection (accession no. 194-7) in Beirut, Lebanon.

HABITAT

Free living.

TYPE LOCALITY

Between Jezzine and Dar-al-Baidhar in southern Lebanon. Beds originate from the Neocomian division of the Early Cretaceous some 130-135 mya.

ETYMOLOGY

The specific epithet is derived from the location of the amber.

Vetus maximus n. sp.

(Figs 132, 133)

This species is represented by 89 specimens in a piece of Dominican amber. The only other inclusion present is a partial beetle. All individuals appear to be adults or developmental stages.

DESCRIPTION

Adults: head attenuated; tail pointed, with blunt tip; length (n = 10) 884 (347-1166) μm; diam. 37 (25-44) μm; a = 23; eggs (free in amber) round to oval, 32-38 μm in greatest diam.

COMMENTS

All developmental stages of this species, including eggs, are present and it is evident that the population was developing on the bark surface when a resin flow swept them away. Based on the general length and diameter as well as the egg shape, *V. maximus* **n. sp.** is probably a member of the Rhabditidae or Diplogastridae.

TYPE MATERIAL

Specimen (accession no. N-3-4) deposited in the Poinar amber collection maintained at Oregon State University.

HABITAT

Free living.

TYPE LOCALITY

Amber mines located between Santiago and Puerto Plata in the Cordillera Septentrional of the northern portion of the Dominican Republic.

ETYMOLOGY

The specific epithet is derived from the Latin 'maximus' = large.

***Vetus microcosmus* n. sp.**
(Figs 240-242)

This species is based on three juveniles in a mass of fungal hyphae together with a pseudoscorpion and several other organisms in a piece of Dominican amber.

DESCRIPTION

Lip region truncate; tail short, but pointed; length (n = 3) 247 (220-294) μm; diam. (n = 3) 16 (11-20) μm; a = 15.

COMMENTS

The nematodes, which appear to be juvenile rhabditoids or diplogastroids, make up only a small portion of this microcosm. Aside from the pseudoscorpion which is the top predator in the system, there are over 500 specimens of tardigrade-like invertebrates as well as oribatid mites and protozoa. It is possible that the nematodes were introduced with the pseudoscorpion since dauer stages of rhabditids are known to attach themselves to pseudoscorpions (Poinar & Ćurčić, 1994; Uri *et al.*, 1996).

TYPE MATERIAL

Specimen (accession no. ON-1-4) deposited in the Poinar amber collection maintained at Oregon State University.

HABITAT

Free-living.

TYPE LOCALITY

Amber mines located between Santiago and Puerto Plata in the Cordillera Septentrional of the northern portion of the Dominican Republic.

ETYMOLOGY

The specific epithet is derived from the Greek words 'micros' = little and 'kosmos' = universe.

Vetus pristinus (Menge, 1866) Taylor, 1935

=*Anguillula pristina* Menge, 1866
(Fig. 48)

This species was described from some 30 specimens in a single piece of amber.

DESCRIPTION (modified from Menge, 1866)

Body fusiform, head rounded; tail pointed; length 139 μm.

COMMENTS

Measurements were given only for a single specimen. This may have been due to the orientation of the nematodes.

TYPE SPECIMEN

Location unknown.

HABITAT

Free in amber.

TYPE LOCALITY

Baltic region some 40-50 mya.

Vetus robustus Poinar, 1977
(Figs 243, 244)

This species is based on nematodes in two separate blocks of Mexican amber. Two size categories of nematodes (large and small) occurred in both blocks. All were considered to represent the same species.

DESCRIPTION (modified from Poinar, 1977a)

Block B-8412-1. Larger forms: six partially fused lips; rounded anterior and posterior ends; tail bearing small mucron; length (n = 4) 650 (580-730) μm; diam. 50 (40-60) μm; a = 13. Smaller forms: pointed or slightly extended tails; length (n = 2) 310-370 μm; diam. 20 μm; a = 17. Block B-7048-5. Larger forms: six partially fused lips; rounded anterior and posterior ends with tail bearing small mucron; length (n = 3) 640 (540-720) μm; diam. 43 (40-50) μm; a = 15. Smaller forms: pointed or slightly extended tails; length (n = 2) 180-200 μm; diam. 10 μm; a = 19.

COMMENTS

The larger forms were considered mature females. Some of the smaller forms resembled dauer juveniles.

TYPE MATERIAL

Holotype female deposited in the University of California Museum of Paleontology (accession no. 14509 on block B-7048-5). Paratype females (accession nos 14510 and 14511 on block B-7048-5) deposited in the University of California Museum of Paleontology, Berkeley.

HABITAT

Free in amber.

TYPE LOCALITY

Vega de La Compana, Rio San Pedro Valley about 7 km north of Simojovel, Chiapas, Mexico, Late Oligocene-Early Miocene, La Quinta formation.

Nematology Monographs & Perspectives, 2011, Vol. 9, 359-366

Appendix II

Key to nematodes presented in this work

1 – Eggs and occasionally developmental stages found in coprolites or ancient cadavers 2
– Nematode developmental stages (rarely accompanied with eggs) in rock deposits, amber or copal 14

2 – Nematode eggs in coprolites 3
– Nematode stages in cadavers 8

3 – Eggs in Cretaceous coprolite 4
– Eggs in Pleistocene or Holocene remains 5

4 – Surface egg coat mammillated . . . . . *Ascarites priscus*
– Surface egg coat not mammillated . . . . . *Ascarites gerus*

5 – Eggs spherical 6
– Eggs oval or elliptical 7

6 – Eggshell thick, surface coat mammillated . . . . . *Ascaris* spp.
– Eggshell thin, surface not mammillated . . . . . *Ancyclostoma* and *Necator*

7 – Eggs flattened on one side; without opercula at the poles . . . . . *Enterobius vermicularis*
– Eggs with plug-like opercula at either pole . . . . . *Trichuris* spp.

8 – Nematode stages in human mummies 9
– Nematode stages in other vertebrate remains 11

9 – Nematodes in alimentary tract . . . . . *Stronglyoides stercoralis*
– Nematodes in body tissues 10

10 – Large (>20 cm in length) worms in body tissues . . . . . *Dracunculus medinensis*
– Smaller (<20 cm in length) adult worms in body tissues . . . . . filarial nematodes

11 – Nematodes in Shasta Ground sloth (*Nothrotheriops shastensis*) dung . . . . . *Agamofilaria oxyura* and *Strongyloides shastensis*
– Nematode stages from other vertebrates 12

12 – Nematodes from Upper Pleistocene horse . . . . . *Strongylus edentatus*
– Nematodes from other animals 13

13 – Nematodes from ground squirrels (*Citellus* sp.) . . . . . *Syphacia* sp.

– Nematodes from the iguaniid, *Tropidarus* sp. .... *Parapharyngodon sceleratus*

14 – Nematodes in marine sedimentary deposits 15
– Nematodes in freshwater or terrestrial (including amber and copal) deposits 17

15 – Nematodes >90 mm in length, in Carboniferous deposits .... *Nemavermes mackeei*
– Nematodes <90 mm in length, in Mesozoic deposits 16

16 – Nematodes >50 mm in length, in Jurassic deposits .... *Eophasma jurasicum*
– Nematodes <50 mm in length, in Cretaceous deposits .... *Palaeoenoploides alessandrellianus*

17 – Nematodes in amber or copal 21
– Nematodes not in amber or copal 18

18 – Nematodes >20 mm in length, associated with insects 19
– Nematodes <20 mm in length, not associated with insects 20

19 – Nematode emerging from a cerambycid beetle in Eocene brown coal .... *Heydenius antiquus*
– Nematode emerging from a tabanid fly in Pliocene deposits .... *Heydenius tabanae*

20 – Nematodes with long body setae, adults <600 μm in length, from Cretaceous swamp .... *Captivonema cretacea*
– Nematodes without long body setae, adults >600 μm in length, in stomatal cavities of Devonian plants (*Aglaophyton*) .... *Palaeonema phyticum*

21 – Nematodes in Cretaceous amber deposits 22
– Nematodes in Tertiary amber or sub-recent copal deposits 30

22 – Nematodes in Early Cretaceous Lebanese amber 23
– Nematodes in Early Cretaceous Burmese amber 24

23 – Nematode >1 mm in length, inside body of chironomid midge .... *Cretacimermis libani*
– Nematodes <1 mm in length, free-living .... *Vetus libani*

24 – Nematodes associated with terrestrial snails .... *Palaeocosmocerca burmanica*
– Nematodes not associated with snails 25

25 – Nematodes mycophagous, <1 mm in length .... *Cretaciaphelenchoides burmensis*
– Nematodes associated with insects, size variable 26

26 – Nematodes >2 mm in length, parasites of Chironomidae and Ceratopogonidae 27
– Nematodes <2 mm in length, parasites or associates of termites, craneflies or beetles 28

27 – Parasites of Chironomidae .... *Cretacimermis chironomae*
– Parasites of Ceratopogonidae .... *Cretacimermis protus*

28 – Dauer juveniles associated with termites .... *Cretodiplogaster termitophilus*
– Parasites of craneflies or beetles 29

29 – Parasites of beetles, nematodes >400 μm in length .... *Proheterorhabditis burmanicus*
– Parasites of Tipuloidea, nematodes <400 μm in length .... *Paleothelastoma tipulae*

30 – Free-living nematodes in Colombian copal .... *Copalonema colombiana*
– Nematodes in Tertiary amber deposits 31

31 – Nematodes in Baltic amber 32
– Nematodes in Dominican or Mexican amber 51

32 – Nematodes free, parasitic in or otherwise associated with arthropods 33
– Free-living mermithids ranging from 2 to 60 mm in length .... *Heydenius quadristriatus*

33 – Nematodes >1 mm in length 34
– Nematodes <1 mm in length 41

34 – Nematodes phoretic on mothflies (Psychodidae) .... *Psychodorhabditis balticus*
– Nematodes parasitic in arthropods 35

35 – Nematode parasites of spiders .... *Heydenius araneus*
– Nematode parasites of insects 36

36 – Nematode parasites of ants .... *Heydenius formicinus*
– Nematode parasites of other insects 37

37 – Nematode parasites of planthoppers (Fulgaroidea) ... *Heydenius brownii*
– Nematode parasites of other insects 38

38 – Nematode parasites of gall gnats (Cecidomyiidae) .... *Heydenius cecidomyae*
– Nematode parasites of other insects 39

39 – Nematode parasites of chironomid midges .... *Heydenius matutinus*
– Nematode parasites of sciarid fungus gnats 40

40 – Parasitic females containing eggs .... *Palaeotetradonema sciarae*

– Parasitic females without eggs . . . . *Heydenius sciarophilus*

41 – Nematodes associated with insects 42
– Nematodes free in the amber or associated with insect frass 46

42 – Nematode parasites of sciarid flies . . . . *Tripius balticus*
– Nematode parasites or associates of other insects 43

43 – Nematode parasites of phorid flies. . . . . *Howardula helenoschini*
– Parasites or associates of other insects 44

44 – Dauer juveniles phoretic on craneflies (Tipuloidea). . . . .
. . . . *Tipulacomis balticus*
– Parasites of other insects 45

45 – Parasites of rove beetles (Coleoptera; Staphylinidae) . . . .
. . . . *Palaeoallantonema baltica*
– Utilising blackflies as intermediate hosts . . . . *Cascofilaria baltica*

46 – Nematodes with stylets (Aphelenchida) . . . .
. . . . *Palaeoaphelenchoides balticus*
– Nematode stages lacking stylets (Rhabditidae & Diplogastridae) 47

47 – Small nematodes <200 μm in length . . . . *Vetus pristinus*
– Larger nematodes >200 μm in length 48

48 – Developing colonies with adults and juveniles 49
– Juvenile stages only 50

49 – Hermaphrodites associated with beetle frass; cuticle not annulated . . . . . .
. . . . *Palaeorhabditis balticus*
– Amphimictic species; cuticle annulated . . . . *Vetus duisburgi*

50 – Cuticle annulated; head knob-like; tail curved ventrally . . . .
. . . . *Oligoplectus succini*
– Cuticle lacking annulations, head not knob-like; tail not curved . . . .
. . . . *Vetus capillaceus*

51 – Nematodes in Dominican amber 52
– Nematodes in Mexican amber 95

52 – Nematodes parasitic in insects, >1.5 mm in length (Mermithidae) 53
– Nematodes with variable habits, <1.5 mm in length. 60

53 – Nematodes >10 mm in length; parasites of Lampyridae (Coleoptera) . . .
. . . . *Heydenius lamprophilus*
– Nematodes <10 mm in length, parasites of Diptera and Hymenoptera . . .
. . . . 54

54 – Parasites of ants (Hymenoptera: Formicidae) . . . .
. . . . *Heydenius myrmecophila*
– Parasites of Diptera 55

55 – Parasites of mothflies (Psychodidae) .... *Heydenius psychodae*
– Parasites of other Diptera 56

56 – Parasites of scavenger flies (Diptera; Scatopsidae) ....
.... *Heydenius scatophilus*
– Parasites of other Diptera 57

57 – Parasites of chironomid midges (Diptera: Chironomidae) ....
.... *Heydenius neotropicus*
– Parasites of other Diptera 58

58 – Parasites of mosquitoes (Diptera: Culicidae) ...... *Heydenius dominicus*
– Parasites of other Diptera 59

59 – Parasites of wood gnats (Diptera: Anisopodidae) ....
.... *Heydenius saprophilus*
– Parasites of cyclorrhaphan flies (Diptera) .... *Heydenius dipterophilus*

60 – Nematodes free in amber, not associated with arthropods or plants 61
– Nematodes associated with arthropods or plants 62

61 – Only one developmental stage present; long, slender nematodes; diam. 10-25 μm; tail rounded .... *Vetus elongatus*
– All developmental stages present; adult diam. 25-44 μm; tail pointed ...
.... *Vetus maximus*

62 – Stylet-bearing nematodes 63
– Non-stylet bearing nematodes 75

63 – Nematodes associated with higher plants 64
– Nematodes associated with arthropods 65

64 – Nematodes developing on rootlets .... *Oligaphelenchoides maximus*
– Nematodes developing in seeds .... *Palaeoanguina dominicana*

65 – Phoretic nematodes associated with platypodid beetles 66
– Parasitic nematodes of insects 67

66 – Female tail digitate, reproducing colony; dauers absent ....
.... *Bursaphelenchus similus*
– Female tail filiform, dauer stages present ....
.... *Cryptaphelenchus dominicus*

67 – Ectoparasitic nematodes on moths .... *Setonema protera*
– Body cavity parasites of other insects 68

68 – Nematode parasites of *Cephalotes* ants 69
– Nematode parasites of other insects 70

69 – Infective stage juveniles inside eggs released from infected ants ....
.... *Myrmeconema antiqua*
– Infective stage juveniles released from haemocoel of infected ants ......

. . . . *Palaeoallantonema cephalotae*

70 – Nematode parasites of mycetophilid fungus gnats (Diptera: Mycetophilidae) . . . . *Paleoiotonchium dominicanum*
– Nematode parasites of other insects 71

71 – Nematode parasites of beetles (Coleoptera) 72
– Nematode parasites of Diptera 73

72 – Nematode parasites of weevils 74
– Nematode parasites of rove beetles (Coleoptera: Staphylinidae) . . . . *Palaeoallantonema dominicana*

73 – Nematode parasites of sand flies (Diptera: Phlebotomidae) . . . . *Palaeoallantonema phlebotomae*
– Nematode parasites of drosophilid flies (Diptera: Drosophilidae) . . . . *Palaeoparasitylenchus dominicanus*

74 – Nematode parasites of *Apion* weevils . . . . *Palaeoallantonema apionae*
– Nematode parasites of zygopine weevils . . . . *Palaeoallantonema zygopinae*

75 – Nematode parasites of insects 76
– Nematodes associated with but not parasites of invertebrates 79

76 – Nematodes parasitic in fig wasps (Hymenoptera: Agaonidae) . . . . *Syconema dominicana*
– Nematodes parasitic in other Hymenoptera or Diptera 77

77 – Nematode parasites of chalcidoid wasps (Hymenoptera: Chalcidoidea) . . . . *Chalcidonema paradoxa*
– Nematodes utilising mosquitoes (Diptera: Culicidae) as intermediate hosts 78

78 – Tail portion stub-like; terminal tail knob absent . . . . *Cascofilaria dominicana*
– Tail portion not stub-like; terminal tail knob present . . . . *Cascofilaria parva*

79 – Nematodes associated with insects 83
– Nematodes associated with invertebrates other than insects 80

80 – Nematodes associated with tardigrade-like invertebrates . . . . *Vetus microcosmus*
– Nematodes associated with myriapods (millipedes or centipedes) or isopods (pill bugs) 81

81 – Nematodes associated with isopods . . . . *Crustafila confina*
– Nematodes associated with myriapods 82

82 – Nematodes associated with millipedes (Diplopoda) . . . *Diplonema minuta*

– Nematodes associated with centipedes (Chilopoda) .... *Chilonema filaforma*

83 – Nematodes phoretic on Hymenoptera 84
– Nematodes free-living or phoretic on other insects 85

84 – Nematodes phoretic on ants (Hymenoptera: Formicidae) .... *Formicodiplogaster myrmenema*
– Nematodes phoretic on stingless bees (Hymenoptera: Apidae) .... *Apifilaria proplebiae*

85 – Nematodes phoretic on termites (Isoptera) 86
– Nematodes phoretic on beetles or flies 88

86 – Nematodes phoretic on lower termites 87
– Nematodes phoretic on *Nausutitermes* spp. (Termitidae); length over 600 μm .... *Prescafilotermes elongatus*

87 – Nematodes <300 μm in length .... *Filotermes minutus*
– Nematodes between 300 and 600 μm in length .... *Prescafilotermes dominicanus*

88 – Nematodes associated with flies 89
– Nematodes associated with beetles and bark lice 91

89 – Nematodes phoretic on Tipuloidea .... *Tipulacomis dominicanus*
– Nematodes phoretic on mothflies 90

90 – Dauer length > 600 μm; tail rounded ... *Psychodorhabditis dominicanus*
– Dauer length < 600 μm; tail pointed ..... *Psychodorhabditis attenuatus*

91 – Nematodes phoretic on bark lice .... *Psocidonema dominicana*
– Nematodes phoretic on beetles 92

92 – Nematodes phoretic on scarab beetles (Coleoptera: Scaraboidea); length < 150 μm .... *Scarabinema minuta*
– Nematodes phoretic on bark (Scolytidae) and pin hole borers (Platypodidae); length > 150 μm 93

93 – Dauer with mouth plug; tail long .... *Panagrolaimus neotropicus*
– Dauer without mouth plug; tail long or short 94

94 – Dauer >200 μm in length; tail long .... *Scolytonema dominicana*
– Dauer <200 μm in length; tail short .... *Scolytonema hispaniolae*

95 – Nematodes free in amber or associated with fungi 96
– Nematodes associated with insects 98

96 – Stylet-bearing mycophagous nematodes ... *Oligaphelenchoides atrebora*
– Non stylet-bearing nematodes 97

97 – Nematodes stout, diam. 40-60 μm; lacking tail filament .... *Vetus robustus*

– Nematodes slender, diam. 20 μm; with long tail filament .... *Oligodiplogaster antiqua*

98 – Nematodes parasitic in sand flies (Diptera: Phlebotomidae) .... *Palaeotetradonema phlebotomae*

– Nematodes phoretic on pinhole borers (Coleoptera: Platypodidae) .... *Scolytonema mexicana*

Appendix III

Techniques for studying nematodes in amber

Most of the newly described nematodes in this book are in amber. While preservation is usually quite good, the specimens are often found several millimetres from the surface and may occur in various positions. This requires that the amber piece be trimmed and re-shaped, then re-sanded and re-polished before detailed examinations can be made. Thin-bladed diamond saws are excellent for trimming away excess amber, and rotary wheels with different grades of sandpaper can be used to re-shape the piece. A soft buffing disk will remove further surface blemishes. Hand polishing may be required when the amber piece is quite small or the amber is cracked.

There are various artefacts in amber that resemble nematodes (see section 10). Plant hairs (trichomes) are one of the most common, but fine bark and wood filaments, insect and mammalian hairs, elongated bubbles and some insect parts, can also appear as nematodes, especially when viewing the piece with a hand lens. While the larger nematodes such as mermithids can be studied under the binocular microscope, a compound microscope is needed for specimens less than 1 mm in length. This requires polishing the piece to within several millimetres of the specimen and shaping the bottom surface of the amber parallel to the top.

Storing amber specimens under the correct conditions is important for maintaining their integrity. Light and air will eventually degrade the amber surface, resulting in cracks that will coalesce and disintegrate the surface layer. The amber can be re-polished to remove these oxidised surface layers; however, if the specimens are near the surface they can be destroyed. Rapid changes in temperature, which may occur when a piece is exposed to certain types of lighting, can also fracture the amber. It is best to store specimens in airtight plastic boxes in a dark room with little temperature variation.

Nematology Monographs & Perspectives, 2011, Vol. 9, 369-393

References

ANDERSON, R.C. (1960). On the development and transmission of *Cosmocercoides dukae* of terrestrial molluscs in Ontario. *Canadian Journal of Zoology* 38, 801-825.

ANDERSON, R.C. (1984). The origins of zooparasitic nematodes. *Canadian Journal of Zoology* 62, 317-328.

ANDERSON, R.C. (1988). Nematode transmission patterns. *Journal of Parasitology* 74, 30-45.

ANDERSON, R.C. (2000). *Nematode parasites of vertebrates*, 2nd edition. Wallingford, UK, CABI Publishing, 650 pp.

ANDERSON, R.V. & LAUMOND, C. (1992). *Noctuidonema daptria* n. sp. (Nematoda: Aphelenchoididae): an ectoparasite of the moth *Lesmone porcia* (Stoll). *Journal of Nematology* 24, 16-22.

ANDRÁSSY, I. (1976). *Evolution as a basis for the systematization of nematodes*. London, UK, Pitman Publishing, 288 pp.

ANDRÁSSY, I. (2005). *Free-living nematodes of Hungary (Nematoda errantia). I*. Csuzdi, C. & Mahunka, S. (Eds). Pedozoologica Hungarica No. 3. Budapest, Hungary, Hungarian Natural History Museum and Systematic Zoology Research Group of the Hungarian Academy of Sciences, 518 pp.

ANDRES, D. (1989). Phosphatisierte Fossilien aus dem Unteren Ordoviz vin Südschweden. *Berliner geowissenschaftliche Abhandlunger Rehe A* 106, 9-19.

ARAÚJO, A.J.G., CONFALONIERI, U.E.C. & FERREIRA, L.F. (1982). Oxyurid (Nematoda) egg from coprolites from Brazil. *Journal of Parasitology* 68, 511-512.

ARDUINI, P., PINNA, G. & TERUZZI, G. (1983). *Eophasma jurasicum* n. g., n. sp., a new fossil nematode of the Sinemurian of Osteno in Lombardy. *Atti della Società Italiana di Scienze Naturali e del Museo Civico di Storia Naturale di Milano* 124, 61-64.

BAIN, O. & CHABAUD, A.G. (1986). Atlas des larves infestantes de filares. *Tropical Medicine and Parasitology* 37, 301-340.

BAIRD, S.E., FITCH, D.H.A. & EMMONS, S.W. (1994). *Caenorhabditis vulgaris* sp. n. (Nematoda: Rhabditidae): a necromenic associate of pill bugs and snails, *Nematologica* 40, 1-11.

BAKER, W.L. (1972). *Eastern forest insects*. Washington, DC, USA, USDA Forest Service Miscellaneous Publication 1175, 614 pp.

BANDS, N. & SNYDER, T.E. (1920). A revision of the Nearctic Termites with notes on biology and geographical distribution. *Bulletin of the United States National Museum* 108, 1-222.

BARRON, G.L. (1977). *The nematode destroying fungi.* Topics in Mycobiology, No. 1. Guelph, Canada, Canadian Biological Publications, 140 pp.

BARUS, V. & KOUBKOVÁ, B. (2002). The first species of *Thelastoma* Leidy, 1849 (Nematoda: Thelastomatidae) parasitising the scorpion *Euscorpius italicus* (Chactidae: Scorpionoidea). *Systematic Parasitology* 53, 141-146.

BEDDING, R.A. (1972a). Biological control of *Sirex noctilio* using the nematode *Deladenus siricidicola.* In: Bedding, R., Akhurst, R. & Kaya, H. (Eds). *Nematodes and the biological control of insects pests.* East Melbourne, Australia, CSIRO Publications, pp. 11-20.

BEDDING, R.A. (1972b). Biology of *Deladenus siricidicola* (Neotylenchidae) an entomophagous-mycetophagous nematode parasitic in siricid woodwasps, *Nematologica* 18, 482-493.

BLEIDORN, C., SCHMIDT-RHAESA, A. & GAREY, J.R. (2002). Systematic relationships of Nematomorpha based on molecular and morphological data. *Invertebrate Biology* 121, 357-364.

BONNICHSEN, R. & SCHNEIDER, A.L. (1999). Breaking the impasse on the peopling of the Americas. In: Bonnichsen, R. & Turnmire, K.L. (Eds). *Ice Age people of North America; environments, origins and adaptations.* Corvallis, Oregon, USA, Oregon State University Press, pp. 497-519.

BOUCHET, F., BAFFIER, D., GIRARD, M., MOREL, P., PAICHELER, J.C. & DAVID, F. (1996). Paléoparasitologie en contexte Pléistocène premières observations à la Grand Grotte d'Arcy-sur-Cure (Yonne), France. *Compte Rendus de l'Académie des Sciences, Paris* 319, 147-151.

BOUCOT, A. (1990). *Evolutionary paleobiology of behaviour and coevolution.* Amsterdam, The Netherlands, Elsevier, 735 pp.

BOUCOT, A. & POINAR JR, G.O. (2010). *Fossil behaviour compendium.* Boca Raton, FL, USA, CRC Press, 424 pp.

BOVIEN, P. (1937). Some types of association between nematodes and insects. *Videnskabelige Meddelelser fra Dansk Naturhistorisk Forening* 101, 114 pp.

BOVIEN, P. (1944). *Proatractonema sciarae* n. g., n. sp., a parasitic nematode from the body cavity of a dipterous larva. *Videnskabelige Meddelelser fra Dansk Naturhistorisk Forening* 108, 1-14.

BRIGHT, D.E. & POINAR JR, G.O. (1994). Scolytidae and Platypodidae (Coleoptera) from Dominican Republic amber. *Annals of the Entomological Society of America* 87, 170-194.

BROWN, B.A., SWIFT, B.L. & MITCHELL, M.J. (1978). Effect of *Oniscus asellus* feeding on bacterial and nematode populations of sewage sludge. *Oikos* 30, 90-94.

BRYAN, C.P. (1931). *The Papyrus Ebers.* New York, NY, USA, Appleton & Co., 167 pp.

CARROLL, R.L. (1988). *Vertebrate paleontology and evolution.* New York, NY, USA, W.H. Freeman & Co., 698 pp.

CHABAUD, A.G. (1971). Evolution of host-parasite adaptation in nematodes of vertebrates. *International Journal for Parasitology* 1, 217-221.

CHAPMAN, C.A., ROTHMAN, J.M. & HODDER, S.A.M. (2009). Can parasite infections be a selective force influencing primate group size? A test with red colobus. In: Huffman, M.A. & Chapman, C.A. (Eds). *Primate parasite ecology, the dynamics and study of host-parasite relationships.* Cambridge, UK, Cambridge University Press, pp. 423-440.

CHITWOOD, B.G. (1933). The systematic position of *Echinonema grayi* Gemmill, 1901. *Journal of Parasitology* 20, 104.

CHITWOOD, B.G. (1950). Nemic relationships. In: Chitwood, B.G. & Chitwood, M.B. (Eds). *Introduction to nematology*. Baltimore, USA, University Park Press, pp. 227-241.

CHITWOOD, B.G. (1951). North American marine nematodes. *Texas Journal of Science* 4, 617-672.

CHOO, H.Y. & KAYA, H.K. (1993). Parasitism of the brown planthopper by the mermithid nematode *Agamermis* in Rice. In: Bedding, R., Akhurst, R. & Kaya, H. (Eds). *Nematodes and the biological control of insect pests.* East Melbourne, Australia, CSIRO Publications, pp. 21-26.

CHRISTIE, J.R. (1933). Notes on some saprophytes or semi-parasitic nematodes infesting plants. *Journal of Parasitology* 20, 63.

CLAUSEN, C.P. (1962). *Entomophagous insects*. New York, NY, USA, Hafner Publishing Company, 688 pp.

COBB, N.A. (1922). *Greeffiella* (*Trichoderma* Greeff, 1869, not *Trichoderma* Steph., 1835). *Journal of the Washington Academy of Science* 12, 299-303.

COCKBURN, A., BARRACO, R.A., REYMAN, T.A. & PECK, W.H. (1975). Autopsy of an Egyptian mummy. *Science* 187, 1155-1160.

COLLINSON, M.E., BOULTER, M.C. & HOLMES, P.L. (1993). Magnoliophyta ('Angiospermae'). In: Benton, M. (Ed.). *The fossil record 2*. London, UK, Chapman & Hall, pp. 809-841.

COPPEL, H.C. & LIANG, M.C. (1987). Rhabditoid nematodes associated with subterranean termites. *University of Wisconsin Forestry Research Notes* 274, 1-4.

COUPLAND, J.B. (1995). Susceptibility of helicid snails to isolates of the nematode *Phasmarhabditis hermaphrodita* from southern France. *Journal of Invertebrate Pathology* 66, 207-208.

CROLL, N.A. (1976). Behavioural coordination of nematodes. In: Croll, N.A. (Ed.). *The organization of nematodes*, London, UK, Academic Press, pp. 343-364.

CROMPTON, D.W.T. (1999). How much human helminthiasis is there in the world? *Journal of Parasitology* 85, 397-403.

CRUICKSHANK, R.D. & KO, K. (2003). Geology of an amber locality in the Hukawng Valley, northern Myanmar. *Journal of Asian Earth Sciences* 21, 441-455.

ĆURČIĆ, B.P.M., DIMITRIJEVIĆ, R.N., MAKAROV, S.E., LUČIĆ, L.R. & ĆURČIĆ, S.B. (1996). Further report on nematode-pseudoscorpion associations. *Acta Arachnologica* 45, 43-46.

DAJOZ, R. (2000). *Insects and forests.* Paris, France, Lavoisier, 668 pp.

DALE, P.S. (1965). Two new nematodes of the genus *Tetleyus* (Thelastomatidae) from insect larvae. *New Zealand Journal of Science* 8, 521-530.

DE LEY, P. & BLAXTER, M.L. (2002). Systematic position and phylogeny. In: Lee, D.L. (Ed.). *The biology of nematodes*. London, UK, Taylor & Francis, pp. 1-30.

DE LEY, P. & BLAXTER, M.L. (2004). A new system for Nematoda: combining morphological characters with molecular trees, and translating clades into ranks and taxa. In: Cook, R. & Hunt, D.J. (Eds). *Proceedings of the Fourth International Congress of Nematology, 8-13 June 2002, Tenerife, Spain.* Nematology Monographs and Perspectives 2. Leiden, The Netherlands, Brill, pp. 633-653.

DISNEY, R.H.L. (1994). *Scuttle flies: the Phoridae*. London, UK, Chapman & Hall, 467 pp.

DITLEVSEN, H. (1919). Marine free-living nematodes from Danish waters. *Videnskabelige Meddelelser fra Dansk Naturhistorisk Forening* 70, 147-214.

DOLLFUS, R.-Ph. (1950). Liste des Némathelminthes connus à l'état fossile. *Compte Rendu Sommaire des Séances de la Société Géologique de France* 5, 82-85.

DOLLFUS, R.-Ph. & THÉODORIDÈS, J. (1950). Mermithide (Nematoda) chez un *Anthaxia quadripunctata* (L. 1758) [Coléop. Buprestidae] d'Europe centrale. *L'Entomologiste* 6, 96-99.

DOMMELIER-ESPEJO, S. (2001). *Contribution à l'étude Paléoparasitologique des sites Néolithique en environment Lacustre dans les Domaines Jurassien et Péri-alpin.* Ph.D. Dissertation, University de Reims, Reims, France, 248 pp.

DOUGHERTY, E.C. (1951). Evolution of zooparasitic groups in the phylum Nematoda, with special reference to host distribution. *Journal of Parasitology* 37, 353-378.

DOYEN, J.T. (1975). Intertidal insects: Order Coleoptera. In: Smith, R.I. & Carlton, J.T. (Eds). *Light's manual: intertidal invertebrates of the central California coast*, 3rd edition. Berkeley, CA, USA, University of California Press, pp. 446-452.

DUBININ, V.B. (1948). [The presence of Pleistocene lice (Anoplura) and nematodes in the cadaver of a fossilised gopher.] *Doklady Akademii Nauk SSSR* 62, 417-420.

DUBININA, M.N. (1972). [The nematode *Alfortia edentatus* (Looss, 1900) from the intestine of an Upper Pleistocene horse.] *Parazitologia* 6, 441-443.

VON DUISBURG, H. (1862). Beitrag zur Bernstein-Fauna. *Schriften der Königlichen Physikalisch-Ökonomischen Gesellschaft zu Königsbergen* 3, 31-36.

DUNN, C.W., HEJNOL, A., MATUS, D.Q., PANG, K., BROWNE, W.E., SMITH, S.A., SEAVER, E., ROUSE, G.W., OBST, M., EDGECOMBE, G.D. *ET AL.* (2008). Broad phylogenomic sampling improves resolution of the animal tree of life. *Nature* 452, 745-749.

DUNN, F.L. & WATKINS, R. (1970). Parasitological examinations of prehistoric human coprolites from Lovelock Cave, Nevada. *Contributions of the University of California Archaeological Research Facility* 10, 178-185.

DUPAIN, J., NELL, C., PETRZELKOVÁ, K.J., GARCIA, P., MODRY, D. & GORDO, F.P. (2009). Gastrointestinal parasites of bonobos in the Lomako Forest, Democratic Republic of Congo. In: Huffman, M.A. & Chapman, C.A. (Eds). *Primate parasite ecology: the dynamics and study of host-parasite relationships.* Cambridge, UK, Cambridge University Press, pp. 297-310.

DURETTE-DESSET, M.-C. (1985). Trichostrongyloid nematodes and their vertebrate hosts: reconstruction of the phylogeny of a parasitic group. *Advances in Parasitology* 24, 239-306.

EBBELL, B. (1937). *The papyrus Ebers, the greatest Egyptian medical document.* Copenhagen, Denmark, Levin & Munksgaard, 135 pp.

ERTELD, C. (1994). Über die Bindung von *Diplogaster* spec. (Nematoda) an sandbienen (*Andrena* spp.). *Sitzüngs-gerichte der Gesellschaft naturforschender Freunde zu Berlin* 33, 143-155.

EVANS, A.A.F. & PERRY, R.N. (1976). Survival strategies in nematodes. In: Croll, N.A. (Ed.). *The organization of nematodes.* London, UK, Academic Press, pp. 381-424.

EVANS, A.C. (1996). Late stone-age coprolite reveals evidence of prehistoric parasitism. *South African Medical Journal* 86, 274-275.

EYUALEM-ABEBE, ANDRÁSSY, I. & TRAUNSPURGER, W. (EDS) (2006). *Freshwater nematodes: ecology and taxonomy.* Wallingford, UK, CABI Publishing, 752 pp.

FEDER, W.A. (1951). Yellow bud blight, a disease of *Vanda Joaquim* in Hawaii. *Hawaiian Agricultural Experiment Station Progress Note* 62, 1-5.

FEDORKO, A. & STANUSZEK, S. (1971). *Pristionchus uniformis* sp. n. (Nematoda, Rhaditida, Diplogasteridae), a facultative parasite of *Leptinotarsa decemlineata* Say and *Melolontha melolontha* L. in Poland. Morphology and Biology. *Acta Parasitologica Polonica* 19, 95-112.

FENCHEL, T.M. & RIEDL, R.J. (1970). The sulfide system: a new biotic community underneath the oxidized layer of marine sand bottoms. *Marine Biology* 7, 255-268.

FERREIRA, L.F., ARAÚJO, A., CONFALONIERI, U. & FILHO, C.M.R. (1987). The finding of hookworm eggs in human coprolites from 7230 ± 80 years BP, from Piauí, Brasil. *Anais da Academia Brasileira de Ciencias* 59, 280-281.

FERREIRA, L.F., ARAÚJO, A., CONFALONIERI, U., CHAME, M. & GOMES, D.G. (1991). *Trichuris* eggs in animal coprolites dated from 30,000 years ago. *Journal of Parasitology* 73, 491-493.

FERREIRA, L.F., ARAÚJO, A. & DUARTE, A.N. (1993). Nematode larvae in fossilized animal coprolites from Lower and Middle Pleistocene sites, central Italy. *Journal of Parasitology* 79, 440-442.

FERRIS, J.M. & FERRIS, V.R. (1966). Observations on *Tetradonema plicans*, an entomoparasitic nematode, with a key to the genera of the family Tetradonematidae (Nematoda: Trichosyringida). *Annals of the Entomological Society of America* 59, 964-971.

FILIPJEV, I.N. (1918). [Free-living marine nematodes of the Sevastopol Area. Issue 1.] *Transactions of the Zoological Laboratory and the Sevastopol Biological Station of the Russian Academy of Sciences*, Series 2, No. 4, 255 pp.

FILIPJEV, I.N. (1934). The classification of the free-living nematodes and their relation to the parasitic nematodes. *Smithsonian Miscellaneous Collections* 89, No. 6, 1-63.

FILIPJEV, I.N. & SCHUURMANS STEKHOVEN JR, J.H. (1959). *A manual of agricultural helminthology.* Leiden, The Netherlands, E.J. Brill, 878 pp.

FOSSEY, D. (1983). *Gorillas in the mist.* Boston, MA, USA, Houghton Mifflin Co., 311 pp.

FORTUNER, R. & MAGGENTI, A.R. (1987). A reappraisal of Tylenchina (Nemata). 4. The Family Anguinidae Nicoll, 1936 (1926). *Revue de Nématologie* 10, 163-176.

FRY, G.F. & HALL, H.J. (1969). Parasitological examination of prehistoric human coprolites from Utah. *Proceedings of the Utah Academy of Science, Arts and Letters* 46, part 2, 102-105.

FRY, G.F. & MOORE, J.G. (1969). *Enterobius vermicularis*: 10,000-year-old human infection. *Science* 166, 1620.

GEMMILL, J.F. & VON LINSTOW, O. (1902). *Ichthyonema grayi* Gemmill & von Linstow. *Archiv für Naturgeschichte* 1, 113-118.

GEORGIS, R., KAYA, H.K. & GAUGLER, R. (1991). Effect of steinernematid and heterorhabditid nematodes (Rhabditida: Steinernematidae and Heterorhabditidae) on nontarget arthropods. *Environmental Entomology* 20, 815-822.

GIBLIN, R.M. & KAYA, H.K. (1984). *Aduncospiculum halicti* n. gen., n. sp. (Diplogasterida: Diplogasteroididae), an associate of bees in the genus *Halictus* (Hymenoptera: Halictidae). *Revue de Nématologie* 7, 189-197.

GIBLIN, R.M., SWAN, J.L. & KAYA, H.K. (1984). *Bursaphelenchus kevini* n. sp. (Aphelenchida: Aphelenchoididae), an associate of bees in the genus *Halictus* (Hymenoptera: Halictidae). *Revue de Nématologie* 7, 177-187.

GONÇALVES, M.L.C., ARAÚJO, A. & FERREIRA, L.F. (2003). Human intestinal parasites in the past: new findings and a review. *Memórias do Instituto Oswaldo Cruz, Supplement 1* 98, 103-118.

GOODEY, J.B. (1963). *Soil and freshwater nematodes*. London, UK, Methuen & Co., 544 pp.

GOODEY, J.B., FRANKLIN, M.T. & HOOPER, D.J. (1965). *T. Goodey's The nematode parasites of plants catalogued under their hosts*. Farnham Royal, UK, Commonwealth Agricultural Bureaux, 214 pp.

GRABENHORST, H. (1985). Eine zweite Bremse (Tabanidae) zusammen mit ihrem Parasiten (Nematoda, Mermithoidae) aus dem oberpliozän von Willershausen, Krs. Osterode. *Osterode Aufschluss* 36, 325-328.

GRANSTEIN, F., OGG, J. & SMITH, A. (2004). *A geologic time scale*. Cambridge, UK, Cambridge University Press, 589 pp.

GRAY, J. (1988). Evolution of the freshwater ecosytem: the fossil record. *Palaeogeography, Palaeoclimatology, Palaeoecology* 62, 1-214.

GRIMALDI, D.A. (2003). Fossil record. In: Resh, V.H. & Cardé, R.T. (Eds). *Encyclopedia of insects*. London, UK, Academic Press, pp. 455-463.

GRIMALDI, D.A., ENGEL, M.S. & NASCIMBENE, P.C. (2002). Fossiliferous Cretaceous amber from Myanmar (Burma): its rediscovery, biotic diversity, and paleontological significance. *American Museum Novitates* 3361, 71 pp.

HAGMEIER, A. (1912). Beiträge zur Kenntnis der Mermithiden. 1. Biologische Notizen und systematische Beschreibung einiger alter und neuer Arten. *Zoologische Jahrbücher Abteilung für Systematik Oekologie und Geographie der Tiere* 32, 521-612.

HALL, H.J. (1972). Analysis of ancient feces: a discussion and annotated bibliography. *Archaeological Research Facility, Department of Anthropology report*. Berkeley, CA, USA, University of California, 47 pp.

HALLETT, S.L., ERSÉUS, C., O'DONOGHUE, P.J. & LESTER, R.J.G. (2001). Parasitic fauna of Australian marine Oligochaetes. *Memoirs of the Queensland Museum* 46, 555-576.

HANTZSCHEL, W. (1975). Trace fossils and problematics. In: Moore, R.C. (Ed.). *Treatise on invertebrate paleontology. Part W. Supplement 1*. Boulder, Colorado, USA, The Geological Society of America, pp. 1-269.

HARWOOD, P.D. (1930). A new species of *Oxysomatium* (Nematoda) with some remarks on the genera *Oxysomatium* and *Aplectana* and observations on the life history. *Journal of Parasitology* 17, 61-73.

HASEGAWA, H. (2009). Methods of collection and identification of minute nematodes from the feces of primates, with special application to coevolutionary studes of pinworms. In: Huffman, M.A. & Chapman, C.A. (Eds). *Pri-*

mate parasite ecology: the dynamics and study of host-parasite relationships. Cambridge, UK, Cambridge University Press, pp. 29-46.

HELDEN, A.J. (2008). First extant records of mermithid nematode parasites of Auchenorrhyncha in Europe. *Journal of Invertebrate Pathology* 99, 351-353.

HERNANDEZ, A., MACINTOSH, A.J. & HUFFMAN, M.A. (2009). Primate parasite ecology: patterns and predictions from an ongoing study of Japanese macaques. In: Huffman, M.A. & Chapman, C.A. (Eds). *Primate parasite ecology: the dynamics and study of host-parasite relationships.* Cambridge, UK, Cambridge University Press, pp. 387-410.

VON HEYDEN, C. (1860). *Mermis antiqua*, ein fossiler Eigeweidewurm. *Stettiner Entomologische Zeitund* 21, 38.

VON HEYDEN, C. (1862). Gliederthiere aus der Braunkohle des Niederrhein's der Wetterau und der Röhn. In: *Palaeontographica. Beiträge zur Naturgeschichte der Vorwelt*, Vol. 10. Cassel, Germany, von Theodor Fischer, pp. 1-83.

HOFMANN, H.J. & AITKEN, J.D. (1979). Precambrian biota from the Little Dal Group, Mackenzie Mountains, northwestern Canada. *Canadian Journal of Earth Sciences* 16, 150-166.

HOPE, W.D. (1977). Gutless nematodes of the deep-sea. *Mikrofauna Meeresboden* 61, 307-308.

HOPE, W.D. & MURPHY, D.G. (1969). *Syringonomus typicus*, new genus, new species (Enoplida: Leptosomatidae) a marine nematode inhabiting arenaceous tubes. *Proceeding of the Biological Society of Washington* 82, 511-518.

HOPE, W.D. & TCHESUNOV, A.V. (1999). *Smithsoninema inaequale* n. g. and n. sp. (Nematoda: Leptolaimidae) inhabiting the test of a foraminiferan. *Invertebrate Biology* 118, 95-108.

HÖRNSCHEMEYER, T., WEDMANN, S. & POINAR, G. O. (2010). How long can insect species exist? Evidence from extant and fossil *Micromalthus* beetles (Insecta: Coleoptera). *Zoological Journal of the Linnean Society* 158, 300-311.

HUFFMAN, M.A., PEBSWORTH, P., BAKUNEETA, C., GOTOH, S. & BARDI, M. (2009). Chimpanzee-parasite ecology at Budongo Forest (Uganda) and the Mahale Mountains (Tanzania): influence of climatic differences on self-medicative behaviour. In: Huffman, M.A. & Chapman, C.A. (Eds). *Primate parasite ecology: the dynamics and study of host-parasite relationships.* Cambridge, UK, Cambridge University Press, pp. 331-350.

HUNT, D.J. (1993). *Aphelenchida, Longidoridae and Trichodoridae: their systematics and bionomics.* Wallingford, UK, CABI Publishing, 352 pp.

HYMAN, L.H. (1951). *The Invertebrates: Acanthocephala, Aschelminthes, and Entoprocta, the pseudocoelomate Bilateria*, Vol. 3. New York, NY, USA, McGraw-Hill Company, Inc., 572 pp.

INGLIS, W.G. (1965). Patterns of evolution in parasitic nematodes. In: Taylor, A.E.R. (Ed.). *Evolution of parasites*. Oxford, UK, Blackwell Scientific Publications, pp. 79-124.

IÑIGUEZ, A.M., REINHARD, K., GONÇALVES, M.L.C., FERREIRA, L.F., ARAÚJO, A. & VICENTE, A.C.P. (2006). SL1 RNA gene recovery from *Enterobius vermicularis* ancient DNA in pre-Columbian human coprolites. *International Journal for Parasitology* 36, 1419-1425.

JANSEN JR, J. & OVER, H.J. (1962). Het voorkomen van parasieten in terpmateriaal uit Noordwest Duitsland. *Tijdschrift voor Diergeneesk* 87, 1377-1379.

JANSEN JR, J. & OVER, H.J. (1966). Observations on helminth infections in a Roman army-camp. *Proceedings of the first International Congress of Parasitology*, Rome, 1964, p. 791.

JANSSON, H.-B. & POINAR JR, G.O. (1986). Some possible fossil nematophagous fungi. *Transactions of the British Mycological Society* 87, 471-474.

JARRY, D.T. (1969). Les Oxyuroïdes de quelques Arthropodes dans le Midi de la France. *Annales de Parasitologie Humaine et Comparée* 39, 381-508.

JERAM, A.J., SELDEN, P.A. & EDWARDS, D. (1990). Land animals in the Silurian: arachnids and myriapods from Shropshire, England. *Science* 250, 658-661.

JONES, G.M. & HAGEN, N.T. (1987). *Echinomermella matsi* sp. n., an endoparasitic nematode from the sea urchin *Strongylocentrotus droebachiensis* in Northern Norway. *Sarsia* 72, 203-212.

KABAT, A.R. (1990). Predatory ecology of natic gastropods with a review of shell boring predation. *Malacologia* 32, 155-193.

KAISER, H. (1986). Über Wechselbeziehungen zwischen Nematoden (Mermithidae) und Ameisen. *Zoologische Anzeiger* 217, 156-177.

KAUR, T. & SINGH, J. (2009). Primate-parasitic zoonoses and anthropozoonoses: a literature review. In: Huffman, M.A. & Chapman, C.A. (Eds). *Primate parasite ecology: the dynamics and study of host-parasite relationships*. Cambridge, UK, Cambridge University Press, pp. 199-230.

KEILIN, M.D. & ROBINSON, V.C. (1933). On the morphology and life history of *Aproctonema entomopagum* Keilin, a parasite in the larva of *Sciara pullula* Winn. (Diptera-Nematocera). *Parasitology* 25, 285-295.

KHAN, M.A. (1957). *Sphaerularia bombi* Duf. (Nematoda: Allantonematidae) infesting bumblebees and *Sphaerularia hastata* sp. nov. infesting bark beetles in Canada. *Canadian Journal of Zoology* 35, 519-523.

KLEIN, M. (1993). Biological control of scarabs with entomopathogenic nematodes. In: Bedding, R., Akhurst, R. & Kaya, H. (Eds). *Nematodes and the biological control of insect pests*. East Melbourne, Australia, CSIRO Publications, pp. 49-57.

KLIKS, M. (1975). *Paleoepidemiological studies on Great Basin coprolites: estimation of dietary fiber intake and evaluation of the ingestion of anthelmintic plant substances.* Contributions of the Archaeological Research Facility, Berkeley, CA, USA, University of California, 1975, 31 pp.

KLIKS, M. (1990). Helminths as heirlooms and souveniers: a review of New World paleoparasitology. *Parasitology Today* 6, 93-100.

KRANTZ, G.W. & POINAR JR, G.O. (2004). Mites, nematodes and the multimillion dollar weevil. *Journal of Natural History* 38, 135-141.

LAM, A.B.Q. & WEBSTER, J.M. (1971). Morphology and biology of *Panagrolaimus tipulae* n. sp. (Panagrolaimidae) and *Rhabditis* (*Rhabitella*) *tipulae* n. sp. (Rhabditidae), from leatherjacket larvae, *Tipula paludosa* (Diptera; Tipulidae). *Nematologica* 17, 201-212.

LAMBERT, J.B. & POINAR JR, G.O. (2002). Amber: the organic gemstone. *Archaeological Chemistry* 35, 628-636.

LARSSON, S.G. (1978). *Baltic amber: a palaeobiological study.* Entomonograph 1. 192 pp.

LEHMANN, U. & HILLMER, G. (1983). *Fossil invertebrates.* Cambridge, UK, Cambridge University Press, 350 pp.

LEIBERSPERGER, E. (1960). *Die Oxyuroidea der europäischen Arthropoden.* Parasitologische Schriftenreihe 11, 150 pp.

LEIBOVITZ, L. & LEWBART, G.A. (2004). Diseases and symbionts: vulnerability despite tough shells. In: Shuster Jr, C.N., Barlow, R.B. & Brockmann, H.J. (Eds). *The American horseshoe crab.* Cambridge, USA, Harvard University Press, pp. 245-275.

LINCOLN, R.J. & BOXSHALL, G.A. (1987). *The Cambridge illustrated dictionary of natural history.* Cambridge, UK, Cambridge University Press, 411 pp.

LLOYD, J. & DAVIES, K.A. (1997). Two new species of *Schistonchus* (Tylenchida: Aphelenchoididae) associated with *Ficus macrophylla* from Australia. *Fundamental and Applied Nematology* 20, 79-86.

LUC, M. (1987). A reappraisal of Tylenchina (Nemata). 7. The family Pratylenchidae Thorne, 1949. *Revue de Nématologie* 10, 203-218.

LUC, M., MAGGENTI, A.R., FORTUNER, R., RASKI, D.J. & GERAERT, E. (1987). A reappraisal of Tylenchina (Nemata) 1. For a new approach to the taxonomy of Tylenchina. *Revue de Nématologie* 10, 127-134.

MAGGENTI, A.R. (1963). Comparative morphology in nemic phylogeny. In: Dougherty, E.C. (Ed.). *The lower Metazoa: comparative biology and phylogeny.* Berkeley, CA, USA, University of California Press, pp. 273-282.

MAGGENTI, A.R. (1971). Nemic relationships and the origins of plant parasitic nematodes. In: Zuckerman, B.M. & Rhode, R.A. (Eds). *Plant parasitic nematodes*, Vol. 1. New York, NY, USA, Academic Press, pp. 65-81.

MAGGENTI, A.R. (1976). Taxonomic position of Nematoda among the pseudocoelomate Bilateria. In: Croll, N.A. (Ed.). *The organization of nematodes.* New York, NY, USA, Academic Press, pp. 1-10.

MALAKHOV, V.V. (1994). *Nematodes.* Washington, DC, USA, Smithsonian Institution Press, 286 pp.

MAMIYA, Y. (1972). Pine wood nematode, *Bursaphelenchus lignicolus* Mamiya & Kiyohara, as a causal agent of pine wilting disease. *Review of Plant Protection Research* 5, 46-60.

MANUM, S.B., BOSE, M.N., SAYER, R.T. & BOSTRÖM, S. (1994). A nematode (*Captivonema cretacea* gen. *et* sp. n.) preserved in a clitellate cocoon wall from the Early Cretaceous. *Zoologica Scripta* 23, 27-31.

MARKIN, G.P. & MCCOY, C.W. (1968). The occurrence of a nematode *Diploscapter lycostoma* in the pharyngeal glands of the Argentine ant, *Iridomyrmex humilis. Annals of the Entomological Society of America* 61, 505-509.

MARTI JR, O.G. & ROGERS, C.E. (1995). *Noctuidonema dibolia* n. sp. (Aphelenchida: Acugutturidae), an ectoparasite of the moth *Mocis latipes* (Lepidoptera: Noctuidae). *Journal of Nematology* 27, 387-394.

MARTI, O.G., SIMMONS, A.M., LALANNE-CASSOU, B., SILVAIN, J.-F. & KERMERREC, A. (2000). Ectoparasitic nematodes (Aphelenchoidoidea: Acugutturidae) of Lepidoptera and Blattodea in Guadeloupe. *Nematology* 2, 669-684.

MASSEY, C.L. (1971). Two new genera of nematodes parasitic in the Eastern subterranean termite, *Reticulitermes flavipes. Journal of Invertebrate Pathology* 17, 238-242.

MASSEY, C.L. (1974). *Biology and taxonomy of nematode parasites and associates of bark beetles in the United States.* Agricultural Handbook No. 446. Washington, DC, USA, 233 pp.

MATTAVELLI, A. & BRACCHI, G. (2008). *Palaeoenoploides alessandrellianus* n. gen. n. sp.: a late Cenomanian 'roundworm' from the Upper Cretaceous of Lebanon (Nematoda: Enoplidae). *Atti della Società Italiana di Scienze Naturali e del Museo Civico di Storia Naturale di Milano* 149, 179-194.

MCCLELLAND, G. (2005). Nematodes (roundworms). In: Rohde, K. (Ed.). *Marine parasitology.* Wallingford, UK, CABI Publishing, pp. 104-115.

MENGE, A. (1866). Ueber ein Rhipidopteron und einige andere im Bernstein eingeschlossene Tiere. *Schriften der Naturforschenden Gesellschaft in Danzig*, Neue Folge 1, 1-8.

MENGE, A. (1872). Ueber eine im Bernstein eingeschlossene *Mermis. Schriften der Naturforschenden Gesellschaft in Danzig*, Neue Folge 3, 1-2.

MENGERT, H. (1953). Nematoden und Schnecken. *Zeitschrift für Morphologie und Ökologie der Tiere* 41, 311-349.

MERRILL, J.H. & FORD, A.L. (1916). Life history and habits of two new nematodes parasitic on insects. *Journal of Agricultural Research* 6, 115-127.

MILLER, L.R. (1989). Some nematode-termite relationships, and the maxillary gland in Isoptera. *Sociobiology* 15, 11-17.

MILLET, N.B., HART, G.D., REYMAN, T.A., ZIMMERMAN, M.R. & LEWIN, P.K. (1980). ROM I: Mummification for the common people. In: Cockburn, A. & Cockburn, E. (Eds). *Mummies, disease, and ancient cultures*. Cambridge, UK, Cambridge University Press, pp. 71-84.

MONTENEGRO, A., ARAUJO, A., EBY, M., FERREIRA, L.F., HETHERINGTON, R. & WEAVER, A.J. (2006). Parasites, paleoclimate, and the peopling of the Americas; using the hookworm to time the Clovis migration. *Current Anthropology* 47, 193-200.

MOUSSA, M.T. (1969). Nematode fossil tracks of Eocene age from Utah. *Nematologica* 15, 376-380.

MYERS, E.H. (1943). Life activities of Foraminifera in relation to marine ecology. *Proceedings of the American Philosophical Society* 86, 439-458.

NICKLE, W.R. & AYRE, G.L. (1966). *Caenorhabditis dolichura* (A. Schneider, 1866) Dougherty (Rhabditidae, Nematoda) in the head glands of the ant, *Camponotus herculeanus* (L.) and *Acanthomyops claviger* (Roger) in Ontario. *Entomological Society of Ontario* 96, 96-98.

NOORDAM, A.P., SAMSON, R.A. & SUDHAUS, W. (1998). Fungi and Nematoda on *Centromerus sylvaticus* (Araneae, Linyphiidae). In: Selden, P.A. (Ed.). *Proceedings of the 17th European colloquium of arachnology, Edinburgh 1997*. British Arachnological Society, Burnham Beeches, Bucks, UK, British Arachnological Society, pp. 343-347.

NUDDS, J.R. & SELDEN, P.A. (2008). *Fossil ecosystems of North America: a guide to the sites and their extraordinary biotas*. Chicago, USA, University of Chicago Press, 288 pp.

OOI, H.K., TENORA, F., ITOH, K. & KAMIYA, M. (1993). Comparative study of *Trichuris trichiura* from nonhuman primates and from man, and their differences with *T. suis*. *Journal of Veterinary Medical Science* 55, 363-366.

OSCHE, G. (1963). Morphological, biological, and ecological considerations in the phylogeny of parasitic nematodes. In: Dougherty, E.C. (Ed.). *The lower Metazoa: comparative biology and phylogeny*. Berkeley, CA, USA, University of California Press, pp. 283-302.

PALM, N.B. (1945). *Sphaerularia bombi* Dufour, en märlig och för Sverige ny hundmask. *Fauna och Flora* 6, 266-274.

PARAMONOV, A.A. (1962). *Principles of plant nematology*. Vol. I. Moscow, Akademii Nauk SSSR, 479 pp. [English translation provided by Israel program for Scientific translations, Jerusalem, 1968, 390 pp.]

PARAMONOV, A.A. (1970). *Studies on phytohelminthology*, Vol. III. Moscow, Akademii Nauk SSSR, 255 pp. [English translation provided by Israel program for Scientific translations, Jerusalem, 1972, 200 pp.]

PATRUCCO, R., TELLO, R. & BONAVIA, D. (1983). Parasitological studies of coprolites of pre-hispanic Peruvian populations. *Current Anthropology* 24, 393-394.

PEMBERTON, C.E. (1928). Nematodes associated with termites in Hawaii, Borneo and Celebes. *Proceedings of the Hawaiian Entomological Society* 7, 148-150.

PEREZ-GELABERT, D.E. (2008). Arthropods of Hispaniola (Dominican Republic and Haiti). A checklist and bibliography. *Zootaxa* 1831, 1-531.

PETERSON, A. (1957). *Larvae of insects. Part II, Coleoptera, Diptera, Neuroptera, Siphonaptera, Mecoptera, Trichoptera.* Ann Arbor, MI, USA, Edwards Brothers, Inc., 416 pp.

PETTER, A.J. (1980). Une nouvelle famille de nématodes parasites d'invertébrés marins, les Benthimermithidae. *Annales de Parasitologie Humaine et Comparée* 55, 209-224.

PETTER, A.J. (1983). Quelques nouvelles espèces du genre *Benthimermis* Petter, 1980 (Benthimermithidae, Nematoda) du Sud de l'Océan Indien. *Systematic Parasitology* 5, 4-15.

PETTER, A.J. (1987). Quelques nouvelles espèces de femelles du genre *Benthimermis* Petter, 1980 (Benthimermithidae, Nematoda) des grands fonds de la mer de Norvige. *Bulletin du Muséum d'Histoire Naturelle Section A* 9, 565-578.

PHILIP, G.K., CREEVEY, C.J. & MCINERNEY, J.O. (2005). The Opistokonta and the Ecdysozoa may not be clades: stronger support for the grouping of plant and animal than for animal and fungi and stronger support for the Coelomata than Ecdysozoa. *Molecular Biology and Evolution* 22, 1175-1184.

PHILIPPE, H., LARTILLOT, N. & BRINKMANN, H. (2005). Multigene analyses of bilaterian animals corroborate the monophyly of Ecdysozoa, Lophotrochozoa and Protostomia. *Molecular Biology and Evolution* 22, 1246-1253.

POINAR JR, G.O. (1964). A new nematode *Orthomermis oedobranchus* gen. n. sp. n. (Mermithidae) parasitising *Smittia* larvae in England. *Nematologica* 10, 501-506.

POINAR JR, G.O. (1965). The bionomics and parasitic development of *Tripus sciarae* (Bovien) (Sphaerulariidae: Aphelenchoidea) a nematode parasite of sciarid flies (Sciaridae: Diptera). *Parasitology* 55, 559-569.

POINAR JR, G.O. (1970). *Oryctonema genitalis* n. gen., n. sp. (Rhabditidae: Nematoda) from the genital system of *Oryctes monoceros* Ol. (Scarabaeidae: Coleoptera) in West Africa. *Journal of Helminthology* 44, 1-10.

POINAR JR, G.O. (1971). *Rhabditis adenobia* n. sp. (Nematoda: Rhabditidae) from the colleterial glands of *Orcytes monoceros* L. and other tropical dynastid beetles (Coleoptera: Scarabaeidae). *Proceedings of the Helminthological Society of Washington* 38, 99-108.

POINAR JR, G.O. (1972). Nematodes as facultative parasites of insects. *Annual Review of Entomology* 17, 103-122.

POINAR JR, G.O. (1975). *Entomogenous nematodes.* Leiden, The Netherlands, Brill, 317 pp.

POINAR JR, G.O. (1976). Description and biology of a new insect parasitic rhabditoid, *Heterorhabditis bacteriophora* n. gen., n. sp. (Heterorhabditidae, n. fam.) (Rhabditida [Oerley]). *Nematologica* 21, 463-470.

POINAR JR, G.O. (1977a). Fossil nematodes from Mexican amber. *Nematologica* 23, 232-238.

POINAR JR, G.O. (1977b). *Empidomermis cozii* n. gen., n. sp. (Mermithidae: Nematoda) a parasite of *Anopheles funestus* (Culicidae: Diptera) in West Africa. *Canadian Journal of Zoology* 55, 1475-1479.

POINAR JR, G.O. (1978a). *Mesidionema praecomasculatis* gen. et sp. n.: Mesidionematidae fam. n. (Drilonematoidea: Rhabditida), a nematode parasite of earthworms. *Proceedings of the Helminthological Society of Washington* 45, 97-102.

POINAR JR, G.O. (1978b). *Thelastoma endoscolicum* sp. n. (Oxyurida: Nematoda) a parasite of earthworms (Oligochaeta: Annelida). *Proceedings of the Helminthological Society of Washington* 45, 92-96.

POINAR JR, G.O. (1979a). *Nematodes for biological control of insects.* Boca Raton, FL, USA, CRC Press, 277 pp.

POINAR JR, G.O. (1979b). *Parasitodiplogaster sycophilon* gen. n., sp. n. (Diplogasteridae: Nematoda), a parasite of *Elisabethiella stuckenbergi* Grandi (Agaionidae: Hymenoptera) in Rhodesia. *Proceedings: Koninklijke Nederlandse Akademie van Wetenschappen* 82, 375-381.

POINAR JR, G.O. (1981). *Thaumamermis cosgrovei* n. gen., n. sp. (Mermithidae: Nematoda) parasitising terrestrial isopods (Isopoda: Oniscoidea). *Systematic Parasitology* 2, 261-266.

POINAR JR, G.O. (1982). Fossil dauer rhabditid nematodes. *Nematologica* 27, 466-467.

POINAR JR, G.O. (1983). *The natural history of nematodes.* Englewood Cliffs, NJ, USA, Prentice Hall, 323 pp.

POINAR JR, G.O. (1984a). *Heydenius dominicus* sp. n. (Nematoda: Mermithidae), a fossil parasite from the Dominican Republic. *Journal of Nematology* 16, 371-375.

POINAR JR, G.O. (1984b). Fossil evidence of nematode parasitism. *Revue de Nématologie* 7, 201-203.

POINAR JR, G.O. (1984c). First fossil record of parasitism by insect-parasitic Tylenchida (Allantonematidae: Nematoda). *Journal of Parasitology* 70, 306-308.

POINAR JR, G.O. (1985). Nematode parasites and infectious diseases of Tabanidae (Diptera). *Myia* 3, 599-616.

POINAR JR, G.O. (1986). *Rhabditis myriophila* sp. n. (Rhabditidae, Rhabditoidea) associated with the millipede, *Oxidis gracilis* (Polydesonida: Diplopoda). *Proceedings of the Helminthological Society of Washington* 53, 232-236.

POINAR JR, G.O. (1987). Nematode parasites of spiders. In: Nentwig, W. (Ed.). *Ecophysiology of spiders.* New York, NY, USA, Springer-Verlag, pp. 299-308.

POINAR JR, G.O. (1988). Morphological and behavioural abnormalities in *Heterorhabditis heliothidis* (Rhabditida) associated with the bacterium, *Alcaligenes faecalis*. *Revue de Nématologie* 11, 381-383.

POINAR JR, G.O. (1990a). Redescription of *Chroniodiplogaster aerivora* (Cobb) gen. n., comb. n. (Rhabditida: Diplogasteridae) from termites. *Journal of the Helminthological Society of Washington* 57, 26-30.

POINAR JR, G.O. (1990b). Taxonomy and biology of Steinernematidae and Heterorhabditidae. In: Gaugler, R. & Kaya, H.K. (Eds). *Entomopathogenic nematodes in biological control.* Boca Raton, FL, USA, CRC Press, pp. 23-61.

POINAR JR, G.O. (1991a). *Hymenaea protera* sp. n. (Leguminosae: Caesalpinoideae) from Dominican amber has African affinities. *Experientia* 47, 1075-1082.

POINAR JR, G.O. (1991b). The mycetophagous and entomophagous stages of *Iotonchium californicum* n. sp. (Iotonchidae: Tylenchida). *Revue de Nématologie* 14, 565-580.

POINAR JR, G.O. (1992). *Life in amber.* Stanford, USA, Stanford University Press, 350 pp.

POINAR JR, G.O. (1993). Origins and phylogenetic relationships of the entomophilic rhabditids, *Heterorhabditis* and *Steinernema*. *Fundamental and Applied Nematology* 16, 333-338.

POINAR JR, G.O. (1996). Older and wiser. *Lapidary Journal* 49, 52-56.

POINAR JR, G.O. (2000). *Heydenius araneus* n. sp. (Nematoda: Mermithidae), a parasite of a fossil spider, with an examination of helminthes from extant spiders (Arachnida: Araneae). *Invertebrate Biology* 119, 388-393.

POINAR JR, G.O. (2001a). *Heydenius brownii* (Nematoda: mermithidae) parasitising a planthopper (Achilidae: Homoptera) in Baltic amber. *Nematology* 3, 753-757.

POINAR JR, G.O. (2001b). Nematoda and Nematomorpha. In: Thorp, J.H. & Covich, A.P. (Eds). *Ecology and classification of North American freshwater invertebrates*. New York, NY, USA, Academic Press, pp. 255-295.

POINAR JR, G.O. (2002). First fossil record of nematode parasitism of ants; a 40 million year tale. *Parasitology* 125, 457-459.

POINAR JR, G.O. (2003a). *Formicitylenchus oregonensis* n. g., n. sp. (Allantonematidae: Nematoda), the first tylenchid parasite of ants, with a review of nematodes described from ants. *Systematic Parasitology* 56, 69-76.

POINAR JR, G.O. (2003b). Trends in the evolution of insect parasitism by nematodes as inferred from fossil evidence. *Journal of Nematology* 35, 129-132.

POINAR JR, G.O. (2003c). Fossil evidence of phorid parasitism (Diptera: Phoridae) by allantonematid nematodes (Tylenchida: Allantonematidae). *Parasitology* 127, 589-592.

POINAR JR, G.O. (2009a). Description of an early Cretaceous termite (Isoptera: Kalotermitidae) and its associated intestinal protozoa, with comments on their coevolution, *Parasites and Vectors* 2, 12. Open access at http://www.parasitesandvectors.com/content/2/1/12

POINAR JR, G.O. (2009b). *Dominibrentus leptus*, n. gen., n. sp. (Curculionoidea, Brentidae, Cyphagoginae, Dominibrentini, n. tribe), a straight-snouted weevil in Dominican amber. *Historical Biology* 21, 1-5.

POINAR JR, G.O. & BENTON JR, C.L.B. (1986). *Aranimermis aptispicula* n. g., n. sp. (Mermithidae: Nematoda), a parasite of spiders (Arachnida: Araneida). *Systematic Parasitology* 8, 56-61.

POINAR JR, G.O. & BOUCOT, A. (2006). Evidence of intestinal parasites of dinosaurs. *Parasitology* 133, 245-249.

POINAR JR, G.O. & BRODZINSKY, J. (1986). Fossil evidence of nematode (Tylenchida) parasitism in Staphylinidae (Coleoptera). *Nematologica* 31, 353-355.

POINAR JR, G.O. & BROWN, A.E. (2002). *Hymenaea mexicana* sp. nov. (Leguminosae: Caesalpinioideae) from Mexican amber indicates Old World connections. *Botanical Journal of the Linnean Society* 139, 125-132.

POINAR JR, G.O. & BUCKLEY, R. (2006). Nematode (Nematoda: Mermithidae) and hairworm (Nematomorpha: Chordodidae) parasites in early Cretaceous amber. *Journal of Invertebrate Pathology* 93, 36-41.

POINAR JR, G.O. & ĆURČIĆ, B.P.M. (1994). Nematode (Nematoda) associates and parasites of pseudoscorpions (Arachnida). *Journal of Invertebrate Pathology* 63, 95-96.

POINAR JR, G.O. & DONCASTER, C.C. (1965). The penetration of *Tripius sciarae* (Bovien) (Sphaerulariidae: Aphelenchoidea) into its insect host, *Bradysia paupera* Tuom. (Mycetophilidae: Diptera). *Nematologica* 11, 73-78.

POINAR JR, G.O. & EARLY, J.W. (1990). *Aranimermis giganteus* n. sp. (Mermithidae: Nematoda), a parasite of New Zealand mygalomorph spiders (Araneae: Arachnida). *Revue de Nématologie* 13, 403-410.

POINAR JR, G.O. & GEORGIS, R. (1990). Characterization and field application of *Heterorhabditis bacteriophora* strain AP88 (Heterorhabditidae: Rhabditida). *Revue de Nématologie* 13, 387-393.

POINAR JR, G.O. & HANSEN, E. (1984). Sex and reproductive modifications in nematodes. *Helminthological Abstracts* 52, 145-163.

POINAR JR, G.O. & HERRE, E.A. (1991). Speciation and adaptive radiation in the fig wasp nematode, *Parasitodiplogaster* (Diplogasteridae: Rhabditida), in Panama. *Revue de Nématologie* 14, 361-374.

POINAR JR, G.O. & HESS, R. (1972). Food uptake by the insect-parasitic nematode, *Sphaerularia bombi* (Tylenchida). *Journal of Nematology* 4, 270-277.

POINAR JR, G.O. & HESS, R. (1977). *Romanomermis culicivorax*: morphological evidence of transcuticular uptake. *Expermental Parasitology* 42, 27-33.

POINAR JR, G.O. & HESS, R. (1985). Preservative qualities of recent and fossil resins: electron micrograph studies on tissue preserved in Baltic amber. *Journal of Baltic Studies* 16, 210-221.

POINAR JR, G.O. & JUDZIEWICZ, E.J. (2005). *Pharus primuncinatus* (Poacae: Pharoideae: Phareae) from Dominican amber. *Sida. Contributions to Botany* 21, 2095-2103.

POINAR JR, G.O. & LANE, R.S. (1978). *Pheromermis myopis* sp. n. (Nematoda: Mermithidae), a parasite of *Tabanus punctifer* (Diptera: Tabanidae). *Journal of Parasitology* 64, 440-444.

POINAR JR, G.O. & LINARES, B. (1985). *Hexamermis dactylocercus* sp. n. (Mermithidae: Nematoda) a parasite of *Aeneolamia varia* (Cercopidae: Homoptera) in Venezuela. *Revue de Nématologie* 8, 109-111.

POINAR JR, G.O. & MALGARIN, B. (2007). Double jeopardy in amber. *Gems and Gemnology* 43, 370.

POINAR JR, G.O. & MILKI, R. (2001). *Lebanese amber*. Corvallis, USA, Oregon State University Press, 96 pp.

POINAR JR, G.O. & MONTEYS, S.I. (2008). Mermithids (Nematoda: Mermithidae) of biting midges (Diptera: Ceratopogonidae): *Heleidomermis cataloniensis* n. sp. from *Culicoides circumscriptus* Kieffer in Spain and a species of *Cretacimermis* Poinar, 2001 from a ceratopogonid in Burmese amber. *Systematic Parasitology* 69, 13-21.

POINAR JR, G.O. & POINAR, R. (1999). *The amber forest*. Princeton, NJ, USA, Princeton University Press, 239 pp.

POINAR JR, G.O. & POINAR, R. (2003). Description and development of *Gastromermis anisotis* sp. n. (Nematoda: Mermithidae) a parasite in a quadritrophic system involving a cyanobacterium, midge and virus. *Nematology* 5, 325-338.

POINAR JR, G.O. & POINAR, R. (2008). *What bugged the dinosaurs?* Princeton, NJ, USA, Princeton University Press, 264 pp.

POINAR JR, G.O. & STOCKWELL, S.A. (1988). A new record of a nematode parasite (Mermithidae) of a scorpion. *Revue de Nématologie* 11, 361-364.

POINAR JR, G.O. & TOURENQ, J.N. (1972). On the occurrence of *Hydromermis contorta* (Kohn) (Nematodea) parasitising midges (Chironomidae) in the Camargue. *Annales de Limnologie* 8, 41-48.

POINAR JR, G.O. & YANOVIAK, S.P. (2008). *Myrmeconema neotropicum* n. g., n. sp., a new tetradonematid nematode parasitising South American populations of *Cephalotes atratus* (Hymenoptera: Formicidae), with the discovery of an apparent parasite-induced host morph. *Systematic Parasitology* 69, 145-153.

POINAR JR, G.O., TRIGGINI, O. & MERRITT, R.W. (1976a). Life history of *Eudiplogaster aphodii* (Rhabditida: Diplogasteridae), a facultative parasite of *Aphodius fimetarius* (Coleoptera: Scarabaeidae). *Nematologica* 22, 79-86.

POINAR JR, G.O., LANE, R.S. & THOMAS, G.M. (1976b). Biology and redescription of *Pheromermis pachysoma* (V. Linstow) n. gen., n. comb., a parasite of yellowjackets. *Nematologica* 22, 360-370.

POINAR JR, G.O., LAUMOND, C. & BONIFASSI, E. (1980). *Howardula apioni* sp. n. (Allantonematidae: Nematoda), a parasite of *Apion carduorum* Kirby (Curculonidae: Coleoptera) in southern France. *Proceedings of the Helminthological Society of Washington* 47, 218-223.

POINAR JR, G.O., KARUNAKAR, G.K. & DAVID, H. (1992). *Heterorhabditis indicus* n. sp. (Rhabtida: Nematoda) from India: separation of *Heterorhabditis* spp. by infective juveniles. *Fundamental and Applied Nematology* 15, 467-472.

POINAR JR, G.O., FERRO, C., MORALES, A. & TESH, R.B. (1993). *Anandranema phlebotophaga* n. gen., n. sp. (Allantonematidae: Tylenchida), a new nematode parasite of phlebotomine sand flies (Psychodidae: Diptera) with notes on experimental infections of these insects with parasitic rhabditoids. *Fundamental and Applied Nematology* 16, 11-16.

POINAR JR, G.O., ACRA, A. & ACRA, F. (1994). Earliest fossil nematode (Mermithidae) in Cretaceous Lebanese amber. *Fundamental and Applied Nematology* 17, 475-477.

POINAR JR, G.O., JAENKE, J. & DOMBECK, I. (1997). *Parasitylenchus nearcticus* sp. n. (Tylenchida: Allantonematidae) parasitising *Drosophila* (Diptera: Drosophilidae) in North America. *Fundamental and Applied Nematology* 20, 187-190.

POINAR JR, G.O., JAENIKE, J. & SHOEMAKER, D.D. (1998a). *Howardula neocosmis* sp. n. parasitising North American Drosophila (Diptera: Drosophilidae) with a listing of the species of *Howardula* Cobb, 1921 (Tylenchida: Allantonematidae). *Fundamental and Applied Nematology* 21, 547-552.

POINAR JR, G.O., ĆURČIĆ, B.P.M. & COKENDOLPHER, J.C. (1998b). Arthro pod phoresy involving pseudoscorpions in the past and present. *Acta Arachnologica* 47, 79-96.

POINAR JR, G.O., LATHAM, D.M. & POULIN, R. (2002). *Thaumamermis zealandica* n. sp. (Mermithidae: Nematoda) parasitising the intertidal marine amphipod *Talorchestia quoyana* (Talitridae: Amphipoda) in New Zealand, with a summary of mermithids infecting amphipods. *Systematic Parasitology* 53, 227-233.

POINAR JR, G.O., JACKSON, T.A., BELL, N.L. & WAHID, M.B. (2003). *Cylindrocorpus inevectus* sp. n. associated with the oil palm weevil, *Elaeidobius kamerunicus* (Faust) (Coleoptera: Curculionidae), with a synopsis of the family Cylindrocorporidae and establishment of Longibuccidae n. fam. (Diplogastroidea: Nematoda). *Nematology* 5, 183-190.

POINAR JR, G.O., MEIKLE, W. & MERCADIER, G. (2006a). *Chroniodiplogaster formosiana* sp. n. (Rhabditida: Diplogastridae) from Chinese populations of *Odontotermes formosanus* Shiraki (Isoptera: Termitidae). *Journal of Nematology* 38, 181-186.

POINAR JR, G.O., LACHAUD, J.-P., CASTILLO, A. & INFANTE, F. (2006b). Recent and fossil nematode parasites (Nematoda: Mermithidae) of Neotropical ants. *Journal of Invertebrate Pathology* 91, 19-26.

POINAR JR, G.O., LAMBERT, J.B. & WU, Y. (2007a). Araucarian source of fossiliferous amber: spectroscopic and anatomical evidence. *Journal of the Botanical Research Institute of Texas* 1, 449-455.

POINAR JR, G.O., PORTER, S.D., TANG, S. & HYMAN, B.C. (2007b). *Allomermis solenopsi* n. sp. (Nematoda: Mermithidae) parasitising the fire ant *Solenopsis invicta* Buren (Hymenoptera: Formicidae) in Argentina. *Systematic Parasitology* 68, 115-128.

POINAR JR, G.O., KERP, H. & HASS, H. (2008). *Palaeonema phyticus* gen. n., sp. n. (Nematoda: Palaeonematidae fam. n.), a Devonian nematode associated with early land plants. *Nematology* 10, 9-14.

POINAR JR, G.O., DURATE, D. & SANTOS, M.J. (2009). *Halomonhystera parasitica* n. sp. (Nematoda: Monhysteridae), a parasite of *Talorchestia brito* (Crustacea: Talitridae) in Portugal. *Systematic Parasitology* 75, 53-58.

POINAR, JR, G. O., LEWIS, S.C., HAGEN, N.T. & HYMAN, B. (2011). Systematic affinity of the sea urchin parasite, *Echinomermella matsi* Jones & Hagen (Enoplida: Echinomermellidae). *Nematology*, in press.

RASNITSYN, A.P. & QUICKE, D.L.J. (2002). *History of insects*. Dordrecht, The Netherlands, Kluwer Academic Publishers, 517 pp.

REBOREDO, G.R. & CAMINO, N.B. (2000). Two new Rhabditida species (Nematoda: Rhabditidae), parasites of *Cyclocephala signaticollis* (Coleoptera: Scaracaeidae) in Argentina. *Journal of Parasitology* 86, 819-821.

REDECKER, D., KODNER, R. & GRAHAM, L.E. (2000). Glomalean fungi from the Ordovician. *Science* 289, 1920-1924.

REINHARD, K.J. (1992). Parasitology as an interpretive tool in archaeology. *American Antiquity* 57, 231-243.

REINHARD, K.J. & ARAÚJO, A. (2008). Archaeoparasitology. In: Pearsall, D.M. (Ed.). *Encyclopedia of archaeology*. Amsterdam, The Netherlands, Elsevier, pp. 494-501.

REMILLET, M. & SILVAIN, J.F. (1988). *Noctuidonema guyanense* n. g., n. sp. (Nematoda: Aphelenchoididae) ectoparasite de noctuelles du genre *Spodoptera* (Lepidoptera: Noctuidae). *Revue de Nématologie* 11, 21-24.

REMILLET, M. & VAN WAEREBEKE, D. (1975). Description et cycle biologique de *Howardula modecassa* n. sp. et *Howardula truncati* n. sp. (Nematoda: Sphaerulariidae), de *Carpophilus* (Coleoptera: Nitidulidae). *Nematologica* 21, 192-206.

RICHARDSON, P.N., HESLING, J.J. & RIDING, I.L. (1977). Life cycle and description of *Howardula husseyi* n. sp. (Tylenchida: Allantonematidae), a nematode parasite of the mushroom phorid, *Megaselia halterata* (Phoridae, Diptera). *Nematologica* 23, 217-231.

RIDING, I. (1970). Microvilli on the outside of a nematode. *Nature* 226, 179-180.

RIEMANN, F. (1993). The mouthless deep-sea nematode *Rhaptothyreus minor* sp. n. (Rhaptothyreidae, mermithoid nematodes of uncertain systematic position). *Zoologica Scripta* 22, 341-346.

RINGUELET, R.A. (1957). Restos de probables huevos de Nematodes en el estiercol del edentado extinguido *Mylodon listae* (Ameghino). *Ameghiniana* 1, 15-16.

ROGER, J. (1965). Némathelminthes fossiles. In: Grassé, P.P. (Ed.). *Traité de Zoologie*, Vol. 4, Part 3. Paris, France, Masson et Cie, 1223 pp.

ROGERS, C.E., SIMMONS, A.M. & MARTI, O.G. (1991). *Noctuidonema guyanense*: an ectoparasitic nematode of fall armyworm adults in the tropical Americas. *Florida Entomologist* 74, 246-257.

ROGERS, C.E., MARTI, O.G. & SIMMONS, A.M. (1993). *Noctuidonema guyanense* (Nematoda: Aphelenchoididae): host range and pathogenicity to the fall armyworm, *Spodoptera frugiperda* (Lepidoptera: Noctuidae). In: Bedding, R., Akhurst, R. & Kaya, H. (Eds). *Nematodes and the biological control of insect pests*. East Melbourne, Australia, CSIRO Publications, pp. 27-32.

ROUBIK, D.W. (1989). *Ecology and natural history of tropical bees*. Cambridge, UK, Cambridge University Press, 514 pp.

RUBSTOV, I.A. (1977). A new genus and species of parasitic nematode, *Ananus asteroideus* (Nematoda, Marimermithidae), from the asteroid *Diplopteraster perigrinator. Bulletin du Muséum National d'Histoire Naturelle* 496, 1113-1117.

RUBSTOV, I.A. (1980). [The new marine parasitic nematode, *Abos bathycola*, from Priapulids and the taxonomic position of the family Marimermithidae in the Class Nematoda.] *Parasitologia* 14, 177-181.

RUBSTOV, I.A. & PLATONOVA, T.A. (1974). Une nouvelle famille de nématodes parasites marins. *Zoologichesky Zhurnal* 53, 1445-1458.

RUHM, W. (1956). Die Nematoden der Ipiden. *Parasitologische Schriftenreihe* 6, 437 pp.

SACHS, H. (1950). Die Nematodenfauna der Rinderexkremente. *Zoologische Jahrbuch Systematische* 79, 209-272.

SANDISON, A.T. & TAPP, E. (1998). Disease in ancient Egypt. In: Cockburn, A., Cockburn, E. & Reyman, T.A. (Eds). *Mummies, disease and ancient cultures*, 2nd edition. Cambridge, UK, Cambridge University Press, pp. 38-58.

SANTIAGO-BLAY, J.A. & POINAR JR, G.O. (1992). Millipedes from Dominican amber, with the description of two new species (Diplopoda: Siphonophoridae) of *Siphonophora. Annals of the Entomological Society of America* 85, 363-369.

SCHLEE, D. (1990). Das Bernstein-Kabinett, Begleitheft zur Bernstein-ausstellung im Museum am Lowentor, Stuttgart. *Stuttgarter Beiträge zur Naturkunde Serie C* 28, 100 pp.

SCHLEE, D. & GLÖCKNER, W. (1978). Bernstein-Bernsteine und Bernstein-Fossilien. *Stuttgarter Beiträge zur Naturkunde Serie C* 28, 72 pp.

SCHLINGER, E.I. (1975). Intertidal insects; Order Diptera. In: Smith, R.I. & Carlton, J.T. (Eds). *Light's manual: intertidal invertebrates of the central California coast*, 3rd edition. Berkeley, CA, USA, University of California Press, pp. 436-446.

SCHMIDT, A.R., DÖRFELT, H. & PERRICHOT, V. (2008). *Palaeoanellus dimorphus* gen. et sp. nov. (Deuteromycotina): a Cretaceous predatory fungus. *American Journal of Botany* 95, 1328-1334.

SCHMIDT, G.D., DUSZYNSKI, D.W. & MARTIN, P.S. (1992). Parasites of the extinct shasta ground sloth, *Nothrotheriops shastensis*, in Rampart Cave, Arizona. *Journal of Parasitology* 78, 811-816.

SCHMIDT-RHAESA, A. (1997). Phylogenetic relationships of the Nematomorpha – a discussion of current hypotheses. *Zoologischer Anzeiger* 236, 203-216.

SCHRAM, F.R. (1973). Pseudocoelomates and a nemertine from the Illinois Pennsylvanian. *Journal of Paleontology* 47, 985-989.

SCHRAM, F.R. (1979). Worms of the Mississippian Bear Gulch Limestone of central Montana, USA. *Transactions of the San Diego Society of Natural History* 19, 107-120.

SCHULTE, F. (1989). The association between *Rhabditis necromena* Sudhaus & Schulte, 1989 (Nematoda: Rhabditidae) and native and introduced millipedes in South Australia. *Nematologica* 35, 82-89.

SCHWENK, J. (1927). Sobre un nematoid parasita de Oniscidae. *Boletim do Biologico* 7, 78-80.

SELDEN, P. & NUDDS, J. (2004). *Evolution of fossil ecosystems.* Chicago, USA, University of Chicago Press, 160 pp.

SHEPARD, M.R.N. (1974). *Arthropods as final hosts of nematodes and nematomorphs.* Technical Communication No. 45. St Albans, UK, Commonwealth Institute of Helminthology, 248 pp.

SHIPLEY, A.E. (1900). On some parasites found in *Echinus esculentus*, L. *The Quarterly Journal of Microscopical Science* 44, 281-290.

SIDDIQI, M.R. (1983). Evolution of plant parasitism in nematodes. In: Stone, A.R., Platt, H.M. & Khalil, L.F. (Eds). *Concepts in nematode systematics.* New York, NY, USA, Academic Press, pp. 113-129.

SIDDIQI, M.R. (2000). *Tylenchida: parasites of plants and insects*, 2nd edition. Wallingford, UK, CABI Publishing, 833 pp.

SKRJABIN, K.I., SCHIKHOBALOVA, N. & LAGODOVSKAYA, E.A. (1966). [Oxyurids of Arthropods, Part 4.] In: Skrjabin, K.I. (Ed.). *Principles of nematology*, Vol. 15. Moscow, USSR, Nauka, 538 pp.

SLITER, W.V. (1971). Predation on benthic foraminifers. *Journal of Foraminifera Research* 1, 20-29.

SMITH, R.L. (1996). *Ecology and field biology*, 5th edition. New York, NY, USA, Harper Collins, 804 pp.

STEINER, G. (1921). *Phlyctainophora lamnae* n. g. n. sp., eine neue parasitische Nematodenform aus *Lamna cornubica* (Heringshai). *Centralblatt für Bakteriologie Originale* 86, 591-595.

STEINER, G. (1927). A new nemic family, Epsilonematidae. *Journal of Parasitology* 14, 65-66.

STØRMER, L. (1963). *Gigantoscorpio willsi*, a new scorpion from the Lower Carboniferous of Scotland and its associated preying microorganisms. *Skrifter Utgitt av Det Norske Vidénskaps-Akademi I Oslo 1. Mat.-Naturv. Klasse,* NY Serie No. 8, 1-171.

STUART, M., PENDERGAST, V., RUMFELT, S., PIERBERG, S., GREENSPAN, L., GLANDER, K. & CLARKE, M. (1998). Parasites of wild howlers (*Alouatta* spp.). *International Journal of Primatology* 19, 493-512.

SUDHAUS, W. (1993). Redescription of *Rhabditis* (*Oscheius*) *tipulae* (Nematoda: Rhabditidae) associated with a leatherjacket larvae of *Tipula paludosa* (Diptera: Tipulidae). *Nematologica* 39, 234-239.

SUDHAUS, W. (2008). Evolution of insect parasitism in rhabditid and diplogastrid nematodes. In: Makarov, S.E. & Dimitrijević, R.N. (Eds). *Advances in arachnology and developmental biology. Papers dedicated to Prof. Dr. Božidar Ćurčić*. Sofia, Institute of Zoology, Belgrade; BAS, Faculty of Life Sciences, Vienna; SASA, Belgrade & UNESCO MAB Serbia.Vienna-Belgrade-Sofia, Monographs, 12, pp. 143-161.

SUDHAUS, W. & KÜHNE, R. (1989). Nematodes associated with Psychodidae: description of *Rhabditis berolina* sp. n. and redescription of *R. dubia* Bovien, 1937 (Nematoda: Rhabditidae), with biological and ecological notes, and a phylogenetic discussion. *Nematologica* 35, 305-320.

SUDHAUS, W. & SCHULTE, F. (1989). *Rhabditis* (*Rhabditis*) *necromena* sp. n. (Nematoda: Rhabditidae) from South Australian Diplopoda with notes on its sibling *R. myriophila* Poinar, 1986 and *R. caulleryi* Maupas, 1919. *Nematologica* 35, 15-24.

SZADZIEWSKI, R. & POINAR JR, G.O. (2005). Additional biting midges (Diptera: Ceratopogonidae) from Burmese amber. *Polskie Pismo Entomologiczne* 74, 349-362.

TANG, Y., HOMINICK, W.M., KILLICK-KENDRICK, R., KILLICK-KENDRICK, M. & PAGE, A.M. (1993). *Didilia ooglypta* n. gen., n. sp. (Tetradonematidae: Mermithoidea: Nematoda), a parasite of phlebotomine sandflies in Afghanistan. *Fundamental and Applied Nematology* 16, 325-331.

TAPP, E. & WILDSMITH, K. (1993). The autopsy and endoscopy of the Leeds mummy. In: David, A.R. & Tapp, E. (Eds). *The mummy's tale*. New York, NY, USA, St Martins Press, pp. 132-153.

TAYLOR, A.L. (1935). A review of the fossil nematodes. *Proceedings of the Helminthological Society of Washington* 2, 47-49.

TAYLOR, A.L. (1955). Helminths in mediaeval remains. *The Veterinary Record* 67, 216-218.

TAYLOR, T.N. & TAYLOR, E.L. (1993). *The biology and evolution of fossil plants*. Englewood Cliffs, NJ, USA, Prentice Hall, 982 pp.

TCHESUNOV, A.V. (1988). A case of nematode parasitism in nematodes. A new find and redescription of a rare species *Benthimermis australis* Petter, 1983 (Nematoda: Marimermithida: Benthimermithidae) in South Atlantic. *Helminthologia* 25, 115-128.

TCHESUNOV, A.V. & HOPE, W.D. (1997). *Thalassomermis megamphis* n. g. and n. sp. (Mermithidae: Nemata) from the deep-sea. *Journal of Nematology* 29, 451-464.

TCHESUNOV, A.V. & SPIRIDONOV, S.E. (1993). *Nematimermis enoplivora* gen. n., sp. n. (Nematoda: Mermithoidea) from marine free-living nematodes *Enoplus* spp. *Russian Journal of Nematology* 1, 7-16.

TEAL, J.M. & WEISER, W. (1966). The distribution and ecology of nematodes in a Georgia salt marsh. *Limnology and Oceanography* 11, 217-222.

THOISY, B.D., VOGEL, I., REYNES, J.M., POULIQUEN, J.F., CARME, B., KAZANJI, M. & VIE, J.C. (2001). Health evaluation of translocated free-ranging primates in French Guiana. *American Journal of Primatology* 54, 1-16.

THORNE, G. (1935). Nemic parasites and associated of the mountain pine beetle (*Dendroctonus monticolae*) in Utah. *Journal of Agricultural Research* 51, 131-144.

THORNE, G. (1961). *Principles of nematology*. New York, NY, USA, McGraw-Hill Book Company, Inc., 553 pp.

TIMM, R.W. & FRANKLIN, M.T. (1969). Two marine species of *Aphelenchoides*. *Nematologica* 15, 370-375.

TRIFFITT, M.J. & OLDHAM, J.N. (1927). Observations on the morphology and bionomics of *Rhabditis coarctata* Leuck. occurring on dung beetles. *Journal of Helminthology* 5, 33-46.

UPENIECE, I. (2001). The unique fossil assemblage from the Lode Quarry (Upper Devonian, Latvia). *Mitteilungen aus dem Museum für Naturkunde in Berlin Geowissenschaftliche Reihe* 4, 101-119.

VANDERBURGH, D.J. & ANDERSON, R.C. (1987). The relationship between nematodes of the genus *Cosmocercoides* Wilkie, 1930 (Nematoda: Cosmocercoidea) in toads (*Bufo americanus*) and slugs (*Deroceras laeve*). *Canadian Journal of Zoology* 65, 1650-1661.

VIDAL, G. (1976). Late Precambrian microfossils from the Visingsö Beds in southern Sweden. *Fossils and Strata* 9, 1-57.

VOIGT, E. (1957). Ein parasitischer Nematode in fossiler Coleopteren-Muskulatur aus der eozänen Braunkohle des Geiseltales bei Halle (Saale). *Paläontologische Zeitschrift* 31, 35-39.

VOJTKOVÁ, L. (1971). Beitrage zur Kenntnis der Helminthofauna der Wasserwirbellosen. 3. Cestoda, Nematoda, Acanthocephala. *Vestnik Cesk Slovenia Společia Zoologie* 35, 146-155.

WACHEK, F. (1955). Die Entoparasitischen Tylenchiden. *Parastiologie Schriftenreihe* 3, 119 pp.

WAHAB, A. (1962). Untersuchungen über Nematoden in den Drüsen des Kopfes der Ameisen (Formicidae). *Zeitschrift für Morphologie und Ökologie der Tiere* 52, 33-92.

WALCOTT, C.D. (1899). Pre-Cambrian fossiliferous formations. *Bulletin of the Geological Society of America* 10, 199-244.

WALOSZEK, D. & MÜLLER, K.J. (1994). Pentastomid parasites from the Lower Palaeozoic of Sweden. *Transactions of the Royal Society of Edinburgh, Earth Sciences* 85, 1-37.

WALTER, M.R., OEHLER, J.H. & OEHLER, D.Z. (1976). Megascopic algae 1300 million years old from the Belt Supergroup, Montana: a reinterpretation of Walcott's *Helminthoidichnites*. *Journal of Paleontology* 50, 872-881.

WARD, H.B. (1933). On *Thalassonema ophioctinis*, a nematode parasitic in the brittle star *Ophiocten amitinum*. *Journal of Parasitology* 19, 262-268.

WASILEWSKA, L. (1979). The structure and function of soil nematode communities in natural ecosystems and agrocenoses. *Polish Ecological Studies* 5, 97-145.

WEI, O. (1973). Internal organs of a 2100-year-old female corpse. *Lancet* 7839, 1198.

WEISER, W. (1953). Free-living marine nematodes. 1. Enoploidea. *Reports of the Lund University Chile Expedition 1948-1949, 10.* Lunds Universitets Årsskrift, N. F., avd 2, 49, 155 pp.

WEISER, W. (1956). Free-living marine nematodes. III. Axonolaimoidea and Monhysteroidea. *Reports of the Lund University Chile Expedition 1948-1949, 10.* Lunds Universitets Årsskrift, N. F., avd 2, 49, 115 pp.

WEITSCHAT, W. & WICHARD, W. (2002). *Atlas of plants and animals in Baltic amber*. Munich, Germany, Verlag Dr Friedrich Pfeil, 256 pp.

WITENBERG, G. & GERICHTER, C. (1944). The morphology and life history of *Foleyella duboisi* with remarks on allied filariids of Amphibia. *Journal of Parasitology* 30, 245-254.

WITTINGTON, H.B. (1985). *The Burgess shale*. New Haven, USA, Yale University Press, 75 pp.

WUNDERLICH, J. (2004). *Fossil spiders in amber and copal*. Beiträge zur Araneologie 3a, Vol. 1, 848 pp.

YAMAGUTI, S. (1961). *Systema Helminthum. The nematode parasites of vertebrates*, Vol. 3. New York, NY, USA, Interscience Publishers, Inc., 1261 pp.

YANOVIAK, S.P., KASPARI, M., DUDLEY, R. & POINAR JR, G.O. (2008). Parasite-induced fruit mimicry in a tropical canopy ant. *The American Naturalist* 171, 536-544.

YEATES, G.W., SPIRIDONOV, S.E. & BLAKEMORE, R. (1998). *Plesioungella kathleenae* gen. n. *et* sp. n. (Nematoda: Drilonematoidea) from the Australian endemic megascocecid earthworm, *Fletcherodrilus unicus* (Fletcher, 1899). *New Zealand Journal of Zoology* 25, 205-212.

ZIMMERMAN, E.C. (1994). *Australian weevils (Coleoptera: Curculionoidea)*, Vol. 2. Canberra, Australia, CSIRO, 755 pp.

Nematology Monographs & Perspectives, 2011, Vol. 9, 395-414

Colour plates

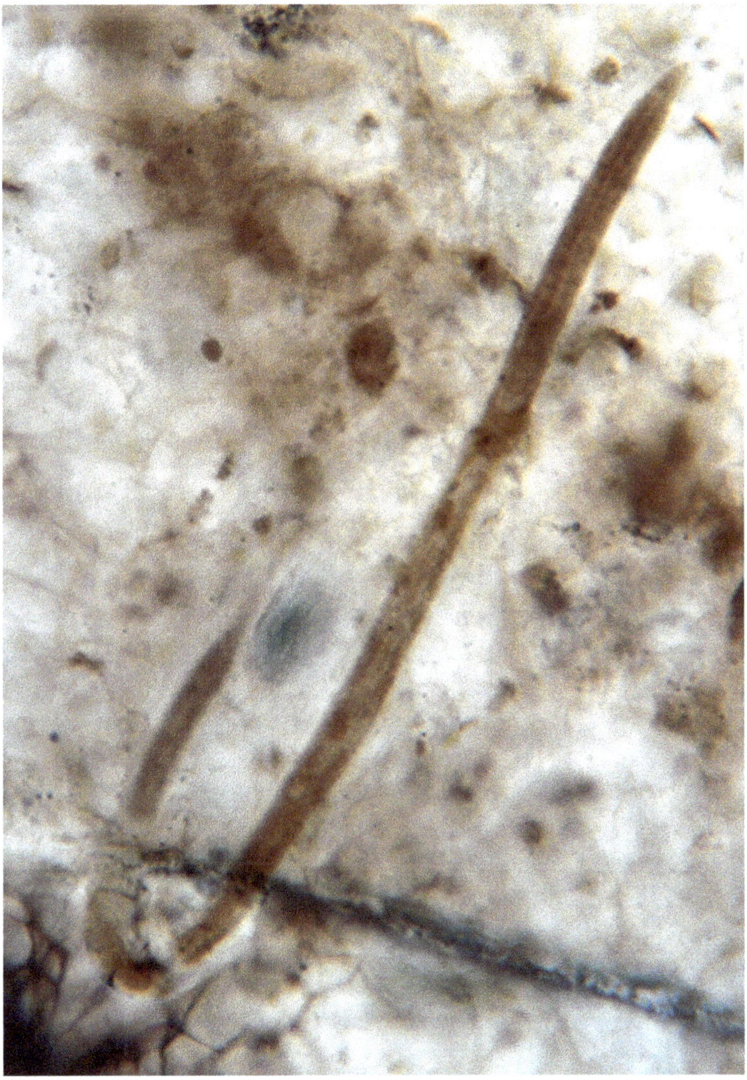

Plate 1. Palaeonema phyticum *Poinar, Kerp & Hass, 2008 in the stomatal chamber of the Early Devonian land plant,* Aglaophyton major.

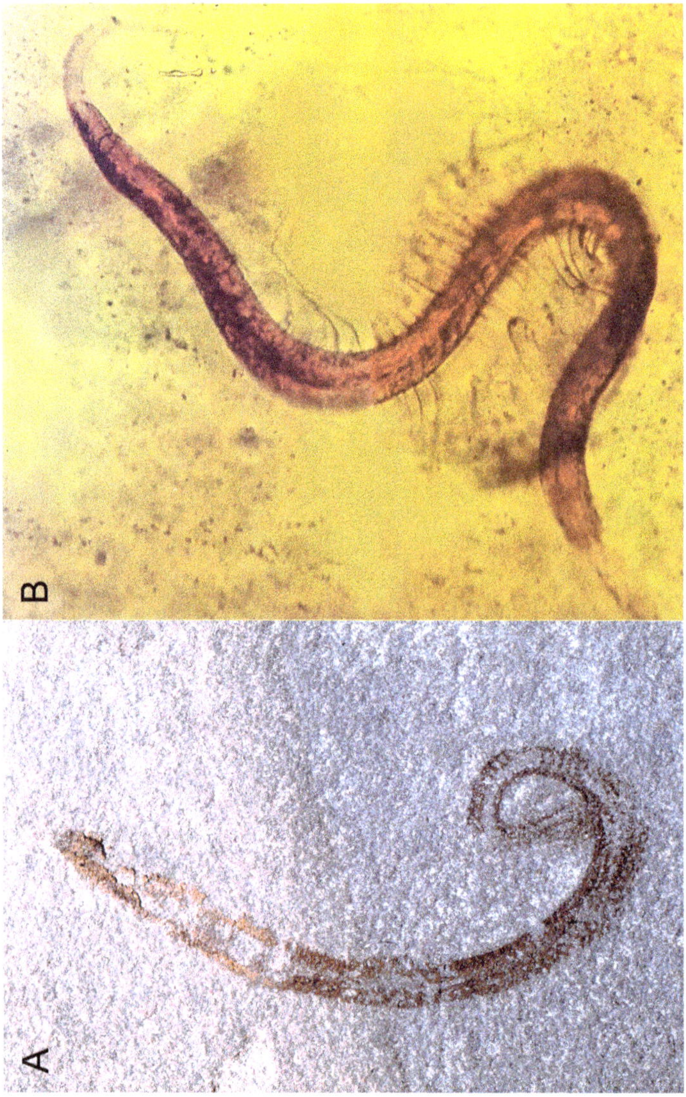

Plate 2. *A*: Eophasma jurasicum *Arduini, Pinna & Teruzzi, 1983 from the Sinemurian deposit of Ostena in Lombardy, Italy; B*: Captivonema cretacea *Boström* in *Manum, Bose, Sayer & Boström, 1994 in the cocoon wall of a clitellate annelid from the Barremian of Spitsbergen, Norway.*

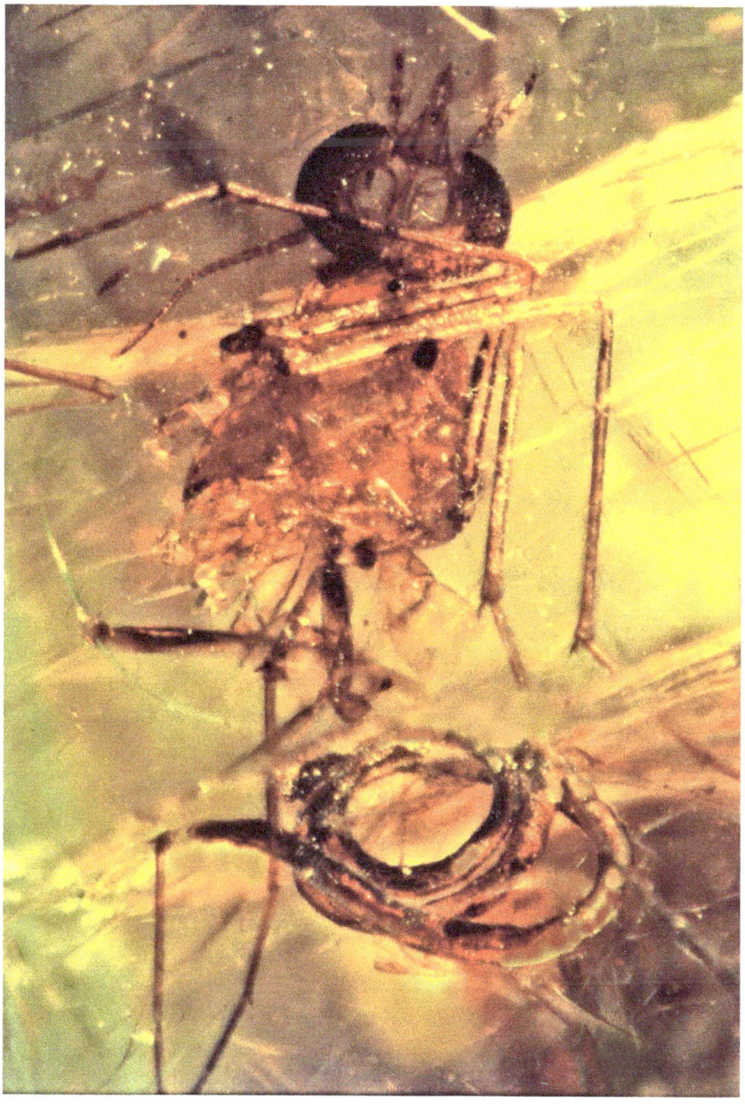

Plate 3. *The mermithid* Cretacimermis libani *(Poinar, Acra & Acra, 1994) parasitising a chironomid midge in Early Cretaceous Lebanese amber.*

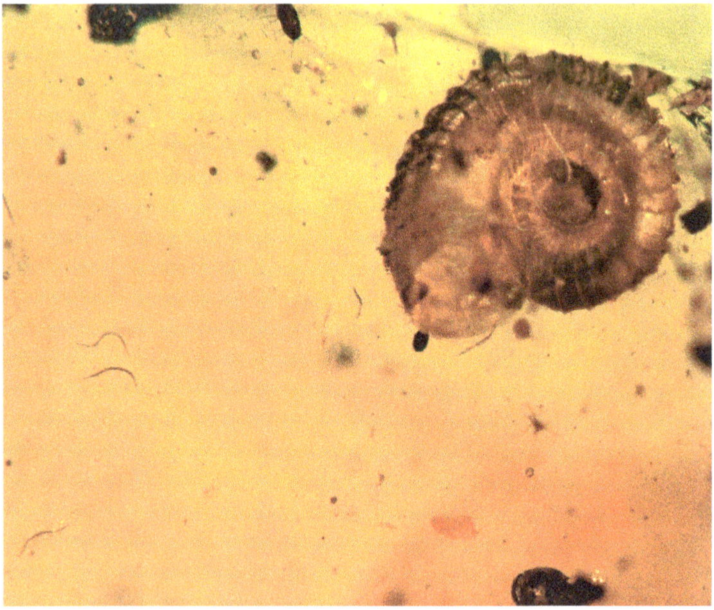

Plate 4. *Juveniles of* Palaeocosmocerca burmanica* *associated with a terrestrial snail in Early Cretaceous Burmese amber.*

Plate 5. *The thelastomatid* Paleothelastoma tipulae* *adjacent to its cranefly host in Early Cretaceous Burmese amber.*

Plate 6. *The mermithid* Cretacimermis protus *Poinar & Buckley, 2006 emerging from a biting midge in Early Cretaceous Burmese amber.*

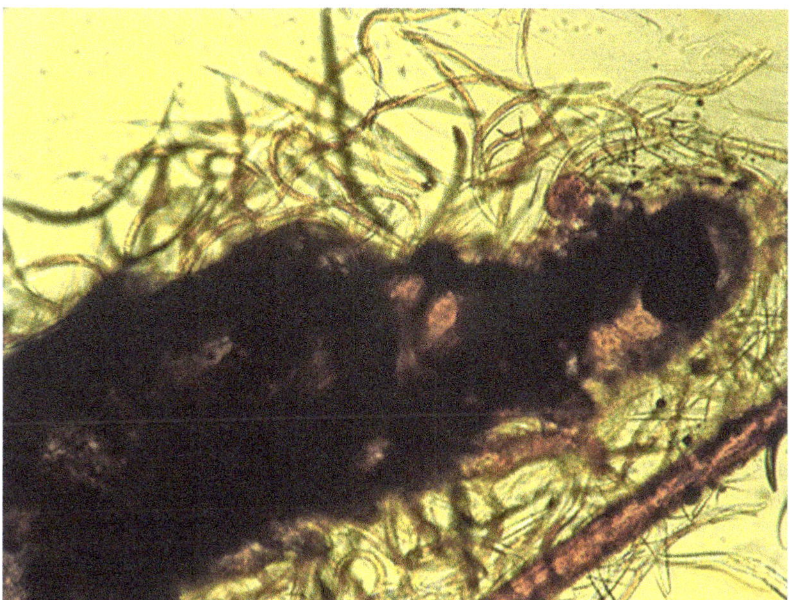

Plate 7. *Dauer juveniles of* Tipulacomis balticus* *leaving their cranefly carrier host in Baltic amber.*

Plate 8. *Two specimens of the mermithid* Heydenius matutinus *(Menge, 1866) with their chironomid midge host in Baltic amber.*

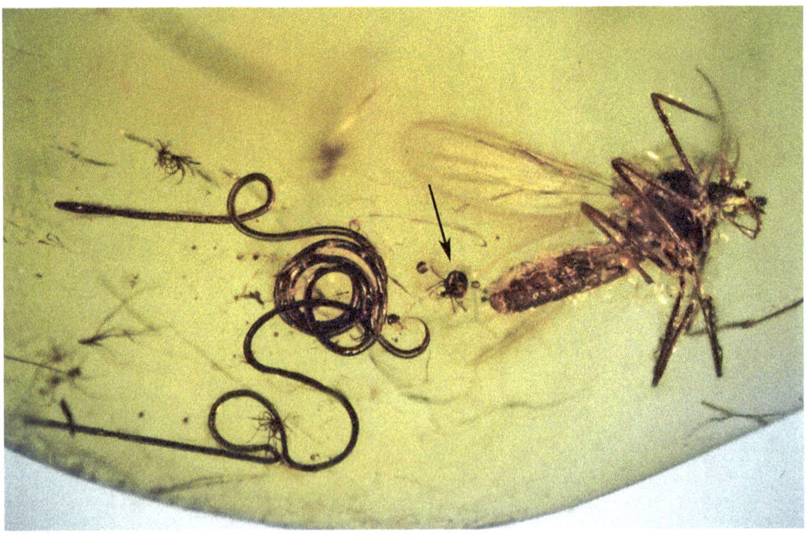

Plate 9. *Two specimens of* Heydenius matutinus *(Menge, 1866) adjacent to their chironomid midge host in Baltic amber. Note the ectoparasitic mite (arrow) adjacent to the midge.*

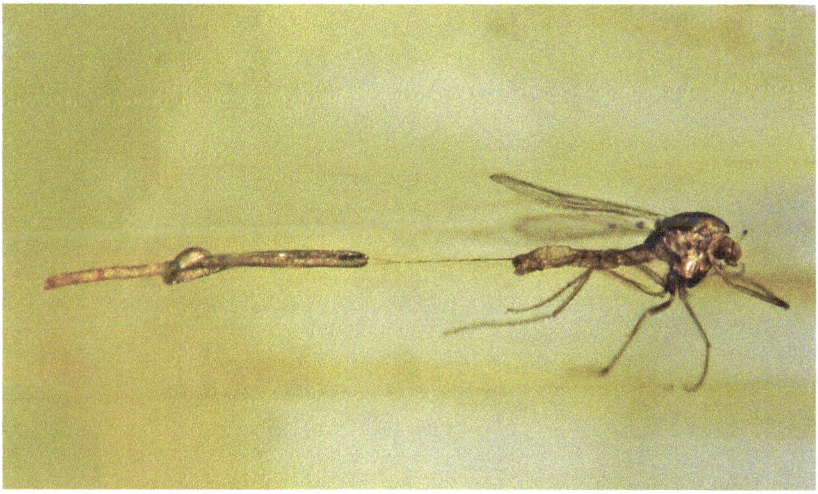

Plate 10. *The mermithid* Heydenius matutinus *(Menge, 1866) connected to its chironomid midge host by a thin strand of congealed blood in Baltic amber.*

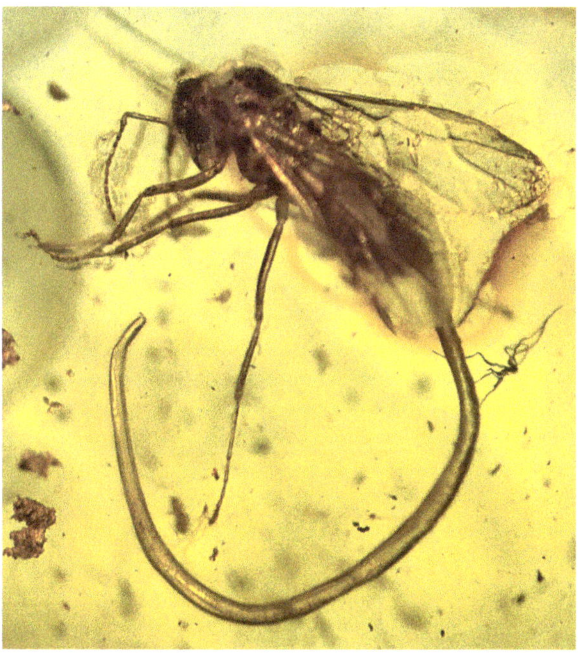

Plate 11. *The mermithid* Heydenius formicinus *Poinar, 2002 emerging from an ant in Baltic amber.*

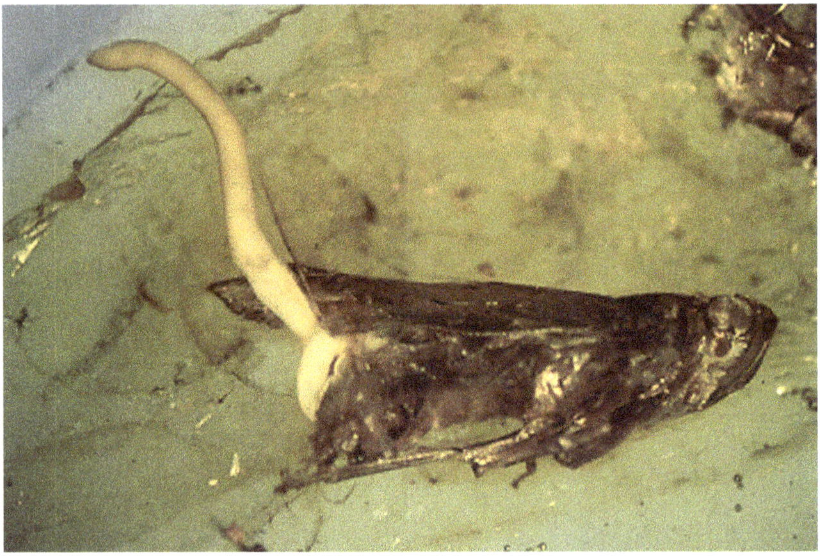

Plate 12. *The mermithid* Heydenius brownii *Poinar, 2001 emerging from an achiliid planthopper in Baltic amber.*

Plate 13. *Four specimens of the mermithid* Heydenius sciarophilus* *with their sciarid host in Baltic amber.*

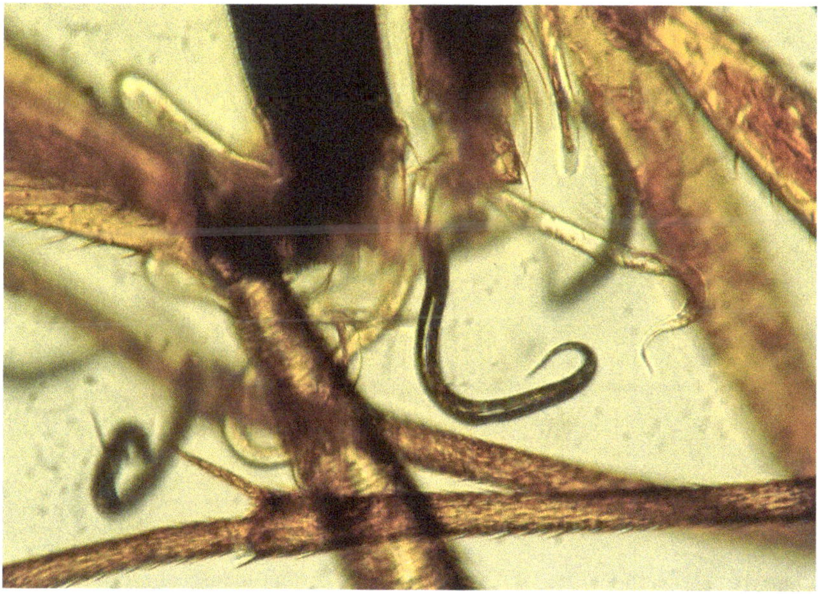

Plate 14. *The sphaerulariid* Tripius balticus* *emerging from the severed legs of a sciarid fly in Baltic amber.*

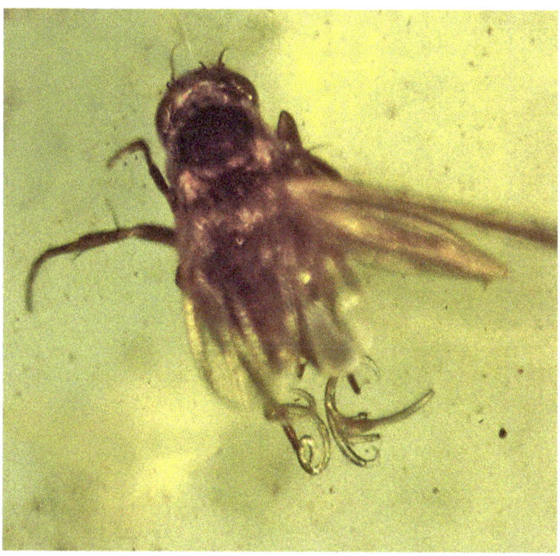

Plate 15. *Parasitic stages of the allantonematid* Howardula helenoschini *Poinar, 2003 emerging from their phorid host in Baltic amber.*

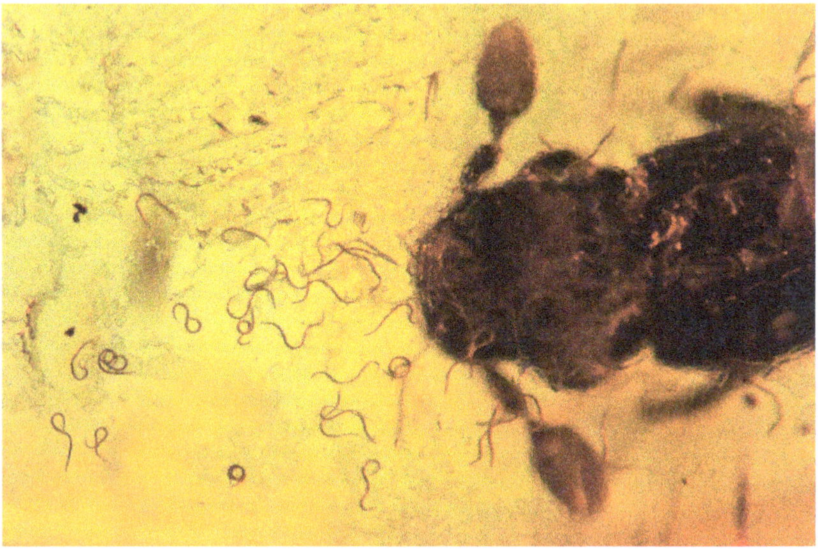

Plate 16. *Dauer juveniles of* Cryptaphelenchus dominicus* *adjacent to their platypodid beetle carrier in Dominican amber.*

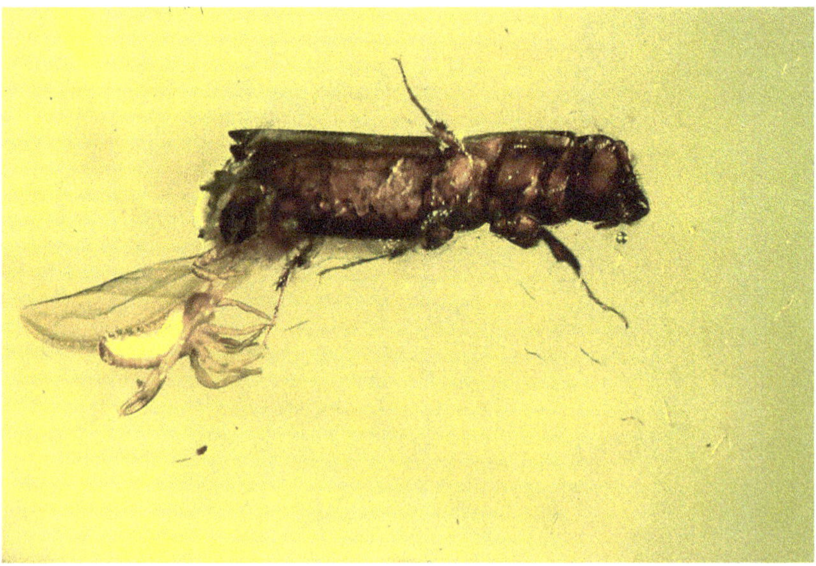

Plate 17. *Dauer juveniles of* Scolytonema dominicana* *and* S. hispaniolae* *beneath their platypodid carrier in Dominican amber. Note pseudoscorpion also being transported by the beetle.*

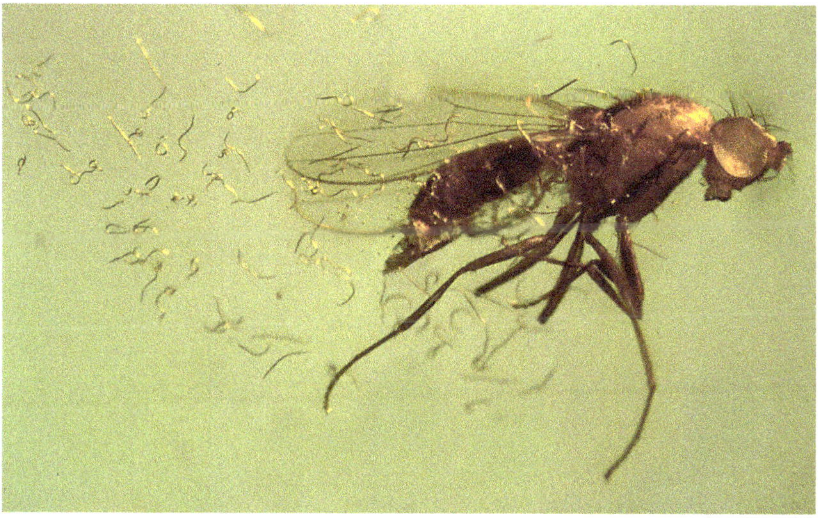

Plate 18. *Parasitic stages of the allantonematid* Palaeoparasitylenchus dominicanus* *emerging from a drosophilid fly in Dominican amber.*

Plate 19. *The mermithid* Heydenius psychodae* *with its mothfly host in Dominican amber.*

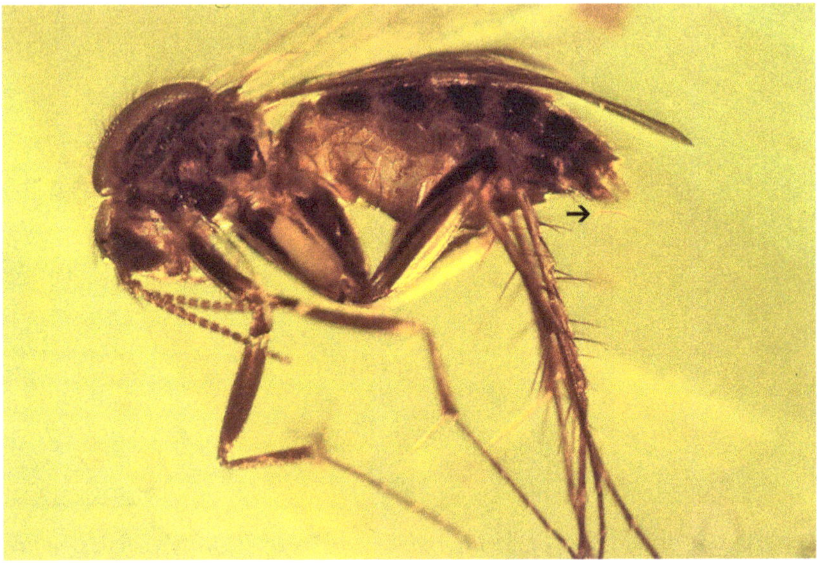

Plate 20. *Juveniles of the iotonchiid* Paleoiotonchium dominicanum* *inside the body cavity of a mycetophilid fungus gnat in Dominican amber. Arrow shows an emerged juvenile.*

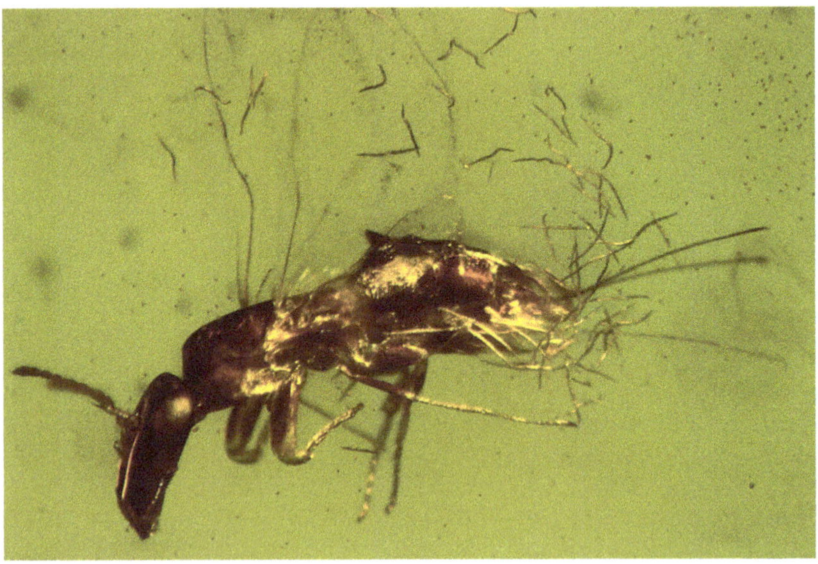

Plate 21. *Juveniles of the diplogastrid* Syconema dominicana* *emerging from a fig wasp in Dominican amber.*

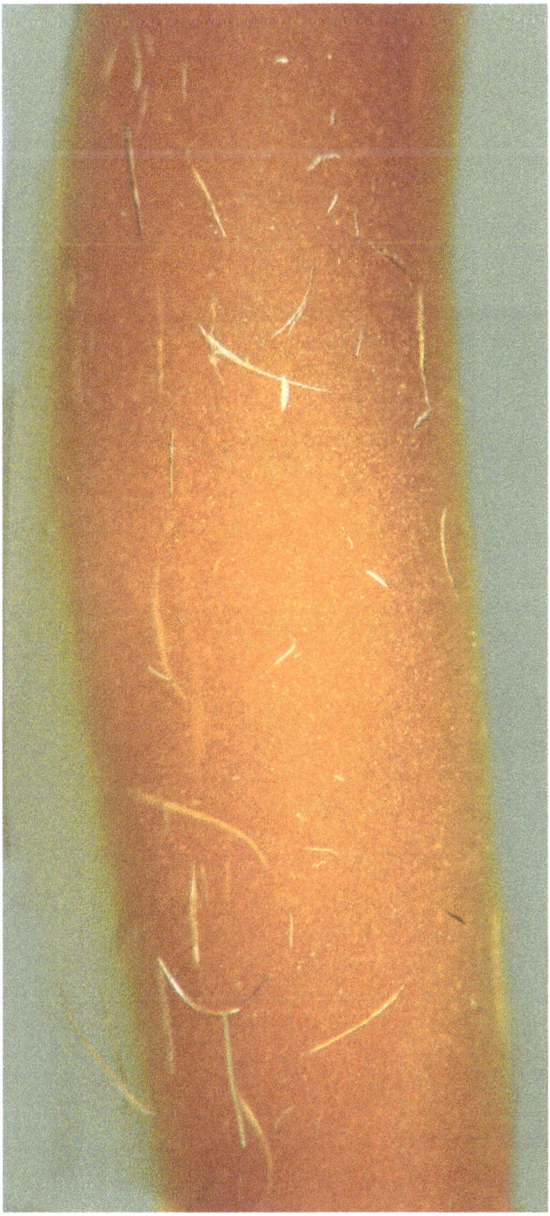

Plate 22. *Population of* Oligaphelenchoides maximus* *developing in a rootlet in Dominican amber.*

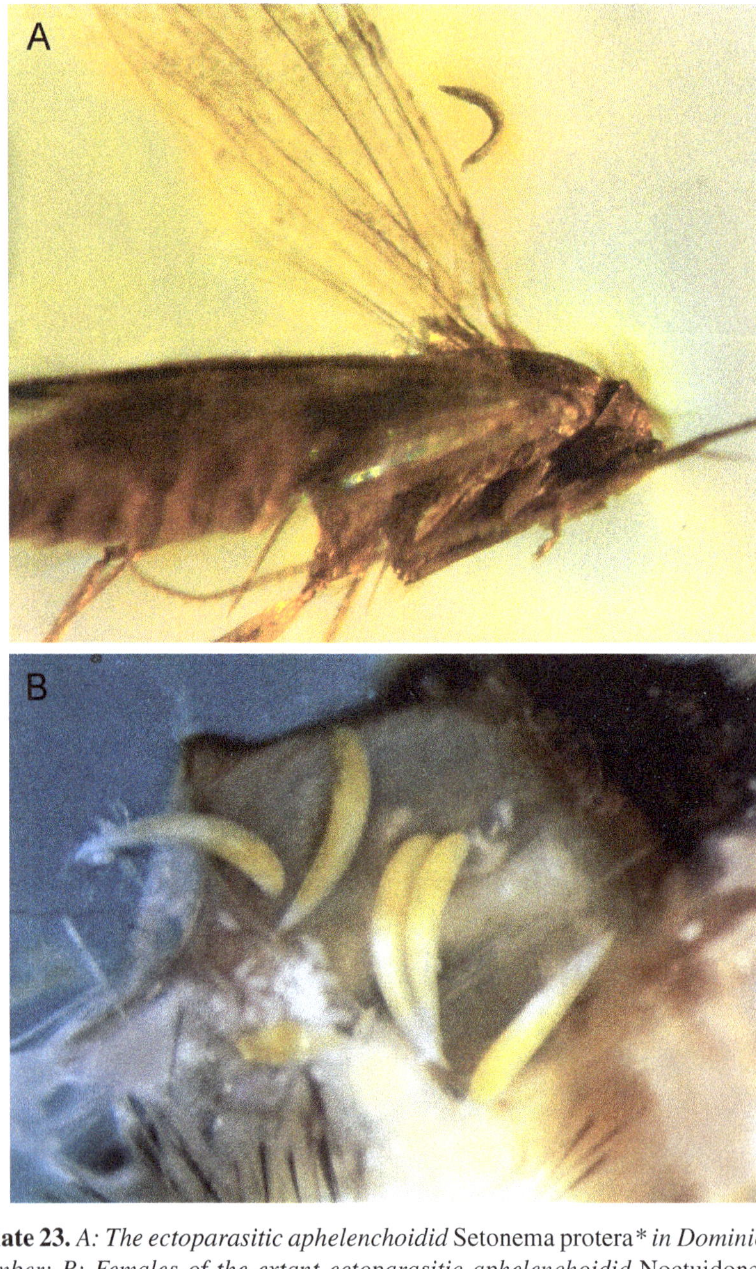

Plate 23. *A: The ectoparasitic aphelenchoidid* Setonema protera* *in Dominican amber; B: Females of the extant ectoparasitic aphelenchoidid* Noctuidonema guyanense *Remillet & Silvain, 1988 on the body of their noctuid moth host.*

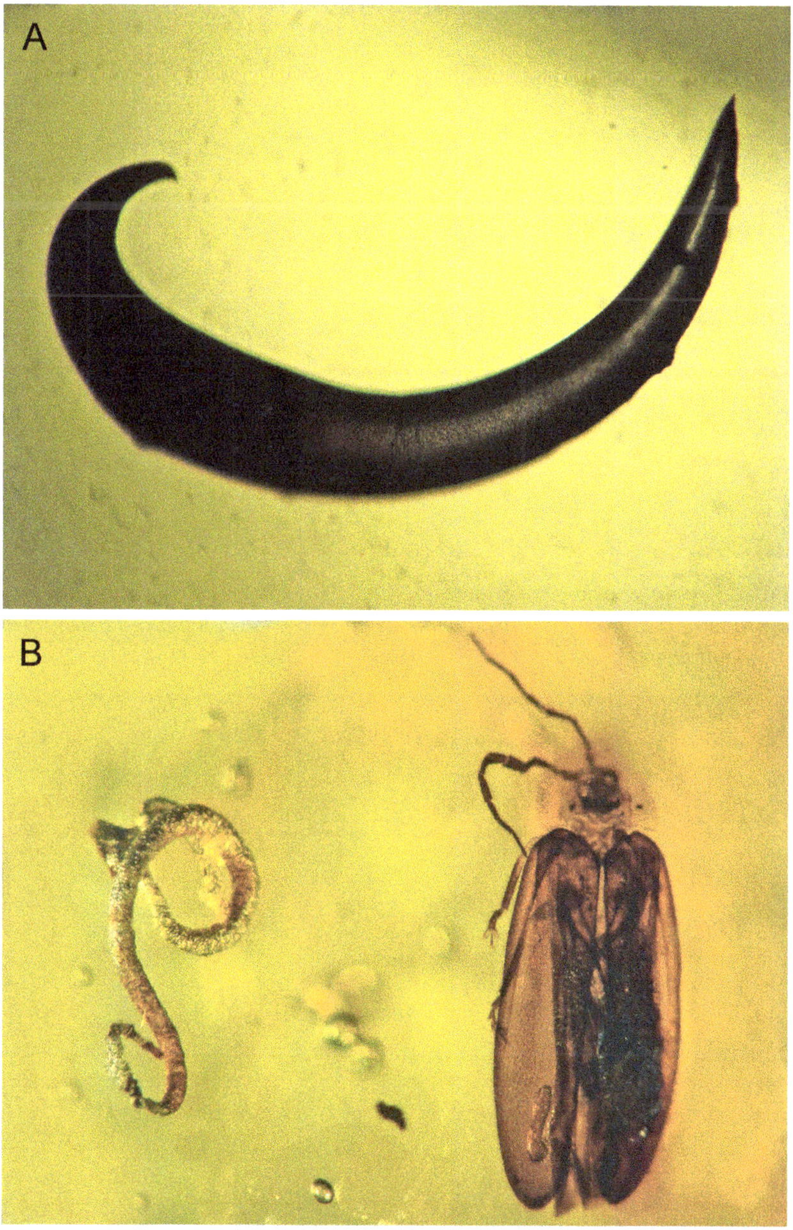

Plate 24. *A: The anguinid* Palaeoanguina dominicana* *in Dominican amber; B: The mermithid* Heydenius lamprophilus* *adjacent to its firefly host in Dominican amber.*

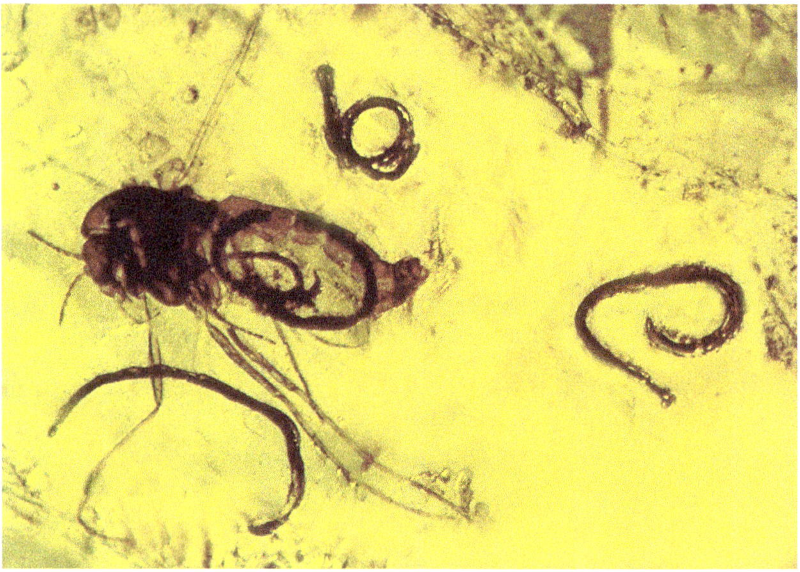

Plate 25. *Four specimens of the mermithid* Heydenius neotropicus *that were parasitising a chironomid in Dominican amber. One individual is still in the host's abdomen.*

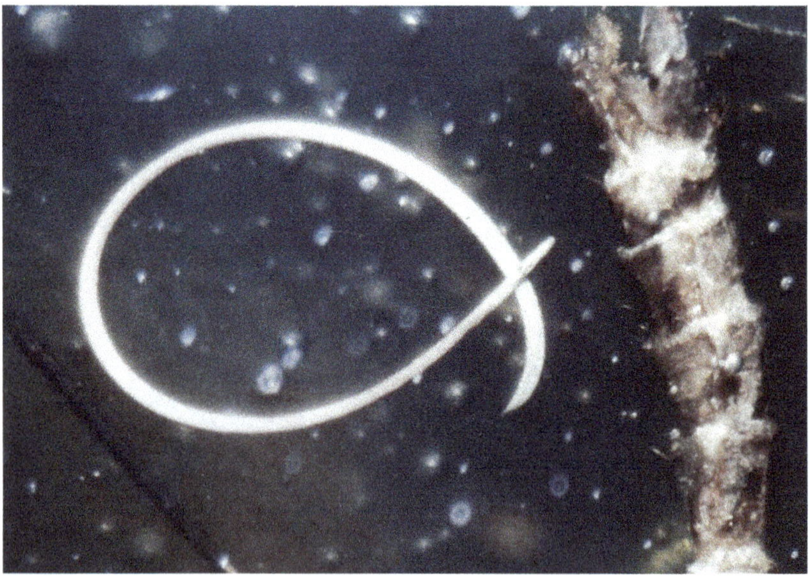

Plate 26. *Parasitic juvenile of the mosquito mermithid* Heydenius dominicus *Poinar, 1984 in Dominican amber.*

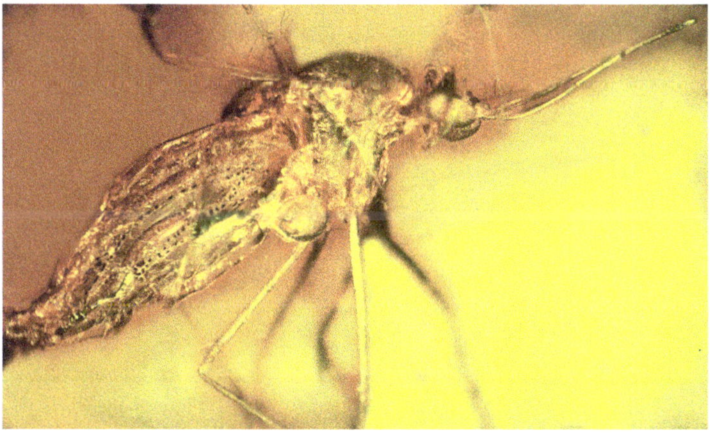

Plate 27. *Specimen of the mermithid* Heydenius dominicus *Poinar, 1984 coiled in the abdomen of a* Culex *mosquito in Dominican amber.*

Plate 28. *Specimen of* Heydenius saprophilus* *adjacent to its wood gnat host in Dominican amber.*

Plate 29. *Parasitic juveniles of* Palaeoallantonema dominicana* *adjacent to their rove beetle host in Dominican amber.*

Plate 30. *A chalcidoid wasp with stages of the aphelenchoidid parasite,* Chalcidonema paradoxa*, *in Dominican amber.*

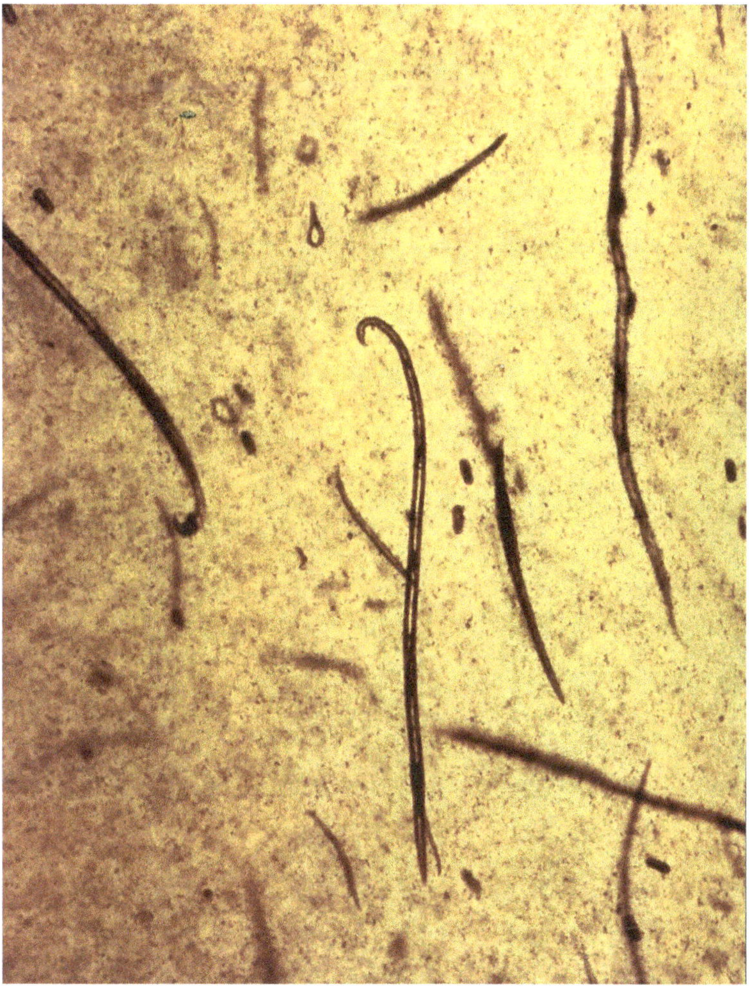

Plate 31. *Population of the mycetophagous* Oligaphelenchoides atrebora *Poinar, 1977 in Mexican amber.*

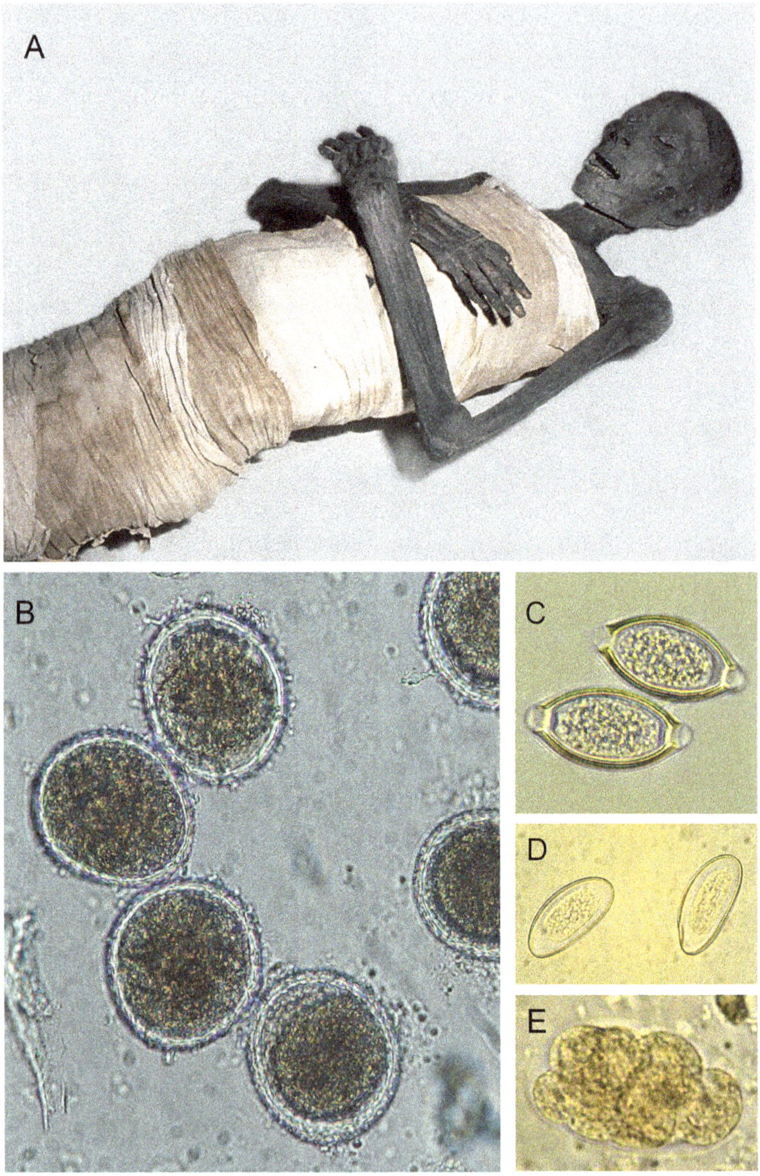

Plate 32. *Human nematode parasites. A: Egyptian mummies such as this one have supplied much information about ancient human nematode parasites; B: Eggs of the human white stomach worm,* Ascaris lumbricoides*; C: Eggs of the human whipworm,* Trichuris trichura*; D: Eggs of the human pinworm,* Enterobius vermicularis*; E: Human hookworm egg. (Not to scale.)*

Nematology Monographs & Perspectives, 2011, Vol. 9, 415-429

Index

www.ingramcontent.com/pod-product-compliance
Lightning Source LLC
Chambersburg PA
CBHW071231230926
51849CB00001B/74
* 9 7 8 9 0 0 4 1 7 5 2 1 1 *